About the Author

Ronald N. Bracewell was born in Australia, received his B.Sc., B.E., and M.E. degrees from the University of Sydney, and earned a Ph.D. from Cambridge University. Currently Professor of Electrical Engineering at Stanford University, Dr. Bracewell has an impressive roster of professional affiliations, awards, and publications to his credit. He is a Fellow of the Royal Astronomical Society, a Fellow of the Institute of Electrical and Electronic Engineers, a Life Member of the Astronomical Society of the Pacific, and a past Councilor of the American Astronomical Society. He has served as an advisor to several government organizations, including the National Science Foundation and the National Academy of Sciences, and in 1952 received the Duddell Premium of the Institute of Electrical Engineers for his contributions to the study of the ionosphere. Among his many achievements, Dr. Bracewell constructed a microwave spectroheliograph which automatically produced in printed form daily temperature maps of the sun, and was used by NASA for support of its first manned landing on the moon.

Fourier's theorem is not only one of the most beautiful results of modern analysis, but it may be said to furnish an indispensable instrument in the treatment of nearly every recondite question in modern physics.

Lord Kelvin

Contents

Note on 1986 Revision

Developments based on Hartley's formulas (page 179) have made it possible to dispense with imaginaries in the computation of Fourier transforms. In fact, in any application where Fourier analysis is practiced it is now possible in numerical work to proceed elegantly and simply using the real Hartley formalism. In computing, a speed advantage, or reduction in the space allocated to memory, results from the fact that multiplying complex numbers takes four times longer than multiplying real numbers; which leads to a net advantage of a factor of 2 in computing with real data when account is taken of the nonredundancy of the Hartley transform. The new ideas and a demonstration computer program are now introduced in Chapters 19 and 20. For a more detailed treatment, see R. N. Bracewell, "The Hartley Transform," Oxford University Press, 1986. For historical interest a chapter on the fascinating life of Joseph Fourier has also been added.

Ronald N. Bracewell

Preface to the Second Edition

The unifying role played by the Fourier transform in linking together the diverse fields mentioned in the original preface has now firmly established transform methods at the heart of the electrical engineering curriculum. Computing and data processing, which have emerged as large curricular segments, though rather different in character from circuits, electronics, and waves, nevertheless do share a common bond through the Fourier transform. Consequently, the subject matter has easily moved into the pivotal role foreseen for it, and faculty members from various specialties have found themselves comfortable teaching it. The course is taken by first-year graduate students, especially students arriving from other universities, and increasingly by students in the last year of their bachelor's degree.

Introduction of the fast Fourier transform (FFT) algorithm has greatly broadened the scope of application of the Fourier transform to data handling and has brought prominence to the discrete Fourier transform (DFT). The technological revolution brought about by these topics, now treated in a new Chapter 18, is only beginning to be felt, but will make an understanding of Fourier notions (such as aliasing, which only aficionados knew about) indispensable to any engineer who handles masses of data. In the future this will mean nearly everyone.

Transforms presented graphically in the Pictorial Dictionary proved to be a useful reference feature and have been added to. Graphical presentation adds a new dimension to the published compilations of integral transforms where it is sometimes frustrating to seek commonly needed entires among the profusion of rare cases. In addition, simple functions that are impulsive, discontinuous, or defined piecewise may not be included or may be hard to recognize.

A good problem assigned at the right stage can be extremely valuable for the student, but a good problem is hard to compose. Among the collection of supplementary problems now included at the end of the book are several that go beyond being mathematical exercises by inclusion of technical background or by asking for opinions.

Notation is a vital adjunct to thinking and I am happy to report that the *sinc function*, which we learned from P. M. Woodward's book, is alive and well and surviving erosion by occasional authors who do not know that "sine x over x" is not the sinc function. The unit rectangle function (unit height and width) $\Pi(x)$, the transform of the sinc function, has also proved extremely useful, especially for blackboard work. In typescript or other circumstances where the Greek letter is less desirable, $\Pi(x)$ may be written "rect x," and it is convenient in any case to pronounce it *rect*. The shah function $\text{III}(x)$ has caught on. It is easy to type and is twice as useful as you might think because it is its own transform. The asterisk for convolution, which was in use a long time ago by Volterra and perhaps earlier, is now in wide use and I recommend $**$ to denote two-dimensional convolution, which has come into extensive use as a result of the explosive growth of image processing.

Early emphasis on convolution in a text on Fourier transforms turned out to be exactly the right way to go. Convolution has changed in a few years from being presented as a rather advanced concept to one that can easily be explained at an early stage as is fitting for an operation that describes all those systems that respond sinusoidally when you shake them sinusoidally.

Ronald N. Bracewell

Preface to the First Edition

Transform methods provide a unifying mathematical approach to the study of electrical networks, devices for energy conversion and control, antennas, and other components of electrical systems, as well as to complete linear systems whether electrical or not. These same methods apply equally to the subjects of electrical communication, radio propagation, and ionized media—which are concerned in the interconnection of electrical systems—and to information theory which, among other things, relates to the acquisition, processing, and presentation of data. Other theoretical techniques are used in handling these basic fields of electrical engineering, but transform methods are virtually indispensable in all of them.

A course on transforms and their applications has formed part of the electrical engineering curriculum at Stanford University for some years, and has been given with no prerequisites which the holder of a bachelor's degree does not normally possess. One objective has been to develop a pivotal course to be taken at an early stage by all graduates, so that in later, more specialized courses, the student would be spared encountering the same basic material over and over again, and the instructor would be able to proceed more directly to his special subject matter.

It is clearly not feasible to give the whole of linear mathematics in a single course, and the choice of material must necessarily remain a matter of judgment. The choice must, however, be sharply defined if later instructors are to rely on it.

An early-level course should be simple, but not trivial; the objective of this book is to simplify the presentation of many key topics that are ordinarily dealt with in advanced contexts by making use of suitable notation and an approach through convolution.

The organization of chapters is as follows:

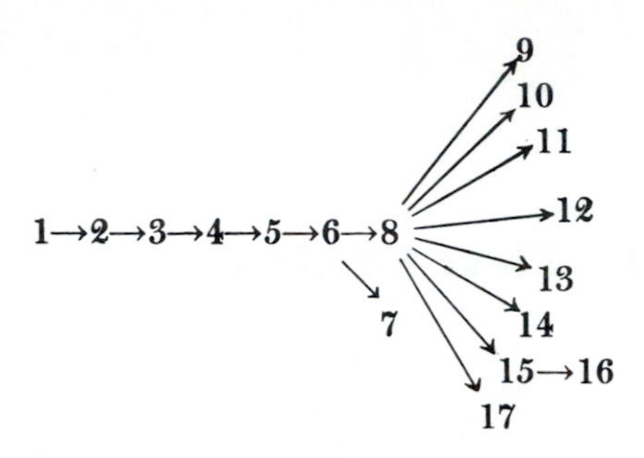

One way of working from the book is to take the chapters in numerical order. This sequence is feasible for students who can read the first half unassisted, or who can be taken through it rapidly in a few lectures, but if the material is approached at a more normal pace, then as a practical teaching matter, it is desirable to illustrate the theorems and concepts by dealing simultaneously with a physical topic, such as waveforms and their spectra (Chapter 9), for which the student already has some feeling.

The amount of material is suitable for one semester, or for one quarter, according to how many of the later chapters on applications are included.

Many fine mathematical texts on the Fourier transform have been published. This book differs in that it is intended for those who are concerned with applying Fourier transforms to physical situations rather than with furthering the mathematical subject as such. The connections of the Fourier transform with other transforms are also explored, and the text has been purposely enriched with condensed information that will suit it for use as a reference source for transform pairs and theorems pertaining to transforms.

My interest in the subject was fired when I was studying analysis from Carslaw's "Fourier Series and Integrals" at the University of Sydney in 1939; more recently I have applied transform methods to various problems arising in connection with directive antennas, a subject that is touched on only briefly in this book, but which may be followed up by reference to the "Encyclopedia of Physics" (vol. 54, S. Flügge, ed., Springer-Verlag, Berlin, 1962). In solving these problems I benefited from the physical approach to the Fourier transformation that I learned from J. A. Ratcliffe at the Cavendish Laboratory, Cambridge.

Ronald N. Bracewell

Chapter 1 Introduction

Linear transforms, especially those named for Fourier and Laplace, are well known as providing techniques for solving problems in linear systems. Characteristically one uses the transformation as a mathematical or physical tool to alter the problem into one that can be solved. This book is intended as a guide to the understanding and use of transform methods in dealing with linear systems.

The subject is approached through the Fourier transform. Hence, when the more general Laplace transform is discussed later, many of its properties will already be familiar and will not distract from the new and essential question of the strip of convergence on the complex plane. In fact, all the other transforms discussed here are greatly illuminated by an approach through the Fourier transform.

Fourier transforms play an important part in the theory of many branches of science. While they may be regarded as purely mathematical functionals, as is customary in the treatment of other transforms, they also assume in many fields just as definite a physical meaning as the functions from which they stem. A waveform—optical, electrical, or acoustical—and its spectrum are appreciated equally as physically picturable and measurable entities: an oscilloscope enables us to see an electrical waveform, and a spectroscope or spectrum analyzer enables us to see optical or electrical spectra. Our acoustical appreciation is even more direct, since the ear hears spectra. Waveforms and spectra are Fourier transforms of each other; the Fourier transformation is thus an eminently physical relationship.

The number of fields in which Fourier transforms appear is surprising. It is a common experience to encounter a concept familiar from one branch of study in a slightly different guise in another. For example, the principle of the phase-contrast microscope is reminiscent of the circuit for detecting frequency modulation, and the explanation of both is conveniently given in terms of transforms along the same lines. Or a problem in statistics may yield to an approach which is familiar from studies of cascaded amplifiers. This is simply a case of the one underlying theorem from Fourier theory assuming different physical embodiments.

It is a great advantage to be able to move from one physical field to another and to carry over the experience already gained, but it is necessary to have the key which interprets the terminology of the new field. It will be evident, from the rich variety of topics coming within its scope, what a pervasive and versatile tool Fourier theory is.

Many scientists know Fourier theory not in terms of mathematics, but as a set of propositions about physical phenomena. Often the physical counterpart of a theorem is a physically obvious fact, and this allows the scientist to be abreast of matters which in the mathematical theory may be quite abstruse. Emphasis on the physical interpretation enables us to deal in an elementary manner with topics which in the normal course of events would be considered advanced.

Although the Fourier transform is vital in so many fields, it is often encountered in formal mathematics courses in the last lecture of a formidable course on Fourier series. As need arises it is introduced ad hoc in later graduate courses but may never develop into a usable tool. If this traditional order of presentation is reversed, the Fourier series then falls into place as an extreme case within the framework of Fourier transform theory, and the special mathematical difficulties with the series are seen to be associated with their extreme nature—which is nonphysical; the handicap imposed on the study of Fourier transforms by the customary approach is thus relieved.

The great generality of Fourier transform methods strongly qualifies the subject for introduction at an early stage, and experience shows that it is quite possible to teach, at this stage, the distilled theorems, which in their diverse applications offer such powerful tools for thinking out physical problems.

The present work began as a pictorial guide to Fourier transforms to complement the standard lists of pairs of transforms expressed mathematically. It quickly became apparent that the commentary would far outweigh the pictorial list in value, but the pictorial dictionary of transforms is nevertheless important, for a study of the entries reinforces the intuition, and many valuable and common types of function are included which, because of their awkwardness when expressed algebraically, do not occur in other lists.

A contribution has been made to the handling of simple but awkward functions by the introduction of compact notation for a few basic functions which are usually defined piecewise. For example, the rectangular pulse, which is at least as simple as a Gaussian pulse, is given the name $\Pi(x)$, which means that it can be handled as a simple function. The picturesque term "gate function," which is in use in electronics, suggests how a gating waveform $\Pi(t)$ opens a valve to let through a segment of a waveform. This is the way we think of the rectangle function mathematically when we use it as a multiplying factor.

Among the special symbols introduced or borrowed are $\text{II}(x)$, the even impulse pair, and $\text{III}(x)$ (pronounced *shah*), the infinite impulse train defined by $\text{III}(x) = \Sigma\, \delta(x - n)$. The first of these two gains importance from its status as the Fourier transform of the cosine function; the second proves indispensable in discussing both regular sampling or tabulation (operations which are equivalent to multiplication by *shah*) and periodic functions (which are expressible as convolutions with *shah*). Since *shah* proves to be its own Fourier transform, it turns out to be twice as useful an entity as might have been expected. Much freedom of expression is gained by the use of these conventions of notation, especially in conjunction with the asterisk notation for convolution. Only a small step is involved in writing $\Pi(x) * f(x)$, or simply $\Pi * f$, instead of

$$\int_{x-\frac{1}{2}}^{x+\frac{1}{2}} f(u)\, du$$

(for example, for the response to a photographic density distribution $f(x)$ on a sound track scanned with a slit). But the disappearance of the dummy variable and the integral sign with limits, and the emergence of the character of the response as a convolution between two profiles Π and f, lead to worthwhile convenience in both algebraic and mental manipulation.

Convolution is used a lot here. Experience shows that it is a fairly tricky concept when it is presented bluntly under its integral definition, but it becomes easy if the concept of a functional is first understood. Numerical practice on serial products confirms the feeling for convolution and incidentally draws attention to the practical character of numerical evaluation: for numerical purposes one normally prefers to have the answer to a problem come out as the convolution of two functions rather than as a Fourier transform.

This is a good place to mention that transform methods do not necessarily involve taking transforms numerically. On the contrary, some of the best methods for handling linear problems do not involve application of the Fourier or Laplace transform to the data at all; but the basis for such methods is often clarified by appeal to the transform domain. Thinking in terms of transforms, we may show how to avoid numerical harmonic analysis or the handling of data on the complex plane.

It is well known that the response of a system to harmonic input is itself harmonic, at the same frequency, under two conditions: linearity and time invariance of the system properties. These conditions are, of course, often met. This is why Fourier analysis is important, why one specifies an amplifier by its *frequency* response, why harmonic variation is ubiquitous. When the conditions for harmonic response to harmonic stimulus break down, as they do in a nonlinear servomechanism, analysis of a stimulus into harmonic components must be reconsidered. Time invariance can often be counted on even when linearity fails, but *space* invariance is by no means as common. Failure of this condition is the reason that bridge deflections are not studied by analyzing the load distribution into sinusoidal components (space harmonics).

The two conditions for harmonic response to harmonic stimulus can be restated as *one* condition: that the response shall be relatable to the stimulus by *convolution*. For work in Fourier analysis, convolution is consequently profoundly important, and such a pervasive phenomenon as convolution does not lack for familiar examples. Good ones are the relation between the distribution on a film sound track or recorder tape and the electrical signal read out through the scanning slit or magnetic head.

Many topics normally considered abstruse or advanced are presented here, and simplification of their presentation is accomplished by minor conveniences of notation and by the use of graphs.

Special care has been given to the presentation and use of the impulse symbol $\delta(x)$, on which, for example, both $\text{II}(x)$ and $\text{III}(x)$ depend. The term "impulse symbol" focuses attention on the status of $\delta(x)$ as something which is not a function; equations or expressions containing it then have to have an interpretation, and this is given in an elementary fashion by recourse to a sequence of pulses (not impulses). The expression containing the impulse symbol thus acquires meaning as a limit which, in many instances, "exists." This commonplace mathematical status of the *complete expression*, in contrast with that of $\delta(x)$ itself, directly reflects the physical situation, where Green's functions, impulse responses, and the like are often accurately producible or observable, within the limits permitted by the resolving power of the measuring equipment, while impulses themselves are fictitious. By deeming all expressions containing $\delta(x)$ to be subject to special rules, we can retain both rigor and the direct procedures of manipulating $\delta(x)$ which have been so successful.

Familiar examples of this physical situation are the moment produced by a point mass resting on a beam and the electric field of a point charge. In the physical approach to such matters, which have long been imbedded in physics, one thinks about the effect produced by smaller and smaller but denser and denser massive objects or charged volumes, and notes whether the effect produced approaches a definite limit. Ways of representing this mathematically have been tidied up to the satisfaction of mathematicians only in recent years, and use of $\delta(x)$ is now licensed if accompanied by an appropriate footnote reference to this recent literature, although in some conservative fields such as statistics $\delta(x)$ is still occasionally avoided in favor of distinctly more awkward Stieltjes integral notation. These developments in mathematical ideas are mentioned in Chapters 5 and 6, the presentation in terms of "generalized functions" following Temple and Lighthill being preferred. The validity of the original physical ideas remains unaffected, and one ought to be able, for example, to discuss the moment of a point mass on a beam by considering those rectangular distributions of pressure that result from restacking the load uniformly on an ever-narrower base to an ever-increasing height; it should not be necessary to limit attention to pressure distributions possessing an infinite number of continuous derivatives merely because in some other problem a derivative of high order is involved. Therefore the subject of impulses is introduced with the aid of the rather simple mathematics of rectangle functions, and in the relatively few cases where the first derivative is wanted, triangle functions are used instead.

Chapter 2 Groundwork

Most of the material in this chapter is stated without proof. This is done because the proofs entail discussions that are lengthy (in fact, they form the bulk of conventional studies in Fourier theory) and remote from the subject matter of the present work.

Omitting the proofs enables us to take the transform formulas and their known conditions as our point of departure. Since suitable notation is an important part of the work, it too is set out in this chapter.

The Fourier transform and Fourier's integral theorem

The Fourier transform of $f(x)$ is defined as

$$\int_{-\infty}^{\infty} f(x)e^{-i2\pi xs}\,dx.$$

This integral, which is a function of s, may be written $F(s)$. Transforming $F(s)$ by the same formula, we have

$$\int_{-\infty}^{\infty} F(s)e^{-i2\pi ws}\,ds.$$

When $f(x)$ is an even function of x, that is, when $f(x) = f(-x)$, the repeated transformation yields $f(w)$, the same function we began with. This is the cyclical property of the Fourier transformation, and since the cycle is of two steps, the reciprocal property is implied: if $F(s)$ is the Fourier transform of $f(x)$, then $f(x)$ is the Fourier transform of $F(s)$.

The cyclical and reciprocal properties are imperfect, however, because when $f(x)$ is odd—that is, when $f(x) = -f(-x)$—the repeated trans ormation yields $f(-w)$. In ge ieral, whether $f(x)$ is even or odd or neither, repeated transformation yields $f(-w)$.

The customary formulas exhibiting the reversibility of the Fourier transformation are

$$F(s) = \int_{-\infty}^{\infty} f(x)e^{-i2\pi xs}\,dx$$
$$f(x) = \int_{-\infty}^{\infty} F(s)e^{i2\pi xs}\,ds.$$

In this form, two successive transformations are made to yield the original function. The second transformation, however, is not exactly the same as the first, and where it is necessary to distinguish between these two sorts of Fourier transform, we shall say that $F(s)$ is the minus-i transform of $f(x)$ and that $f(x)$ is the plus-i transform of $F(s)$.

Writing the two successive transformations as a repeated integral, we obtain the usual statement of Fourier's integral theorem:

$$f(x) = \int_{-\infty}^{\infty} \left[\int_{-\infty}^{\infty} f(x)e^{-i2\pi xs}\,dx \right] e^{i2\pi xs}\,ds.$$

The conditions under which this is true are given in the next section, but it must be stated at once that where $f(x)$ is discontinuous the left-hand side should be replaced by $\frac{1}{2}[f(x+) + f(x-)]$, that is, by the mean of the unequal limits of $f(x)$ as x is approached from above and below.

The factor 2π appearing in the transform formulas may be lumped with s to yield the following version (system 2):

$$F(s) = \int_{-\infty}^{\infty} f(x)e^{-ixs}\,dx$$
$$f(x) = \frac{1}{2\pi} \int_{-\infty}^{\infty} F(s)e^{ixs}\,ds.$$

And for the sake of symmetry, authors occasionally write (system 3):

$$F(s) = \frac{1}{(2\pi)^{\frac{1}{2}}} \int_{-\infty}^{\infty} f(x)e^{-ixs}\,dx$$
$$f(x) = \frac{1}{(2\pi)^{\frac{1}{2}}} \int_{-\infty}^{\infty} F(s)e^{ixs}\,ds.$$

All three versions are in common use, but here we shall keep the 2π in the exponent (system 1). If $f(x)$ and $F(s)$ are a transform pair in system 1, then $f(x)$ and $F(s/2\pi)$ are a transform pair in system 2, and $f[x/(2\pi)^{\frac{1}{2}}]$ and $F[s/(2\pi)^{\frac{1}{2}}]$ are a transform pair in system 3. An example of a transform pair in each of the three systems follows.

System 1		*System 2*		*System 3*	
$f(x)$	$F(s)$	$f(x)$	$F(s)$	$f(x)$	$F(s)$
$e^{-\pi x^2}$	$e^{-\pi s^2}$	$e^{-\pi x^2}$	$e^{-s^2/4\pi}$	$e^{-\frac{1}{2}x^2}$	$e^{-\frac{1}{2}s^2}$

An excellent notation which may be used as an alternative to $F(s)$ is $\bar{f}(s)$. Various advantages and disadvantages are found in both notations.

The bar notation leads to compact expression, including some convenient manuscript or blackboard forms which are not very suitable for typesetting. Consider, for example, these versions of the convolution theorem,

$$\overline{FG} = \bar{F} * \bar{G}$$
$$\overline{F * G} = \bar{F}\bar{G}$$
$$\overline{\bar{F}\bar{G}} = F * G$$
$$\overline{\bar{F} * \bar{G}} = FG,$$

which display shades of distinction not expressible at all in the capital notation. See Chapter 6 for illustrations of the freedom of expression permitted by the bar notation. A certain awkwardness sets in, however, when complex conjugates or primes representing derivatives have to be handled; this awkwardness does not afflict the capital notation. Therefore we have departed from the usual custom of adopting a single notation. In the early mathematical sections, where f and g are nearly the only symbols for functions, the capital notation is mainly used. In the physical sections, preemption of capitals such as E and H for the representation of physical quantities leads more naturally to bars.

Neither of the above two notations lends itself to symbolic statements equivalent to "the Fourier transform of exp $(-\pi x^2)$ is exp $(-\pi s^2)$"; however, we can write

$$\mathfrak{F}e^{-\pi x^2} = e^{-\pi s^2}$$

or

$$e^{-\pi x^2} \supset e^{-\pi s^2}.$$

In the first of these, $\mathfrak{F}$ can be regarded as a functional operator which converts a function into its transform. It may be applied wherever the bar and capital notations are employed, but will be found most appropriate in connection with specific functions such as those above. It lends itself to the use of affixes, (for example, $\mathfrak{F}_s$, $\mathfrak{F}_c$, ${}^2\mathfrak{F}$) and mixes with symbols for other transforms. It can also distinguish between the minus-i and plus-i transforms through the use of $\mathfrak{F}^{-1}$ for the inverse of $\mathfrak{F}$; or like the bar notation, but not the capital notation, it can remain discreetly silent. The properties of this notation make it indispensable, and it is adopted in suitable places in the sequel.

The sign $\supset$ or equivalents, several of which are in use, is not as versatile as $\mathfrak{F}$ but is useful for algebraic work. It is also often used to denote the Laplace transform.

Conditions for the existence of Fourier transforms

A circuit expert finds it obvious that every waveform has a spectrum, and the antenna designer is confident that every antenna has a radiation pattern. It sometimes comes as a surprise to those whose acquaintance with Fourier transforms is through physical experience rather than mathematics that there are some functions without Fourier transforms. Nevertheless, we may be confident that no one can generate a waveform without a spectrum or construct an antenna without a radiation pattern.

The question of the existence of transforms may safely be ignored when the function to be transformed is an accurately specified description of a physical quantity. Physical possibility is a valid sufficient condition for the existence of a transform. Sometimes, however, it is convenient to substitute a simple mathematical expression for a physical quantity. It is very common, for example, to consider the waveforms

$\sin t$ (harmonic wave, pure alternating current)
$H(t)$ (step)
$\delta(t)$ (impulse).

It turns out that none of these three has, strictly speaking, a Fourier transform. Of course, none of them is physically possible, for a waveform $\sin t$ would have to have been switched on an infinite time ago, a step $H(t)$ would have to be maintained steady for an infinite time, and an impulse $\delta(t)$ would have to be infinitely large for an infinitely short time. However, in a given situation we can often achieve an approximation so close that any further improvement would be immaterial, and we use the simple mathematical expressions because they are less cumbersome than various slightly different but realizable functions. Nevertheless, the above functions do not have Fourier transforms; that is, the Fourier integral does not converge for all s. It is therefore of practical importance to consider the conditions for the existence of transforms.

Transforming and retransforming a single-valued function $f(x)$, we have the repeated integral

$$\int_{-\infty}^{\infty} e^{i2\pi sx} \left[\int_{-\infty}^{\infty} f(x) e^{-i2\pi sx}\, dx \right] ds.$$

This expression is equal to $f(x)$ (or to $\frac{1}{2}[f(x+) + f(x-)]$ where $f(x)$ is discontinuous), provided that

1. The integral of $|f(x)|$ from $-\infty$ to ∞ exists
2. Any discontinuities in $f(x)$ are finite

A further but less important condition is mentioned below. In physical circumstances these conditions are violated[1] when there is infinite energy, and a kind of duality between the two conditions is often noted. For instance, absolutely steady direct current which has always been

[1] Exceptions are provided by finite-energy waveforms such as $(1 + |x|)^{-\frac{3}{4}}$ and $x^{-1} \sin x$, which nevertheless do not have absolutely convergent infinite integrals.

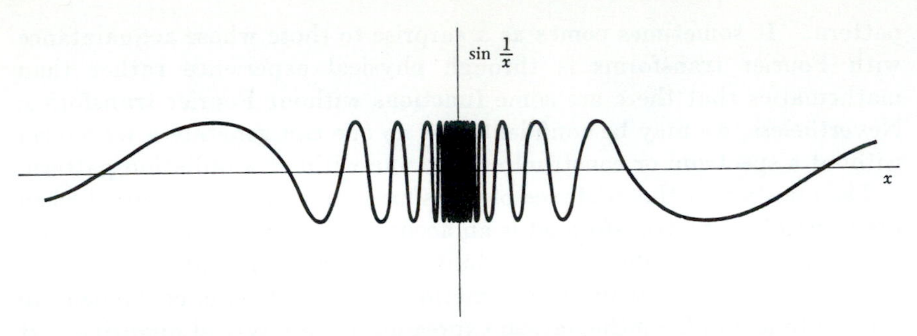

Fig. 2.1 A function with an infinite number of maxima.

flowing and always will flow represents infinite energy and violates the first condition. The distribution of energy with frequency would have to show infinite energy concentrated entirely at zero frequency and would violate the second condition. The same applies to harmonic waves.

It is sometimes stated that an infinite number of maxima and minima in a finite interval disqualifies a function from possessing a Fourier transform, the stock example being $\sin x^{-1}$ (see Fig. 2.1), which oscillates with ever-increasing frequency as x approaches zero. This kind of behavior is not important in real life, even as an approximation. We therefore record for general interest that some functions with infinite numbers of maxima and minima in a finite interval do have transforms. This is allowed for when the further condition is given as *bounded variation.*[2] Again, however, there are transformable functions with an infinite number of maxima and with unbounded variation in a finite interval, a circumstance which may be covered by requiring $f(x)$ to satisfy a Lipschitz condition.[3] Then there is a more relaxed condition used by Dini. This is a fascinating topic in Fourier theory, but it is not immediately relevant to our branch of it, which is physical applications. Furthermore, we by no means propose to abandon useful functions which do not possess Fourier transforms in the ordinary sense. On the contrary, we include them equally, by means of a generalization to Fourier transforms in the

[2] A function $f(x)$ has bounded variation over the interval $x = a$ to $x = b$ if there is a number M such that

$$|f(x_1) - f(a)| + |f(x_2) - f(x_1)| + \ . \ . \ . + |f(b) - f(x_{n-1})| \leqslant M$$

for every method of subdivision $a < x_1 < x_2 < \ . \ . \ . < x_{n-1} < b$. Any function having an absolutely integrable derivative will have bounded variation.

[3] A function $f(x)$ satisfies a Lipschitz condition of order α at $x = 0$ if

$$|f(h) - f(0)| \leqslant B|h|^\beta$$

for all $|h| < \epsilon$, where B and β are independent of h, β is positive, and α is the upper bound of all β for which finite B exists.

limit. Conditions for the existence of Fourier transforms now merely distinguish, where distinction is desired, between those transforms which are ordinary and those which are transforms in the limit.

Transforms in the limit

Although a periodic function does not have a Fourier transform, as may be verified by reference to the conditions for existence, it is nevertheless considered in physics to have a spectrum, a "line spectrum." The line spectrum may be identified with the coefficients of the Fourier series for the periodic function, or we may broaden the mathematical concept of the Fourier transform to bring it into harmony with the physical standpoint. This is what we shall do here, taking the periodic function as one example among others which we would like Fourier transform theory to embrace.

Let $P(x)$ be a periodic function of x. Then

$$\int_{-\infty}^{\infty} |P(x)|\, dx$$

does not exist, but if we modify $P(x)$ slightly by multiplication with a factor such as $\exp(-\alpha x^2)$, where α is a small positive number, then the modified version may have a transform, for

$$\int_{-\infty}^{\infty} |e^{-\alpha x^2} P(x)|\, dx$$

may exist. Of course, any infinite discontinuities in $P(x)$ will still disqualify it, but let us select $P(x)$ so that $\exp(-\alpha x^2)P(x)$ possesses a Fourier transform. Then as α approaches zero, the modifying factor for each value of x approaches unity, and the modified functions of the sequence generated as α approaches zero thus approach $P(x)$ in the limit. Since each modified function possesses a transform, a corresponding sequence of transforms is generated; now, as α approaches zero, does this sequence of transforms also approach a limit? We already know that it does not, at least not for all s; we content ourselves with saying that the sequence of regular transforms defines or constitutes an entity that shall be called a generalized function. The periodic function and the generalized function form a Fourier transform pair in the limit.

The idea of dealing with things that are not functions but are describable in terms of sequences of functions is well established in physics in connection with the impulse symbol $\delta(x)$. In this case a progression of ever-stronger and ever-narrower unit-area pulses is an appropriate sequence, and a little later in the chapter we go into this idea more fully. We use the term "generalized function" to cover impulses and their like.

Periodic functions fail to have Fourier transforms because their infinite integral is not absolutely convergent; failure may also be due to the infinite discontinuities associated with impulses. In this case we replace any impulse by a sequence of functions that do have transforms; then the sequence of corresponding transforms may approach a limit, and again we have a Fourier transform pair in the limit. As before, only one member of the pair is a generalized function involving impulses.

It may also happen that the sequence of transforms does not approach a limit. This would be so if we began with something that was both impulsive and periodic; then the members of the transform pair in the limit would both be generalized functions involving impulses.

At this point we might proceed to lay the groundwork leading to the definition of a generalized function. Instead we defer the rather severe

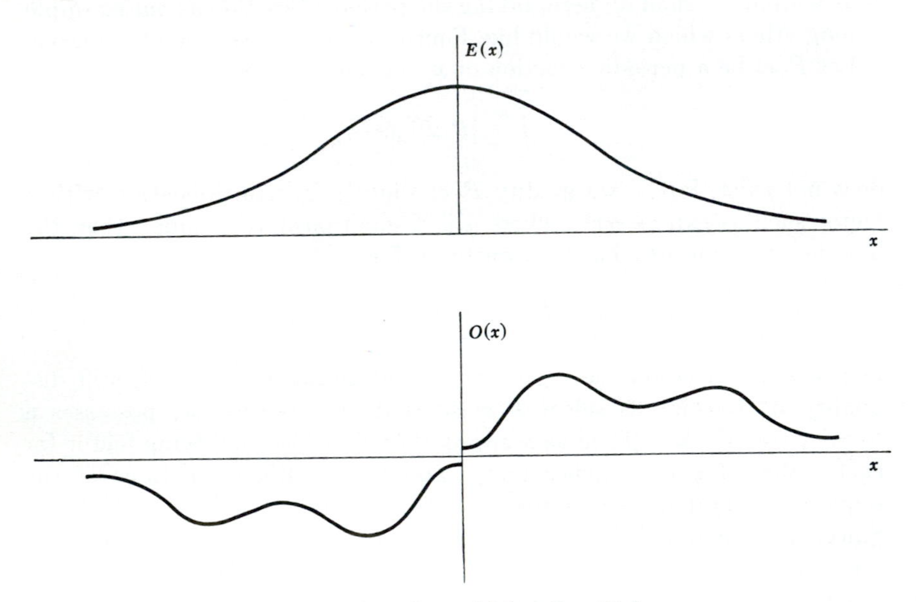

Fig. 2.2 An even function $E(x)$ and an odd function $O(x)$.

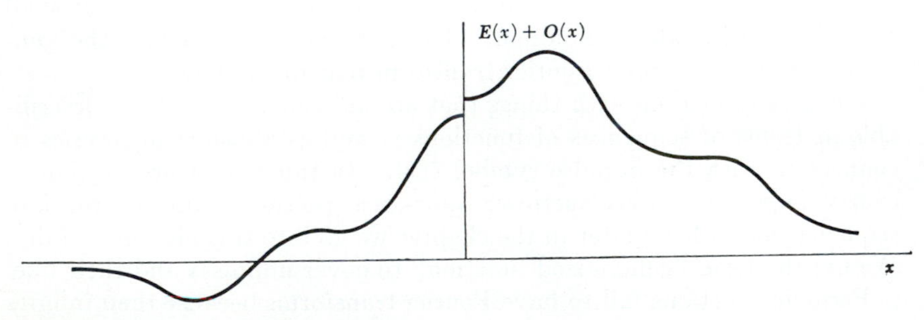

Fig. 2.3 The sum of $E(x)$ and $O(x)$.

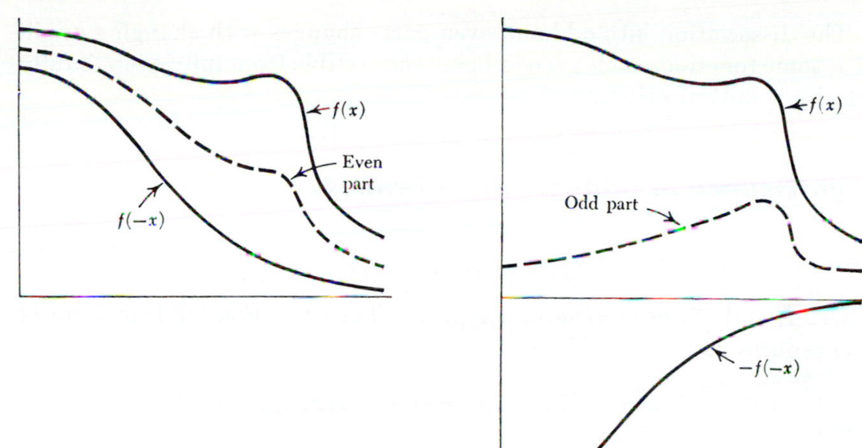

Fig. 2.4 Constructions for the even and odd parts of a given function $f(x)$.

general discussion to a much later stage, since it can be read with more profit after facility has been acquired in handling the impulse symbol $\delta(x)$.

Oddness and evenness

Symmetry properties play an important role in Fourier theory. Arguments from symmetry to show directly that certain integrals vanish, without the need of evaluating them, are familiar and perhaps often seem trivial in print. More alertness is needed, however, to ensure full exploitation in one's own reasoning of symmetry restrictions and the corresponding restrictive properties generated under Fourier transformation. Some simple terminology is recalled here.

A function $E(x)$ such that $E(-x) = E(x)$ is a symmetrical, or even, function. A function $O(x)$ such that $O(-x) = -O(x)$ is an antisymmetrical, or odd, function (see Fig. 2.2). The sum of even and odd functions is in general neither even nor odd, as illustrated in Fig. 2.3, which shows the sum of the previously chosen examples.

Any function $f(x)$ can be split unambiguously into odd and even parts. For if $f(x) = E_1(x) + O_1(x) = E_2(x) + O_2(x)$, then $E_1 - E_2 = O_2 - O_1$; but $E_1 - E_2$ is even and $O_2 - O_1$ odd, hence $E_1 - E_2$ must be zero.

The even part of a given function is the mean of the function and its reflection in the vertical axis, and the odd part is the mean of the function and its negative reflection (see Fig. 2.4). Thus

$$E(x) = \tfrac{1}{2}[f(x) + f(-x)]$$

and

$$O(x) = \tfrac{1}{2}[f(x) - f(-x)].$$

The dissociation into odd and even parts changes with changing origin of x, some functions such as $\cos x$ being convertible from fully even to fully odd by a shift of origin.

Significance of oddness and evenness

Let

$$f(x) = E(x) + O(x),$$

where E and O are in general complex. Then the Fourier transform of $f(x)$ reduces to

$$2\int_0^\infty E(x)\cos(2\pi xs)\,dx - 2i\int_0^\infty O(x)\sin(2\pi xs)\,dx.$$

It follows that if a function is even, its transform is even, and if it is odd, its transform is odd. Full results are

Real and even	Real and even
Real and odd	Imaginary and odd
Imaginary and even	Imaginary and even
Complex and even	Complex and even
Complex and odd	Complex and odd
Real and asymmetrical	Complex and hermitian
Imaginary and asymmetrical	Complex and antihermitian
Real even plus imaginary odd	Real
Real odd plus imaginary even	Imaginary
Even	Even
Odd	Odd

These properties are summarized in the following diagram:

$$f(x) = o(x) + e(x) = \mathrm{Re}\ o(x) + i\ \mathrm{Im}\ o(x) + \mathrm{Re}\ e(x) + i\ \mathrm{Im}\ e(x)$$

$$F(s) = O(s) + E(s) = \mathrm{Re}\ O(s) + i\ \mathrm{Im}\ O(s) + \mathrm{Re}\ E(s) + i\ \mathrm{Im}\ E(s).$$

Figure 2.5, which records the phenomena in another way, is also valuable for revealing at a glance the "relative sense of oddness": when $f(x)$ is real and odd with a *positive* moment, the odd part of $F(s)$ has i times a *negative* moment; and when $f(x)$ is real but not necessarily odd, we also find opposite senses of oddness. However, inverting the procedure—that is, going from $F(s)$ to $f(x)$, or taking $f(x)$ to be imaginary—produces the same sense of oddness.

Real even functions play a special part in this work because both they and their transforms may easily be graphed. Imaginary odd, real odd, and imaginary even functions are also important in this respect.

Another special kind of symmetry is possessed by a function $f(x)$ whose

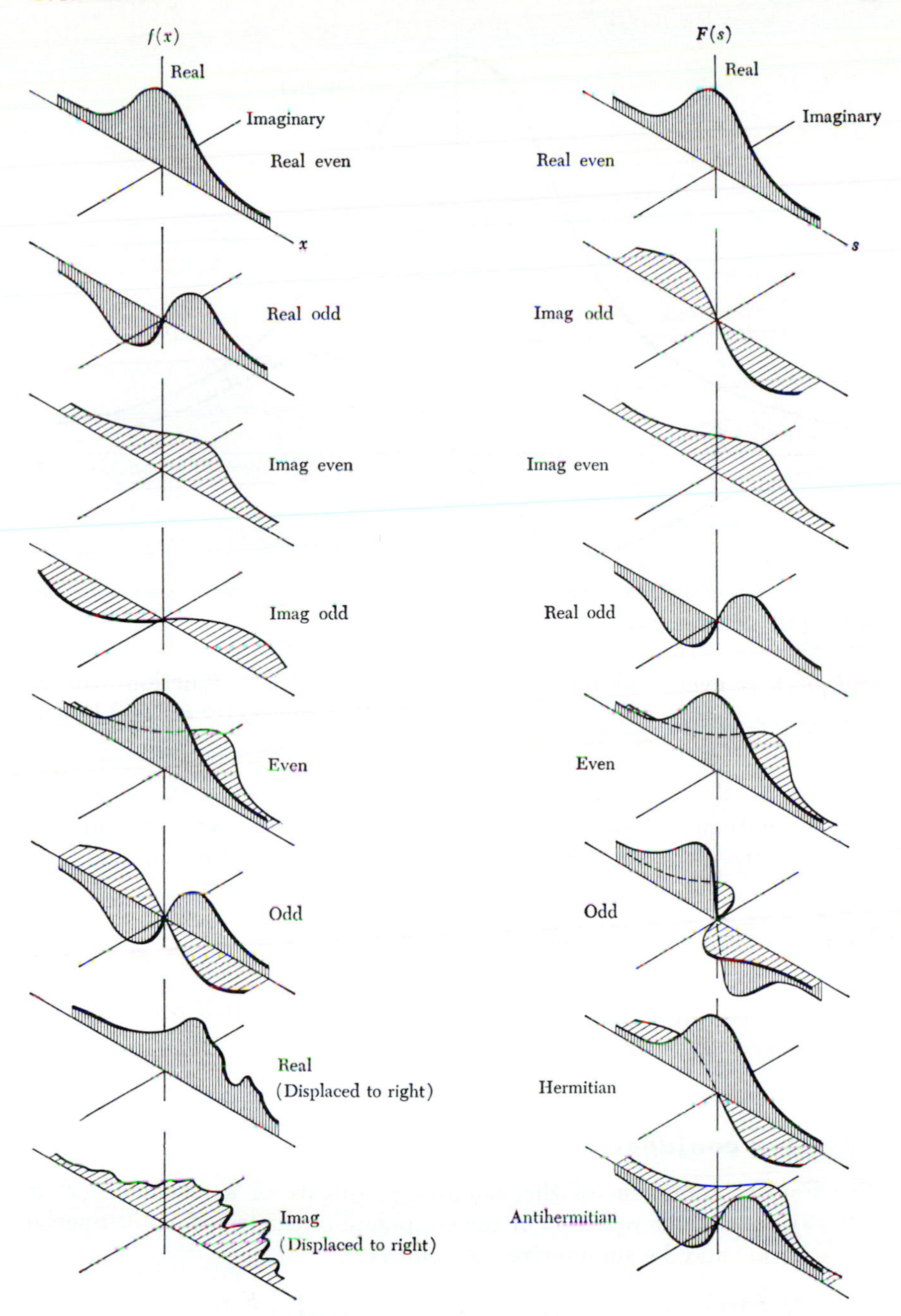

Fig. 2.5 Symmetry properties of a function and its Fourier transform.

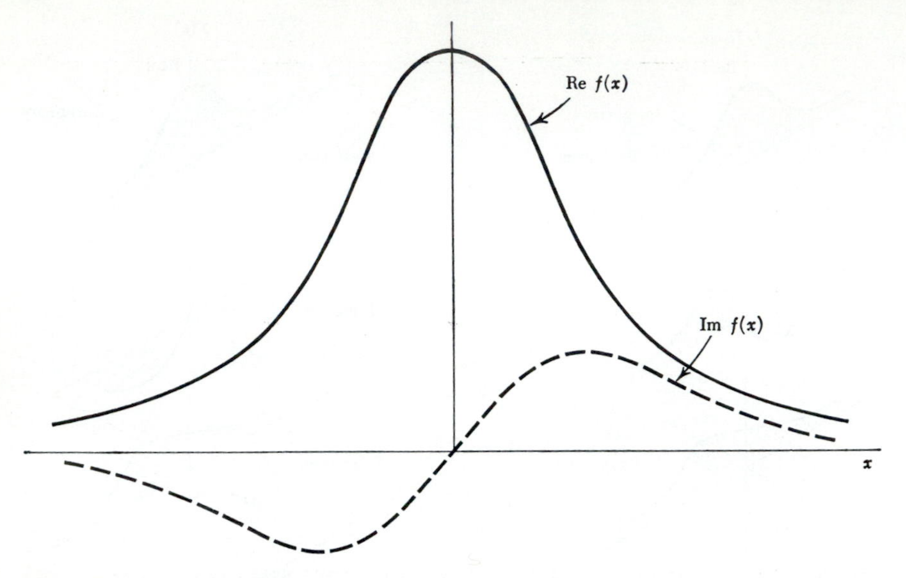

Fig. 2.6 Hermitian functions have their real part even and their imaginary part odd. Their Fourier transform is pure real.

real part is even and imaginary part odd. Such a function will be described as hermitian (see Fig. 2.6); it is often succinctly defined by the property

$$f(x) = f^*(-x),$$

and as mentioned above its Fourier transform is real. As an example of algebraic procedure for handling matters of this kind, consider that

$$f(x) = E + O + i\hat{E} + i\hat{O}.$$

Then
$$f(-x) = E - O + i\hat{E} - i\hat{O}$$
and
$$f^*(-x) = E - O - i\hat{E} + i\hat{O}.$$

If we now require that $f(x) = f^*(-x)$ we must have $O = 0$ and $\hat{E} = 0$. Hence $f(x) = E + i\hat{O}$.

Complex conjugates

The Fourier transform of the complex conjugate of a function $f(x)$ is $F^*(-s)$, that is, the *reflection* of the conjugate of the transform. Special cases of this may be summarized as follows:

$$\text{If } f(x) \text{ is } \begin{cases} \text{real} \\ \text{imaginary} \\ \text{even} \\ \text{odd} \end{cases} \text{ the transform of } f^*(x) \text{ is } \begin{cases} F(s) \\ -F(s) \\ F^*(s) \\ -F^*(s) \end{cases} = F^*(-s).$$

Related statements are tabulated for reference.

$$
\begin{aligned}
f(x) &\supset F(s)\\
f^*(x) &\supset F^*(-s)\\
f^*(-x) &\supset F^*(s)\\
f(-x) &\supset F(-s)\\
2\,\mathrm{Re}\, f(x) &\supset F(s) + F^*(-s)\\
2\,\mathrm{Im}\, f(x) &\supset F(s) - F^*(-s)\\
f(x) + f^*(-x) &\supset 2\,\mathrm{Re}\, F(s)\\
f(x) - f^*(-x) &\supset 2\,\mathrm{Im}\, F(s)
\end{aligned}
$$

Cosine and sine transforms

The cosine transform of a function $f(x)$ is defined as

$$2 \int_0^\infty f(x) \cos 2\pi sx \, dx.$$

The cosine transform is the same as the Fourier transform if $f(x)$ is an even function. In general the even part of the Fourier transform of $f(x)$ is the cosine transform of the even part of $f(x)$.

It will be noted that the cosine transform, as defined, takes no account of $f(x)$ to the left of the origin.

Let $F_c(s)$ represent the cosine transform of $f(x)$. Then the cosine transformation and the reverse transformation by which $f(x)$ is obtained from $F_c(s)$ are identical. Thus

$$
\begin{aligned}
F_c(s) &= 2 \int_0^\infty f(x) \cos 2\pi sx \, dx\\
f(x) &= 2 \int_0^\infty F_c(s) \cos 2\pi sx \, ds.
\end{aligned}
$$

The sine transform of $f(x)$ is defined by

$$F_s(s) = 2 \int_0^\infty f(x) \sin 2\pi sx \, dx.$$

This transformation is also identical with its reverse; thus

$$f(x) = 2 \int_0^\infty F_s(s) \sin 2\pi sx \, dx.$$

We may say that i times the odd part of the Fourier transform of $f(x)$ is the sine transform of the odd part of $f(x)$.

Combining the sine and cosine transforms of the even and odd parts leads to the Fourier transform of the whole of $f(x)$:

$$\mathfrak{F} f(x) = \mathfrak{F}_c E(x) - i \mathfrak{F}_s O(x),$$

where the operators $\mathfrak{F}$, $\mathfrak{F}_c$, and $\mathfrak{F}_s$ stand for the minus-i Fourier, cosine, and sine transformations, respectively. It will be clear that the terms on the right are the even and odd parts of $F(s)$, not the real and imaginary parts.

If $f(x)$ is zero to the left of the origin, then

$$F(s) = \tfrac{1}{2}F_c(s) - \tfrac{1}{2}iF_s(s),$$

or, to restate this property for any $f(x)$,

$$\tfrac{1}{2}F_c(s) - \tfrac{1}{2}iF_s(s) = \mathfrak{F}f(x)\ H(x),$$

where $H(x)$ is the unit step function (unity when x is positive and zero when x is negative).

In this text the terms "cosine transform" and "sine transform" will be avoided, but they are commonly used elsewhere. The definitions other than the above which may, however, be encountered are

$$\int_0^\infty f(x) \begin{matrix}\cos \\ \sin\end{matrix}\ sx\, dx$$

and

$$\left(\frac{2}{\pi}\right)^{\frac{1}{2}} \int_0^\infty f(x) \begin{matrix}\cos \\ \sin\end{matrix}\ sx\, dx.$$

Some advantage would accrue from the adoption of

$$\int_{-\infty}^{\infty} f(x) \cos 2\pi sx\, dx.$$

Interpretation of the formulas

Habitués in Fourier analysis undoubtedly are conscious of graphical interpretations of the Fourier integral. Since the integral contains a complex factor, probably the simpler cosine and sine versions are more often pictured. Thus, given $f(x)$, we picture $f(x) \cos 2\pi sx$ as an oscillation (see Fig. 2.7*a*), lying within the envelope $f(x)$ and $-f(x)$. Twice the

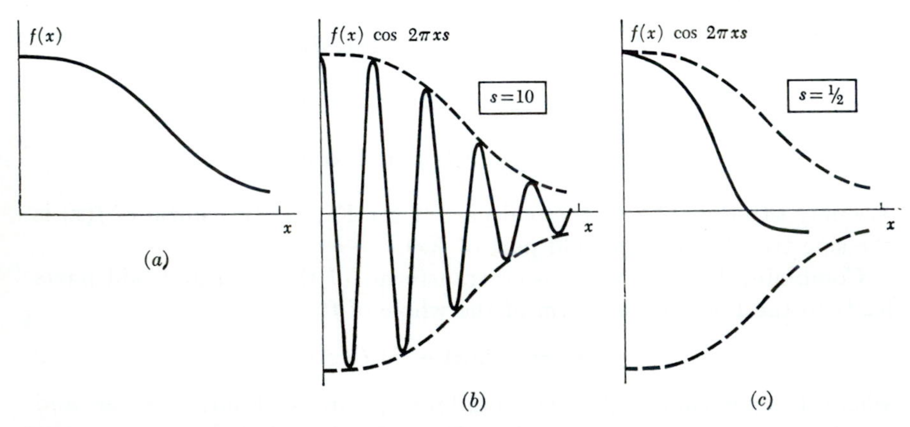

Fig. 2.7 The product of $f(x)$ with $\cos 2\pi xs$, as a function of x.

area under $f(x)$ cos $2\pi sx$ is then $F_c(s)$, for $F_c(s) = 2 \int_0^\infty f(x) \cos 2\pi sx \, dx$. In Fig. 2.7*b* this area is virtually zero, but a rather high value of s is implied. Figure 2.7*c* is for a low value of s.

The Fourier integral is thus visualized for discrete values of s. The

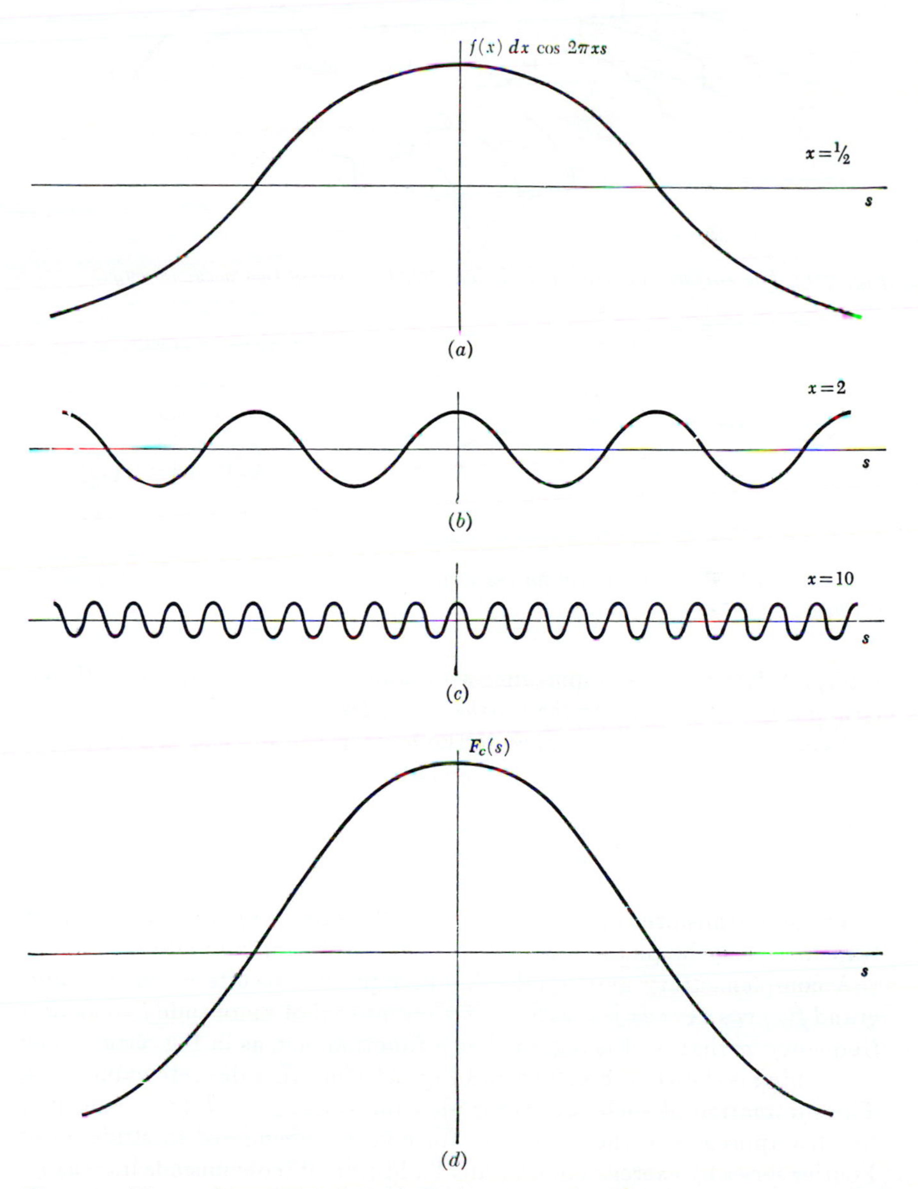

Fig. 2.8 The product of $f(x)$ dx with cos $2\pi xs$, as a function of s.

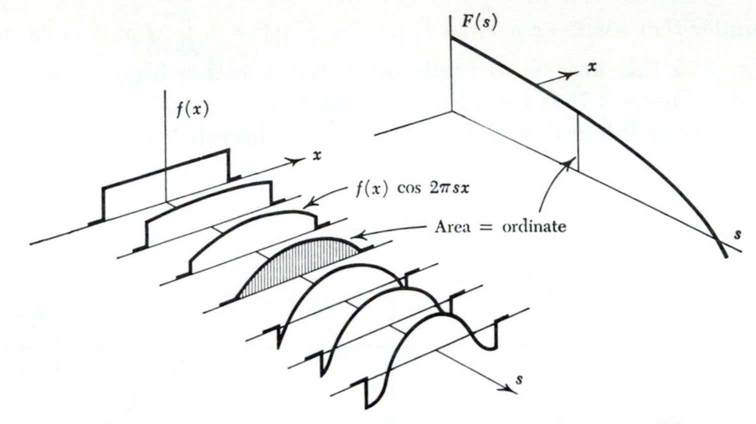

Fig. 2.9 The surface $f(x)$ cos $2\pi sx$ shown sliced in one of two possible ways.

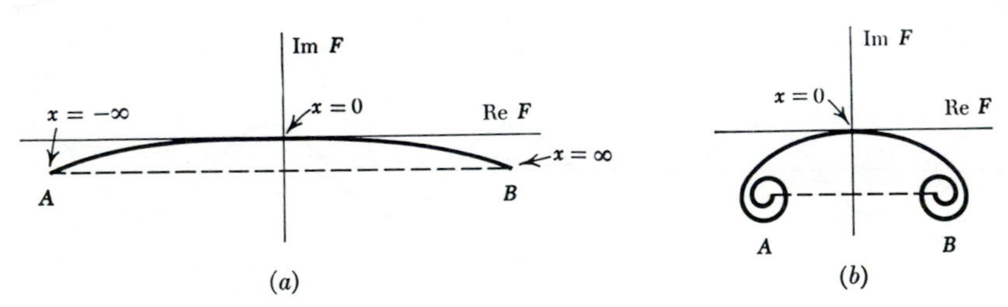

Fig. 2.10 The Fourier integral on the complex plane of $F(s)$ when s is (a) small, (b) large.

interpretation of s is important: s characterizes the frequency of the cosinusoid and is equal to *the number of cycles per unit of x.*

As an exercise in this approach to the matter, contemplate the graphical interpretation of the algebraic statements

$$F_c(s) \begin{cases} \to 0 & s \to \infty \\ = 2 \int_0^\infty f(x)\, dx & s = 0. \end{cases}$$

The sine transform may be pictured in the same way, and the complex transform may be pictured as a combination of even and odd parts.

A complementary and equally familiar picture results when the integrand $f(x) \cos 2\pi xs\, dx$ is regarded as a cosinusoid of amplitude $f(x)\, dx$ and frequency x; that is, it is regarded as a function of s, as in Fig. 2.8*a*. The same thing is shown in Fig. 2.8*b* and Fig. 2.8*c* for other discrete values of x. The summation of such curves for all values of x gives $F_c(s)$. A feeling for this approach to the transform formula is engendered in students of Fourier series by exercises in graphical addition of (co)sinusoids increasing arithmetically in frequency.

Each of the foregoing points of view has dual aspects, according as one ponders the analysis of a function into components or its synthesis from components. The curious fact that whether you analyze or synthesize you do the same thing simply reflects the reciprocal property of the Fourier transform. Figure 2.9 illustrates the first view of the matter. If we visualize the surface represented by the slices shown for particular values of s, and then imagine it to be sliced for particular values of x, we perceive the second view.

In a further point of view we think on the complex plane (see Fig. 2.10), taking s to be fixed. The vector $f(x)\,dx$ is rotated through an angle $2\pi sx$ by the factor $\exp(-i2\pi sx)$. As $x \to \pm\infty$, the integrand $f(x)\,dx \exp(-i2\pi sx)$ shrinks in amplitude and rotates indefinitely in angle, causing the integral to spiral into two limiting points, A and B. The vector AB represents the infinite integral $F(s)$ as in Fig. 2.10*a*. In Fig. 2.10*b* the more rapid coiling-up for a larger value of s is shown. The behavior of $F(s)$ as $s \to \infty$ and when $s = 0$ is readily perceived.

This kind of diagram, of which Cornu's spiral is an example, is familiar from optical diffraction. It is known as a useful tool both for qualitative thinking and for numerical work in optics. It arises in the propagation of radio waves, and neatly summarizes the behavior of radio echoes reflected from ionized meteor trails as they form. Probably it would be illuminating in fields where its use is not customary.

Problems

1 What condition must $F(s)$ satisfy in order that $f(x) \to 0$ as $x \to \pm\infty$?

2 Prove that $|F(s)|^2$ is an even function if $f(x)$ is real.

3 The Fourier transform in the limit of sgn x is $(i\pi s)^{-1}$. What conditions for the existence of Fourier transforms are violated by these two functions? (sgn x equals 1 when x is positive and -1 when x is negative.)

4 Show that all periodic functions violate a condition for the existence of a Fourier transform.

5 Verify that the function $\cos x$ violates one of the conditions for existence of a Fourier transform. Prove that $\exp(-\alpha x^2)\cos x$ meets this condition for any positive value of α.

6 Give the odd and even parts of $H(x)$, e^{ix}, $e^{-x}H(x)$, where $H(x)$ is unity for positive x and zero for negative x.

7 Graph the odd and even parts of $[1 + (x - 1)^2]^{-1}$.

8 Show that the even part of the product of two functions is equal to the product of the odd parts plus the product of the even parts.

9 Investigate the relationship of $\mathfrak{F}\mathfrak{F}f$ to f when f is neither even nor odd.

10 Show that $\mathfrak{F}\mathfrak{F}\mathfrak{F}\mathfrak{F}f = f$.

11 It is asserted that the odd part of log x is a constant. Could this be correct?

12 Is an odd function of an odd function an odd function? What can be said about odd functions of even functions and even functions of odd functions?

13 Prove that the Fourier transform of a real odd function is imaginary and odd. Does it matter whether the transform is the plus-i or minus-i type?

14 An antihermitian function is one for which $f(x) = -f^*(-x)$. Prove that its real part is odd and its imaginary part even and thus that its Fourier transform is imaginary.

15 Point out the fallacy in the following reasoning. "Let $f(x)$ be an odd function. Then the value of $f(-a)$ must be $-f(a)$; but this is not the same as $f(a)$. Therefore an odd function cannot be even."

16 Let the odd and even parts of a function $f(x)$ be $o(x)$ and $e(x)$. Show that, irrespective of shifts of the origin of x,

$$\int_{-\infty}^{\infty} |o(x)|^2\,dx + \int_{-\infty}^{\infty} |e(x)|^2\,dx = \text{const.}$$

17 Note that the odd and even parts into which a function is analyzed depend upon the choice of the origin of abscissas. Yet the sum of the integrals of the squares of the odd and even parts is a constant that is independent of the choice of origin. What is the constant?

18 Let axes of symmetry of a real function $f(x)$ be defined by values of a such that if o and e are the odd and even parts of $f(x - a)$, then

$$\frac{\int o^2\,dx - \int e^2\,dx}{\int o^2\,dx + \int e^2\,dx}$$

has a maximum or minimum with respect to variation of a. Show that all functions have at least one axis of symmetry. If there is more than one axis of symmetry, can there be arbitrary numbers of each of the two kinds of axis?

19 Note that cos x is fully even and has no odd part, and that shift of origin causes the even part to diminish and the odd part to grow until in due course the function becomes fully odd. In fact the even part of any periodic function will wax and wane relative to the odd part as the origin shifts. Consider means of assigning "abscissas of symmetry" and quantitative measures of "degree of symmetry" that will be independent of the origin of x. Test the reasonableness of your conclusions—for example, on the functions of period 2 which in the range $-1 < x < 1$ are given by $\Lambda(x)$, $\Lambda(x - \frac{1}{2})$, $\Lambda(x - \frac{1}{4})$. See Chapter 4 for triangle-function notation $\Lambda(x)$.

20 The function $f(x)$ is equal to unity when x lies between $-\frac{1}{2}$ and $\frac{1}{2}$ and is zero outside. Draw accurate loci on the complex plane of $F(s)$ from which values of $F(0)$, $F(\frac{1}{2})$, $F(1)$, $F(1\frac{1}{2})$, and $F(2)$ can be measured.

21 The function $f(x)$ is equal to 100 when x differs by less than 0.01 from 1, 2, 3, 4, or 5 and is zero elsewhere. Draw a locus on the complex plane of $F(s)$ from which $F(0.05)$ can be measured in amplitude and phase.

Bibliography

Fourier integral The following are standard texts on the Fourier integral.

Beurling, A.: "Sur les intégrales de Fourier absolument convergentes et leur application à une transformation fonctionnelle" (Neuvième congrès des mathématiciens scandinaves), Helsingfors, Finland, 1939.

Bochner, S.: "Vorlesungen über Fouriersche Integrale," Chelsea Publishing Company, New York, 1948.

Bochner, S., and K. Chandrasekharan: "Fourier Transforms," Princeton University Press, Princeton, N.J., 1949.

Carleman, T.: "L'Intégrale de Fourier et questions qui s'y rattachent," Almqvist and Wiksells, Uppsala, Sweden, 1944.

Carslaw, H. S.: "Introduction to the Theory of Fourier's Series and Integrals," Dover Publications, New York, 1930.

Lighthill, M. J.: "An Introduction to Fourier Analysis and Generalised Functions," Cambridge University Press, Cambridge, England, 1958.

Paley, R. E. A. C., and N. Wiener: "Fourier Transforms in the Complex Domain," vol. 19, American Mathematical Society, Colloquium Publications, New York, 1934.

Sneddon, I. N.: "Fourier Transforms," McGraw-Hill Book Company, New York, 1951.

Titchmarsh, E. C.: "Introduction to the Theory of Fourier Integrals," Oxford University Press, Oxford, England, 1937.

Wiener, N.: "The Fourier Integral and Certain of Its Applications," Cambridge University Press, Cambridge, England, 1933.

Tables of Fourier integrals Since tables of the Laplace transform, properly interpreted, are sources of Fourier integrals, some are included here. The following are the most lengthy compilations.

Campbell, G. A., and R. M. Foster: "Fourier Integrals for Practical Applications," Van Nostrand Company, Princeton, N.J., 1948.

Doetsch, G., H. Kniess, and D. Voelker: "Tabellen zur Laplace Transformation," Springer-Verlag, Berlin, 1947.

Erdélyi, A.: "Tables of Integral Transforms," vol. 1, McGraw-Hill Book Company, New York, 1954.

McLachlan, N. W., and P. Humbert: "Formulaire pour le calcul symbolique," 2d ed., Gauthier-Villars, Paris, 1950.

Tables of the one-sided Laplace transformation (lower limit of integration zero) are necessarily sources of Fourier transforms only of functions that are zero for negative arguments. A table of the double-sided Laplace transform is given in the following work.

Van der Pol, B., and H. Bremmer: "Operational Calculus Based on the Two-sided Laplace Integral," 2d ed., Cambridge University Press, Cambridge, England. 1955.

Chapter 3 Convolution

The *idea* of the convolution of two functions occurs widely, as witnessed by the multiplicity of its aliases. The *word* "convolution" is coming into more general use as awareness of its oneness spreads into various branches of science. The German term *Faltung* is widely used, as is the term "composition product," adapted from the French. Terms encountered in special fields include superposition integral, Duhamel integral, Borel's theorem, (weighted) running mean, cross-correlation function, smoothing, blurring, scanning, and smearing.

As some of these last terms indicate, convolution describes the action of an observing instrument when it takes a weighted mean of some physical quantity over a narrow range of some variable. When, as very often happens, the form of the weighting function does not change appreciably as the central value of the variable changes, the observed quantity is a value of the convolution of the distribution of the desired quantity with the weighting function, rather than a value of the desired quantity itself. All physical observations are limited in this way by the resolving power of instruments, and for this reason alone convolution is ubiquitous. Later we show that the appearance of convolution is coterminous with linearity plus time or space invariance, and also with sinusoidal response to sinusoidal stimulus.

Not only is convolution widely significant as a physical concept, but because of a powerful theorem encountered below, it also offers an advantageous starting point for theoretical developments. Conversely, because of its adaptability to computing, it is an advantageous terminal point for numerical work.

The convolution of two functions $f(x)$ and $g(x)$ is

$$\int_{-\infty}^{\infty} f(u)g(x-u)\,du,$$

or briefly,

$$f(x) * g(x).$$

The convolution itself is also a function of x, let us say $h(x)$.

Various ways of looking at the convolution integral suggest themselves. For example, suppose that $g(x)$ is given. Then for every function $f(x)$ for which the integral exists there will be an $h(x)$. Following Volterra, we may say that $h(x)$ is a *functional* of the function $f(x)$. Note that to calculate $h(x_1)$ we need to know $f(x)$ for a whole range of x, whereas to calculate a *function* of the function $f(x)$ at $x = x_1$, we need only know $f(x_1)$.

In Fig. 3.1 the product $f(u)g(x-u)$ is shown shaded, and the ordinate $h(x)$ is equal to the shaded area.

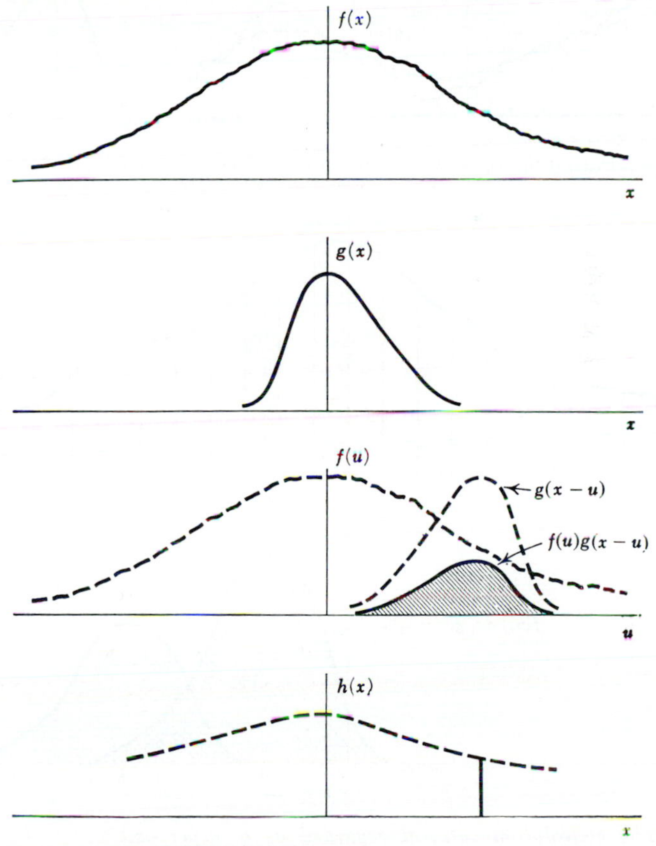

Fig. 3.1 The convolution integral $h(x) = f(x) * g(x)$ represented by a shaded area.

A second example with a different $f(x)$ but the same $g(x)$ is shown in Fig. 3.2 to illustrate the general features relating the functional $h(x)$ to $f(x)$. It will be seen that $h(x)$ is smoother in detail than $f(x)$, is more spread out, and has less total variation.

In another approach, $f(x)$ is resolved into infinitesimal columns (see Fig. 3.3). Each column is regarded as melted out into heaps having the form of $g(x)$ but centered at its original value of x. Just two of these melted-down columns are shown in the figure, but all are to be so pictured. Then $h(x)$ is equal to the sum of the contributions at the point x made by all the heaps; that is,

$$h(x) = \int_{-\infty}^{\infty} f(x_1)g(x - x_1)\,dx_1.$$

A further view is illustrated in Fig. 3.4, where $g(u)$ is shown folded back on

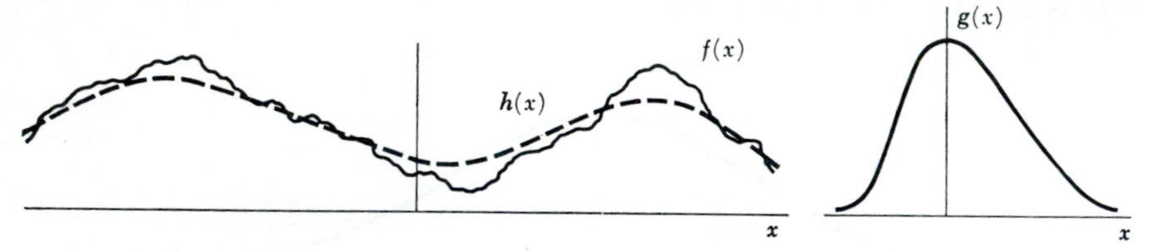

Fig. 3.2 Illustrating the smoothing effect of convolution ($h = f * g$).

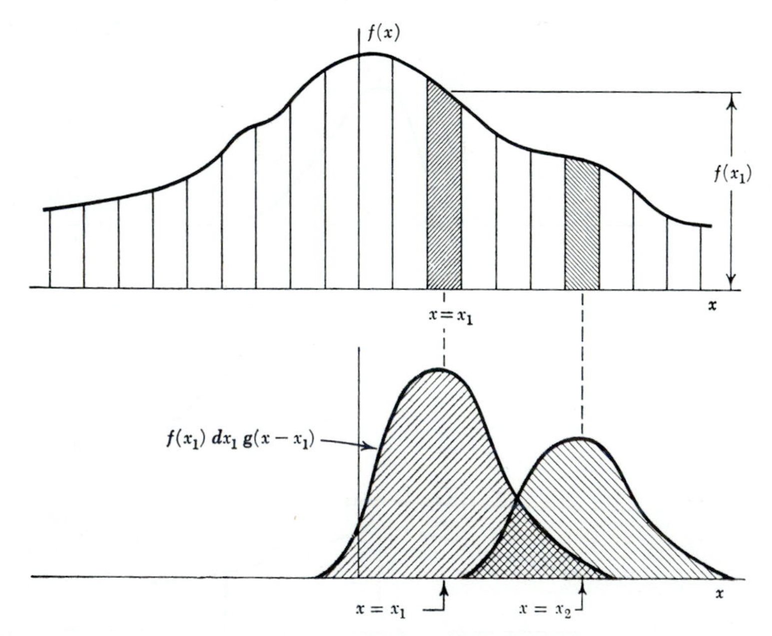

Fig. 3.3 The convolution integral regarded as a superposition of characteristic contributions.

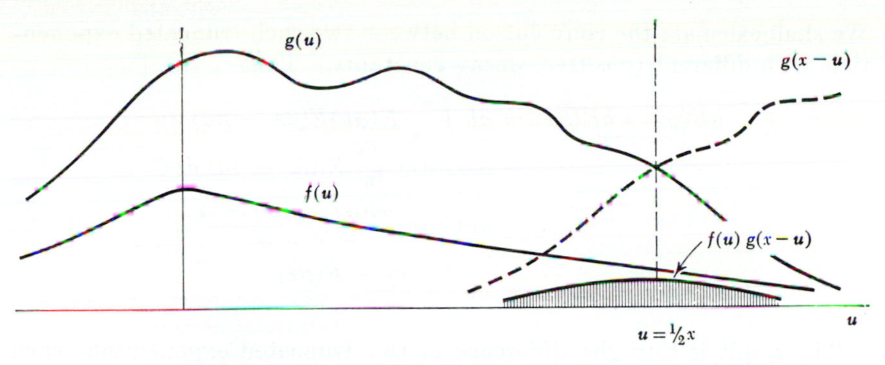

Fig. 3.4 Convolution from the Faltung *standpoint.*

itself about the line $u = \frac{1}{2}x$. As before, the area under the product curve $f(u)g(x - u)$ is the convolution $h(x)$, and its dependence on the position of the line of folding (German *Faltung*) may often be visualized from this standpoint.

These suggestions do not exhaust the possible interpretations of the convolution integral; others will appear later.

It should be noticed that before the operations of multiplication and integration take place $g(x)$ must be reversed. It is possible to dispense with the reversal, and in some subjects this is more natural. A consequence of the reversal is, however, that convolution is commutative; that is,

$$f * g = g * f,$$

or

$$\int_{-\infty}^{\infty} f(u)g(x - u)\, du = \int_{-\infty}^{\infty} g(u)f(x - u)\, du.$$

Convolution is also associative (provided that all the convolution integrals exist),

$$f * (g * h) = (f * g) * h,$$

and distributive over addition,

$$f * (g + h) = f * g + f * h.$$

The abbreviated notation with asterisks (*) thus proves very convenient in formal manipulation, since the asterisks behave like multiplication signs.

Examples of convolution

Consider the truncated exponential function

$$E(x) = \begin{cases} e^{-x} & x > 0 \\ 0 & x < 0. \end{cases}$$

We shall calculate the convolution between two such truncated exponentials with different (positive) decay constants. Thus

$$
\begin{aligned}
aE(\alpha x) * bE(\beta x) &= ab \int_{-\infty}^{\infty} E(\alpha u)E(\beta x - \beta u)\, du \\
&= abE(\beta x) \int_0^x E(\alpha u - \beta u)\, du \\
&= abE(\beta x) \frac{E(\alpha x - \beta x) - 1}{\beta - \alpha} \\
&= ab \frac{E(\alpha x) - E(\beta x)}{\beta - \alpha}.
\end{aligned}
$$

The result is thus the difference of two truncated exponentials, each with the same amplitude, as illustrated in Fig. 3.5. This function occurs commonly; for instance, it describes the concentration of a radioactive isotope which decays with a constant α while simultaneously being replenished as the decay product of a parent isotope which decays with a constant β. From the commutative property of convolution we see that the result is the same if α and β are interchanged, and as $t \to \infty$ one of the terms dies out, leaving a simple exponential with constant α or β, whichever describes the slower decay. As a special case we calculate $E(\alpha x) * E(\alpha x)$ by taking the limit as $\beta - \alpha \to 0$. Thus we find

$$
\begin{aligned}
E(\alpha x) * E(\alpha x) &= \lim_{\beta - \alpha \to 0} \frac{E(\alpha x) - E(\beta x)}{\beta - \alpha} \\
&= -\frac{d}{d\alpha} E(\alpha x) \\
&= xE(\alpha x).
\end{aligned}
$$

This particular function describes the response of a critically damped resonator, such as a dead-beat galvanometer, to an impulsive disturbance.

As a further example consider $E(-\alpha x) * E(\beta x)$. This gives an entirely

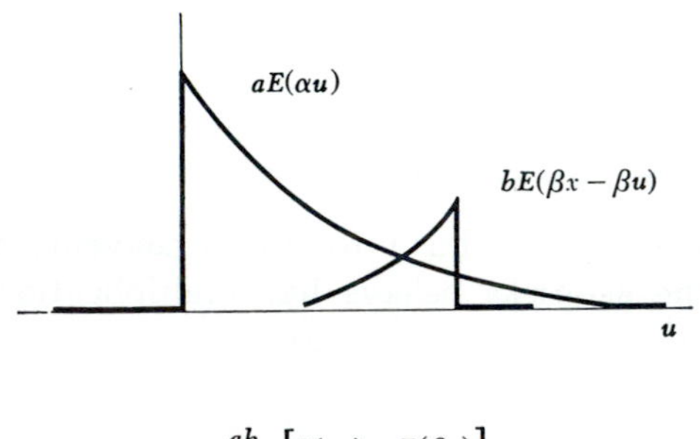

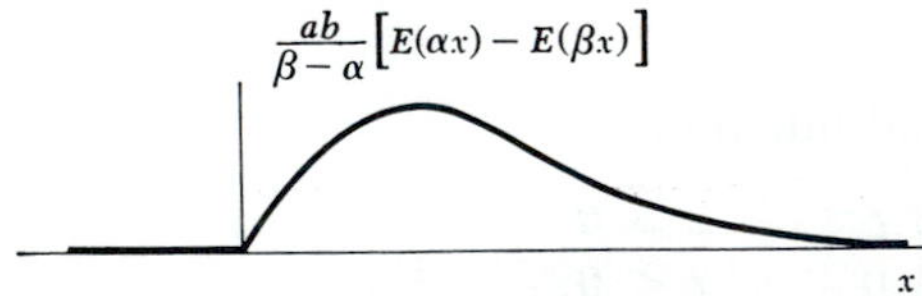

Fig. 3.5 Convolution of two truncated exponentials.

different type of result. Thus

$$E(-\alpha x) * E(\beta x) = \int_{-\infty}^{\infty} E(-\alpha x + \alpha u)E(\beta u)\,du$$

$$= \begin{cases} \int_{x}^{\infty} e^{-\alpha u + \alpha x} e^{-\beta u}\,du & x > 0 \\ \int_{0}^{\infty} e^{-\alpha u + \alpha x} e^{-\beta u}\,du & x < 0 \end{cases}$$

$$= \frac{E(-\alpha x) + E(\beta x)}{\alpha + \beta}.$$

Here we have a function that is peaked at the origin and dies away with a constant α to the left and with a constant β to the right.

In calculations of this kind care is required in fixing the limits of the integrals and checking signs, because the ordinary sort of algebraic error can make a radical change in the result. The following graphical construction for convolution is useful as a check. Plot one of the functions entering into the convolution backward on a movable piece of paper as shown in Fig. 3.6, and slide it along in the direction of the axis of abscissas.

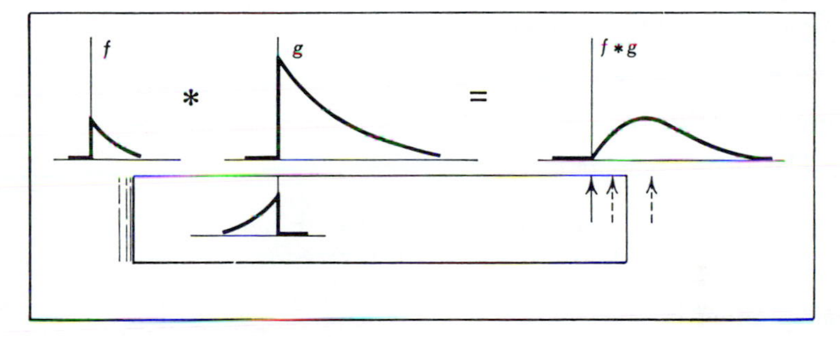

Fig. 3.6 Graphical construction for convolution. The movable piece of paper has a graph of one of the functions plotted backward.

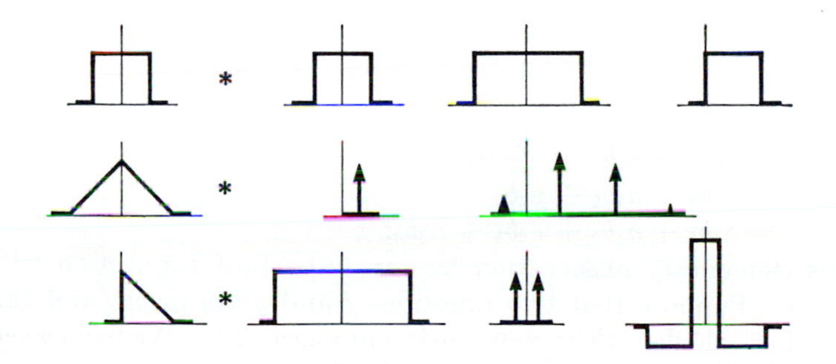

Fig. 3.7 Examples for practicing graphical convolution. The arrows represent impulses.

When the movable piece is to the left of the position shown, the product of g with f reversed is zero. By marking an arrow in some convenient position, we can keep track of this. Then suddenly, at the position shown, the integral of the product begins to assume nonzero values. By moving the paper a little farther along, as indicated by a broken outline, we find that the convolution will be positive and increasing from zero approximately linearly with displacement. Farther along still, we see that a maximum will occur beyond which the convolution dies away.

In the second example given above, two oppositely directed exponential tails, the absence of a zero stretch, and the presence of a cusp at the origin are immediately apparent from the construction. In addition to qualitative conclusions such as these, certain quantitative results are yielded by this construction in many cases, particularly where the functions break naturally into parts in successive ranges of the abscissa.

It is well worth while to carry out this moving construction until it is thoroughly familiar. There is no doubt that experts do this geometrical construction in their heads all the time, and a little practice soon enables the beginner also to dispense with the actual piece of paper. Examples for practice are given in Fig. 3.7.

Serial products

Consider two polynomials

$$a_0 + a_1x + a_2x^2 + a_3x^3 + \cdots$$

and

$$b_0 + b_1x + b_2x^2 + b_3x^3 + \cdots.$$

Their product is

$$a_0b_0 + (a_0b_1 + a_1b_0)x + (a_0b_2 + a_1b_1 + a_2b_0)x^2 + (a_0b_3 + a_1b_2 + a_2b_1 + a_3b_0)x^3 + \cdots,$$

which we may call

$$c_0 + c_1x + c_2x^2 + c_3x^3 + \cdots,$$

where

$$\begin{aligned} c_0 &= a_0b_0 \\ c_1 &= a_0b_1 + a_1b_0 \\ c_2 &= a_0b_2 + a_1b_2 + a_2b_0 \\ c_3 &= a_0b_3 + a_1b_2 + a_2b_1 + a_3b_0. \end{aligned}$$

This elementary observation has an important connection with convolution. Suppose that two functions f and g are given, and that it is required to calculate their convolution numerically. We form a sequence of values of f at short regular intervals of width w,

$$\{f_0\ f_1\ f_2\ f_3\ \cdots\ f_m\},$$

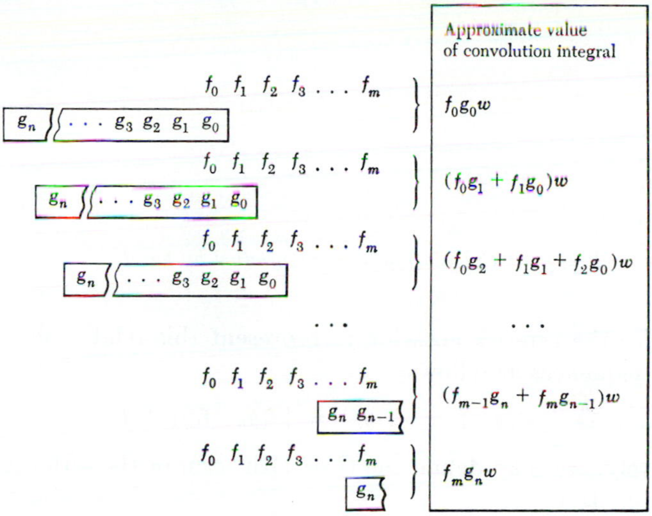

Approximate value of convolution integral

$f_0 \; f_1 \; f_2 \; f_3 \; \ldots \; f_m$

$g_n \; \ldots \; g_3 \; g_2 \; g_1 \; g_0$ — $f_0 g_0 w$

$f_0 \; f_1 \; f_2 \; f_3 \; \ldots \; f_m$

$g_n \; \ldots \; g_3 \; g_2 \; g_1 \; g_0$ — $(f_0 g_1 + f_1 g_0)w$

$f_0 \; f_1 \; f_2 \; f_3 \; \ldots \; f_m$

$g_n \; \ldots \; g_3 \; g_2 \; g_1 \; g_0$ — $(f_0 g_2 + f_1 g_1 + f_2 g_0)w$

$\ldots \qquad \ldots$

$f_0 \; f_1 \; f_2 \; f_3 \; \ldots \; f_m$

$g_n \; g_{n-1}$ — $(f_{m-1} g_n + f_m g_{n-1})w$

$f_0 \; f_1 \; f_2 \; f_3 \; \ldots \; f_m$

g_n — $f_m g_n w$

Fig. 3.8 Explaining the serial product.

and a corresponding sequence of values of g,

$$\{g_0 \; g_1 \; g_2 \; g_3 \; . \; . \; . \; g_n\}.$$

We then approximate the convolution integral

$$\int_{-\infty}^{\infty} f(x')g(x - x')\, dx'$$

by summing products of corresponding values of f and g, taking different discrete values of x one by one. It is convenient to write the g sequence on a movable strip of paper which can be slid into successive positions relative to the f sequence for each successive value. The first few stages of this approach are shown in Fig. 3.8. It will be noticed that the g sequence has been written in reverse as required by the formula. Since $f * g = g * f$, the f sequence could have been written in reverse, in which case *it* would have been written on the movable strip.

It will be seen that this procedure generates the same expressions that occur in the multiplication of series, and we therefore introduce the term "serial product" to describe the sequence of numbers

$$\{f_0 g_0 \quad f_0 g_1 + f_1 g_0 \quad f_0 g_2 + f_1 g_1 + f_2 g_0 \quad . \; . \; . \;\}$$

derived from the two sequences

$$\{f_0 \; f_1 \; f_2 \; f_3 \; . \; . \; . \;\} \qquad \text{and} \qquad \{g_0 \; g_1 \; g_2 \; g_3 \; . \; . \; . \;\}.$$

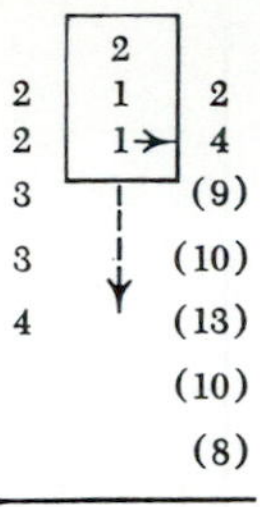

Fig. 3.9 Calculating a serial product by hand.

We transfer the asterisk notation to represent this relationship between the three sequences as follows:

$$\{f_0\ f_1\ .\ .\ .\ f_m\} * \{g_0\ g_1\ .\ .\ .\ g_n\} = \{f_0g_0\quad f_0g_1 + f_1g_0\quad .\ .\ .\ f_mg_n\}.$$

Alternatively, we may define the $(i + 1)$th term of the serial product of $\{f_i\}$ and $\{g_i\}$ to be

$$\sum_j f_j g_{i-j}.$$

In practice the calculation of serial products is an entirely feasible procedure (for example, on a hand calculator by allowing successive products to accumulate in the product register). The two sequences are most conveniently written in vertical columns, and the answers are written opposite an arrow marked in a convenient place on the movable strip. Figure 3.9 shows an early stage in the calculation of $\{2\ 2\ 3\ 3\ 4\} * \{1\ 1\ 2\}$. The value of 4, shown opposite the arrow, has just been calculated, and the values still to be calculated as the movable strip is taken downward are shown in parentheses. Note that the sequence on the *moving* strip has been written in *reverse* (upward).

It will be seen that the serial product is a longer sequence than either of the component sequences, the number of terms being one less than the sum of the numbers of terms in the components:

$$\{2\ 2\ 3\ 3\ 4\} * \{1\ 1\ 2\} = \{2\ 4\ 9\ 10\ 13\ 10\ 8\}.$$

Furthermore the sum of the terms of the serial product is the product of the sums of the component sequences, a fact which allows a very valuable check on numerical work. As a special case, if the sum of one of the component sequences is unity, then the sum of the serial product will be the same as the sum of the other component. These properties are analogous to properties of the convolution integral.

A semi-infinite sequence is one such as

$$\{f_0\ f_1\ f_2\ .\ .\ .\},$$

which has an end member in one direction but runs on without end in the other. If we take the serial product of such a sequence with a finite sequence, the result is also semi-infinite; for example,

$$\{1\ 1\} * \{1\ 2\ 3\ 4\ .\ .\ .\} = \{1\ 3\ 5\ 7\ 9\ .\ .\ .\}.$$

A semi-infinite sequence may or may not have a finite sum, but this does not lead to problems in defining the serial product because each member is the sum of only a finite number of terms. Of course, the numerical check, according to which the sum of the members of the serial product equals the product of the sums, breaks down if any one of the sums does not exist.

The serial product of two semi-infinite sequences presents no problems when both run on indefinitely in the same direction; thus

$$\{1\ 2\ 3\ 4\ .\ .\ .\} * \{1\ 2\ 3\ 4\ \ .\ .\ .\} = \{1\ 4\ 10\ 20\ .\ .\ .\},$$

but if they run on in opposite directions then each member of the serial product is the sum of an infinite number of terms. This sum may very well not exist; thus

$$\{.\ .\ .\ 4\ 3\ 2\ 1\} * \{1\ 2\ 3\ 4\ .\ .\ .\}$$

does not lead to convergent series.

Two-sided sequences are often convenient to deal with and call for no special comment other than that one must specify the origin explicitly, for example, by an arrow as in

$$\{.\ .\ .\ 0.1\ 0.2\ 0.4\ \underset{\uparrow}{0.9}\ 0.8\ 0.7\ 0.6\ .\ .\ .\}.$$

The sequence

$$\{.\ .\ .\ 0\ 0\ 0\ \underset{\uparrow}{1}\ 0\ 0\ 0\ .\ .\ .\} = \{J\}$$

plays an important role analogous to that of the impulse symbol $\delta(x)$. It has the property that

$$\{J\} * \{f\} = \{f\}$$

for all sequences $\{f\}$. This property is, of course, also possessed by the one-member sequence $\{1\}$ and by other sequences such as $\{1\ \ 0\ \ 0\}$.

Serial multiplication by the sequence

$$\{.\ .\ .\ 0\ 0\ 0\ 1\ -1\ 0\ 0\ 0\ .\ .\ .\} \qquad \text{or} \qquad \{1\ \ -1\}$$

is equivalent to taking the first finite difference; that is,

$$\begin{aligned}\{1\ \ -1\} * \{.\ .\ .\ f_{-2}\ f_{-1}\ f_0\ f_1\ f_2\ .\ .\ .\} \\ = \{.\ .\ .\ \ f_{-1} - f_{-2}\ \ f_0 - f_{-1}\ \ f_1 - f_0\ \ f_2 - f_1\ \ .\ .\ .\}.\end{aligned}$$

Where it is important to distinguish between the central difference

$f_{n+\frac{1}{2}} - f_{n-\frac{1}{2}}$ and the forward difference $f_{n+1} - f_n$, it is necessary to specify the origins of the sequences appropriately. Apart from a shift in the indexing of the terms, however, the resulting sequence is the same.

The sequence

$$\left\{\frac{1}{n}\ \frac{1}{n}\ \ldots\ (n\text{ terms})\ \ldots\ \frac{1}{n}\right\}$$

generates the running mean over n terms, a familiar process for smoothing sequences of meteorological data. It may be written, as in the case of weekly running means of daily values, as

$$\frac{1}{7}\{1\ 1\ 1\ 1\ 1\ 1\ 1\}.$$

The sequence $\{1\ 1\ 1\ 1\ 1\ 1\ 1\}$ itself generates running sums, in this case over seven terms.

The semi-infinite sequence

$$\{1\ 1\ 1\ 1\ \ldots\}$$

enters into a serial product with another sequence to generate a familiar result. Thus if $\{f\}$ is a sequence and S_n is the sum of the first n terms, then the sequence $\{S\}$ defined by

$$\{S\} = \{S_1\ S_2\ S_3\ \ldots\}$$

may be expressed in the form

$$\{1\ 1\ 1\ 1\ \ldots\} * \{f\} = \{S\}.$$

Since each term of the sequence $\{f\}$ is the difference between two successive terms of $\{S\}$, a converse relationship can also be written:

$$\{1\ -1\} * \{S\} = \{f\}.$$

Substituting in the previous equation, we have

$$\{1\ 1\ 1\ 1\ \ldots\} * \{1\ -1\} * \{S\} = \{S\};$$

in other words the two operations neutralize each other. We may express this reciprocal relationship by writing

$$\{1\ 1\ 1\ 1\ \ldots\} * \{1\ -1\} = \{1\ 0\ 0\ 0\ \ldots\}$$

or, in a two-sided version,

$$\{\ldots\ 0\ 0\ 0\ \underset{\uparrow}{1}\ 1\ 1\ 1\ \ldots\} * \{\ldots\ 0\ 0\ 0\ \underset{\uparrow}{1}\ -1\ 0\ 0\ \ldots\}$$
$$= \{\ldots\ 0\ 0\ 0\ 1\ 0\ 0\ 0\ \ldots\}.$$

Another example of a reciprocal pair of sequences is

$$\{1\ 2\ 3\ 4\ .\ .\ .\} \qquad \text{and} \qquad \{1\ -2\ 1\}.$$

This is an important relationship because the reciprocal sequence represents the solution to the problem of finding $\{g\}$ when $\{f\}$ and $\{h\}$ are given and it is known that

$$\{f\} * \{g\} = \{h\}.$$

The answer is

$$\{g\} = \{f\}^{-1} * \{h\},$$

where $\{f\}^{-1}$ is the reciprocal of $\{f\}$, that is, the sequence with the property that

$$\{f\}^{-1} * \{f\} = \{.\ .\ .\ 0\ 0\ 1\ 0\ 0\ .\ .\ .\}.$$

Inversion of serial multiplication If $\{f\} * \{g\} = \{h\}$, then $\{h\}$ is called the serial product of $\{f\}$ and $\{g\}$, because the sequence $\{h\}$ comprises the coefficients of the polynomial which is the product of the polynomials represented by $\{f\}$ and $\{g\}$.

Conversely, the process of finding $\{g\}$ when $\{f\}$ and $\{h\}$ are given may be called serial division, and in fact one could carry out the solution of such a problem by actually doing long division of polynomials.

For example, to solve the problem

$$\{1\ 1\} * \{g_0\ g_1\ g_2\ .\ .\ .\} = \{1\ 3\ 3\ 1\}$$

one could write

$$g_0 + g_1x + g_2x^2 + \ .\ .\ .\ = \frac{1 + 3x + 3x^2 + x^3}{1 + x}.$$

The long division would then proceed as follows:

$$\begin{array}{r}
1 + 2x\ + x^2 \\
1 + x\overline{)1 + 3x + 3x^2 + x^3} \\
\underline{1 + \ \ x} \qquad\qquad\quad \\
2x + 3x^2 \qquad \\
\underline{2x + 2x^2} \qquad \\
x^2 + x^3 \\
\underline{x^2 + x^3}
\end{array}$$

Thus

$$g_0 + g_1x + g_2x^2 + \ .\ .\ .\ = 1 + 2x + x^2$$

or

$$\{g_0\ g_1\ g_2\ .\ .\ .\} = \{1\ 2\ 1\}.$$

This, of course, is recognizable immediately as the correct answer.

However, it is not necessary to organize the calculation at such length. A convenient way of doing the job on a hand calculator may be shown by going back to an example in the previous section, where $\{2\ 2\ 3\ 3\ 4\} *$

{1 1 2} was calculated. The situation at the end of the calculation is as shown in Fig. 3.10*a*, the serial product being in the right-hand column. Now, suppose that the left-hand column $\{f\}$ and the right-hand column $\{h\}$ were given. To find $\{g\}$, write the sequence upward on the movable strip of paper, and place it in position for beginning the calculation of the serial product (see Fig. 3.10*b*). Clearly g_0 is immediately deducible, for it has to be such that when it is multiplied by the 2 on the left the result is 2. Hence in the space labeled g_0 one may write 1 and then move the paper down to the next position (see Fig. 3.10*c*). Then, when g_1 is multiplied by 2 and added to the product of 2 and 1, the result must be 4. This gives g_1, and so on.

With a hand calculator one forms the products of the numbers in the left-hand column with such numbers on the moving strip as are already known and accumulates the sum of the products. Subtract this sum from the current value of $\{h\}$ opposite the arrow and divide by the first member of $\{f\}$ at the top of the left column to obtain the next value of $\{g\}$. Enter this value on the paper strip, which may then be moved down one more place. The process is started by entering the first value of $\{g\}$, which is h_0/f_0. This inversion of serial multiplication is as easy as the direct process and takes very little longer.

Compared with long division, this method is superior, since it involves writing down no numbers other than the data and the answer. Furthermore it is as readily applicable when the coefficients are large numbers or fractions. Long division might, however, be considered for short sequences of small integers.

Table 3.1 lists a few common sequences and their inverses. Most of

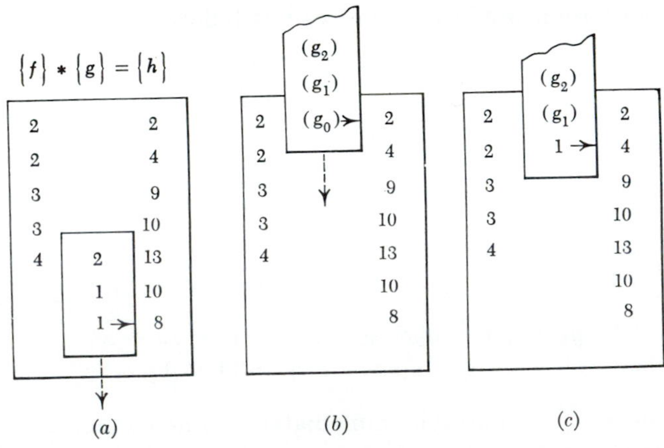

Fig. 3.10 (a) Completion of serial multiplication of {2 2 3 3 4} by *{1 1 2}; (b) and (c) first steps in the inversion of the process.*

Table 3.1 Some sequences and their inverses

Sequence	*Inverse*
$\{1\ 1\}$	$\{1\ -1\ 1\ -1\ 1\ -1\ .\ .\ .\}$
$\{1\ -1\}$	$\{1\ 1\ 1\ 1\ 1\ 1\ .\ .\ .\}$
$\{1\ 0\ 1\}$	$\{1\ 0\ -1\ 0\ 1\ 0\ .\ .\ .\}$
$\{1\ 0\ -1\}$	$\{1\ 0\ 1\ 0\ 1\ 0\ .\ .\ .\}$
$\{1\ 1\ 1\}$	$\{1\ -1\ 0\ 1\ -1\ 0\ .\ .\ .\}$
$\{1\ 2\ 1\}$	$\{1\ -2\ 3\ -4\ 5\ -6\ .\ .\ .\}$
$\{n+1\} = \{1\ 2\ 3\ 4\ .\ .\ .\}$	$\{1\ -1\}^{*2} = \{1\ -2\ 1\}$
$\{(n+1)^2\} = \{1\ 4\ 9\ 16\ .\ .\ .\}$	$\{1\ -1\}^{*3} * \{1\ 1\}^{-1}$
$\{(n+1)^3\} = \{1\ 8\ 27\ .\ .\ .\}$	$\{1\ -1)^{*4} * \{1\ 4\ 1\}^{-1}$
$\{1\ -a\}$	$\{1\ a\ a^2\ a^3\ .\ .\ .\}$
$\{1\ -a\}^{*2}$	$\{1\ 2a\ 3a^2\ 4a^3\ .\ .\ .\}$
$\{e^{-\alpha n}\} = \{1\ e^{-\alpha}\ e^{-2\alpha}\ .\ .\ .\}$	$\{1\ -e^{-\alpha}\}$
$\{1 - e^{-\alpha(n+1)}\}$	$\{1\ -1\} * \{1\ -e^{-\alpha}\}/(1 - e^{-\alpha})$
$\{\sin \omega(n+1)\}$	$\{1\ -2 \cos \omega\ 1\}/\sin \omega$
$\{\cos \omega n\}$	$\{1\ -2 \cos \omega\ 1\} * \{1\ - \cos \omega\}^{-1}$
$\{e^{-\alpha(n+1)} \sin \omega(n+1)\}$	$\{1\ -2e^{-\alpha} \cos \omega\ e^{-2\alpha}\}/e^{-\alpha} \sin \omega$
$\{e^{-\alpha n} \cos \omega n\}$	$\{1\ -2e^{-\alpha} \cos \omega\ e^{-2\alpha}\} * \{1\ -e^{-\alpha} \cos \omega\}^{-1}$

the entries may be verified by serial multiplication of corresponding entries, whereupon the result will be found to be $\{1\ 0\ 0\ 0\ .\ .\ .\}$. The table is precisely equivalent to a list of polynomial pairs beginning as follows:

$$
\begin{array}{ll}
1 + x & 1 - x + x^2 - x^3 + \ldots \\
1 - x & 1 + x + x^2 + x^3 + \ldots \\
1 + x^2 & 1 - x^2 + x^4 + \ldots \\
\ldots & \ldots
\end{array}
$$

In this case the property of corresponding entries would be that their product was unity.

Some of the later entries in Table 3.1 represent exponential and other variations reminiscent of the natural behavior of circuits, and will prove more difficult, though not impossible, to verify at this stage. They are readily derivable by transform methods.

The serial product in matrix notation Let the sequence $\{h\}$ be the serial product of two sequences $\{f\}$ and $\{g\}$, where

$$
\begin{aligned}
\{f\} &= \{f_0\ f_1\ f_2\ .\ .\ .\ f_m\} \\
\{g\} &= \{g_0\ g_1\ g_2\ .\ .\ .\ \ g_n\} \\
\{h\} &= \{h_0\ h_1\ h_2\ .\ .\ .\ \ \ h_{m+n}\}.
\end{aligned}
$$

Evidently all three sequences can be expressed as single-row or single-column matrices, but since the sequences have different numbers of members, matrix notation might not immediately spring to mind. The relationship between $\{f\}$, $\{g\}$, and $\{h\}$ can, however, be expressed in terms of matrix multiplication as follows.

First we recall the special case of a matrix product where the first factor has n rows and n columns, and the second has n rows and one column only. The product is a single-column matrix of n elements. Thus

$$\begin{bmatrix} a_{11} & a_{12} & \cdots & a_{1n} \\ a_{21} & a_{22} & \cdots & a_{2n} \\ \cdots & \cdots & \cdots & \cdots \\ a_{n1} & a_{n2} & \cdots & a_{nn} \end{bmatrix} \begin{bmatrix} x_1 \\ x_2 \\ \cdots \\ x_n \end{bmatrix} = \begin{bmatrix} a_{11}x_1 + a_{12}x_2 + \cdots + a_{1n}x_n \\ a_{21}x_1 + a_{22}x_2 + \cdots + a_{2n}x_n \\ \cdots \qquad \cdots \qquad \cdots \\ a_{n1}x_1 + a_{n2}x_2 + \cdots + a_{nn}x_n \end{bmatrix}$$

Now we form a column matrix $[y]$ whose elements are equal one by one to the members of the sequence $\{h\}$; we also form a column matrix $[x]$, whose early members are equal one by one to the members of the sequence $\{g\}$ as far as they go, and beyond that are zeros until there are as many elements in $[x]$ as in $[y]$. Since each member of $\{h\}$ is a linear combination of members of $\{g\}$ with a suitable set of coefficients selected from $\{f\}$, we can arrange the successive sets of coefficients, row by row, to form a square matrix such that the rules of matrix multiplication will generate the desired result. Thus, to illustrate a case where $\{f\}$ has five members, $\{g\}$ has three, and $\{h\}$ has seven, we can write:

$$\begin{bmatrix} f_0 & 0 & 0 & 0 & 0 & 0 & 0 \\ f_1 & f_0 & 0 & 0 & 0 & 0 & 0 \\ f_2 & f_1 & f_0 & 0 & 0 & 0 & 0 \\ f_3 & f_2 & f_1 & f_0 & 0 & 0 & 0 \\ f_4 & f_3 & f_2 & f_1 & f_0 & 0 & 0 \\ 0 & f_4 & f_3 & f_2 & f_1 & f_0 & 0 \\ 0 & 0 & f_4 & f_3 & f_2 & f_1 & f_0 \end{bmatrix} \begin{bmatrix} g_0 \\ g_1 \\ g_2 \\ 0 \\ 0 \\ 0 \\ 0 \end{bmatrix} = \begin{bmatrix} f_0g_0 \\ f_1g_0 + f_0g_1 \\ f_2g_0 + f_1g_1 + f_0g_2 \\ f_3g_0 + f_2g_1 + f_1g_2 \\ f_4g_0 + f_3g_1 + f_2g_2 \\ f_4g_1 + f_3g_2 \\ f_4g_2 \end{bmatrix} = \begin{bmatrix} h_0 \\ h_1 \\ h_2 \\ h_3 \\ h_4 \\ h_5 \\ h_6 \end{bmatrix}$$

The elements in the southeast quadrant of the f matrix could be replaced by zeros without affecting the result.

By listing parts of the sequence $\{f\}$ row by row with progressive shift as required, we have succeeded in forcing the serial product into matrix notation. This may appear to be a labored and awkward exercise, and one that sacrifices the elegant commutative property. In fact, however, matrix representation for serial products is widely used in control-system engineering and elsewhere because of the rich possibilities offered by the theory of matrices, especially infinite matrices.

Sequences as vectors Just as the short sequence $\{x_1 \; x_2 \; x_3\}$ may be regarded as the representation of a certain vector in three-dimensional space in terms of three orthogonal components, so a sequence of n members may be regarded as representing a vector in n-dimensional space. As a rule we do not expend much effort trying to visualize the n mutually perpendicular axes along which the vector is to be resolved, but simply handle the members of the sequence according to algebraic rules which are natural extensions of those for handling vector components in three-dimensional space. Nevertheless, we borrow a whole vocabulary of terms and expressions from geometry, which lends a certain picturesqueness to linear algebra.

If the relationship between two sequences $\{x\}$ and $\{y\}$ is

$$\{a\} * \{x\} = \{y\},$$

where $\{a\}$ is some other sequence, then an equivalent statement is

$$T\mathbf{X} = \mathbf{Y},$$

where $\mathbf{X}$ and $\mathbf{Y}$ are the vectors whose components are $\{x\}$ and $\{y\}$, and T is an operator representing the transformation that converts $\mathbf{X}$ to $\mathbf{Y}$. The transformation consists of forming linear combinations of the components of $\mathbf{X}$ in accordance with the rules that are so concisely expressed by the serial product.

It will be recalled that the number of members of the sequence $\{y\}$ exceeds the number of members of $\{x\}$ if $\{x\}$ has a finite number of members. To avoid the awkwardness that arises if $\mathbf{X}$ and $\mathbf{Y}$ are vectors in spaces with different numbers of dimensions, the vector interpretation is usually applied to infinite sequences $\{x\}$ and $\{y\}$. If the sequences arising in a given problem are in actual fact not infinite, it is often permissible to convert them to infinite sequences by including an infinite number of extra members all of which are zero.

As in the case of representation of sequences by column matrices, the commutative property $\{a\} * \{x\} = \{x\} * \{a\}$ is abandoned. One of the sequences entering into the serial product is interpreted as a vector, and one in an entirely different way. To reverse the roles of $\{x\}$ and $\{a\}$—for example, to talk of the vector $\mathbf{A}$—would be unthinkable in physical fields where the vector interpretation is used. This apparent rigidification of the notation is compensated by greater generality. To illustrate this we may note first that, to the extent that matrix methods are used for dealing with vectors, there is no distinction between the representations of sequences as column matrices and as vectors. Now, examination of the square matrix in the preceding section that was built up from members of the sequence $\{f\}$ reveals that in each row the sequence $\{f\}$ was written backward, the sequence shifting one step from each row to the next. This reflects the mode of calculation of the serial product, where the sequence $\{f\}$ would be written on a movable strip of paper alongside a sequence $\{g\}$ and shifted one step after each cycle of forming products and adding. In the procedure for taking serial products there is no provision for changing the sequence $\{f\}$ from step to step. Indeed it is the essence of convolution, and of serial multiplication, that no such change occurs. But in the square-matrix formulation, it is just as convenient to express such changes as not. Thus the matrix notation allows for a general situation of which the serial product is a special case.

This more general situation is simply the general linear transformation as contrasted with the general shift-invariant linear transformation. If the indexing of a sequence is with respect to time, then serial multiplication is the most general time-invariant linear transformation, and would for example be applied in problems concerned with linear filters whose elements did not change as time went by. If, on the other hand, one were dealing with the passage of a signal through a filter containing time-varying linear elements such as motor-driven potentiometers, then the relationship between output and input could not be expressed in the form of a serial product. In such circumstances, where convolution is inapplicable, the property referred to loosely as "harmonic response to harmonic excitation" also breaks down.

The autocorrelation function

The self-convolution of a function $f(x)$ is given by

$$f * f = \int_{-\infty}^{\infty} f(u)f(x - u)\, du.$$

Suppose, however, that prior to multiplication and integration we do not reverse one of the two component factors; then we have the integral

$$\int_{-\infty}^{\infty} f(u)f(u - x)\, du,$$

which may be denoted by $f \star f$. A single value of $f \star f$ is represented by

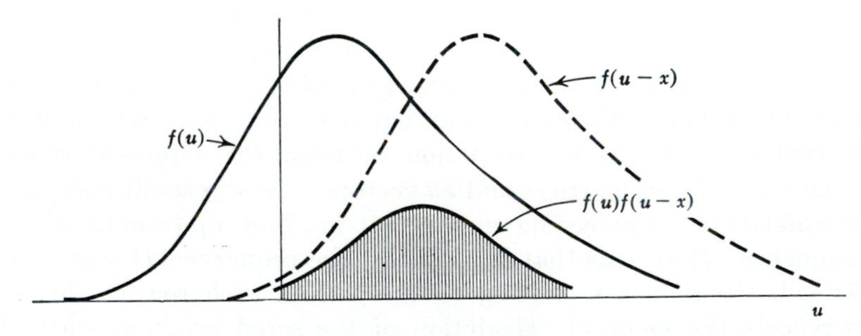

Fig. 3.11 The autocorrelation function represented by an area (shown shaded).

the shaded area in Fig. 3.11. A moment's thought will show that if the function f is to be displaced relative to itself by an amount x (without reversal), then the integral of the product will be the same whether x is positive or negative. In other words, if $f(x)$ is a real function, then $f \star f$ is an even function, a fact which is not true in general of the convolution integral. It follows that

$$f \star f = \int_{-\infty}^{\infty} f(u)f(u - x)\, du = \int_{-\infty}^{\infty} f(u)f(u + x)\, du,$$

which, of course, is deducible from the previous expression by substitution of $w = u - x$.

It is shown in the appendix to this chapter that the value of $f \star f$ at its origin is a maximum; that is, as soon as some shift is introduced, the integral of the product falls off.

When $f(x)$ is complex it is customary to use the expression

$$\int_{-\infty}^{\infty} f(u)f^*(u - x)\, du$$

or

$$\int_{-\infty}^{\infty} f^*(u)f(u + x)\, du.$$

(Note that it is possible to place the asterisk in the wrong place and thus obtain the conjugate of the standard version.)

It is often convenient to normalize by dividing by the central value. Then we may define a quantity $\gamma(x)$ given by

$$\gamma(x) = \frac{\int_{-\infty}^{\infty} f^*(u)f(u + x)\, du}{\int_{-\infty}^{\infty} f(u)f^*(u)\, du}$$

and it is clear that

$$\gamma(0) = 1.$$

We shall refer to $\gamma(x)$ as the autocorrelation function of $f(x)$. However, it often happens that the question of normalization is unimportant in a particular application, and the character of the autocorrelation is of more interest than the magnitude; then the nonnormalized form is referred to as the autocorrelation function.

As an example, take the function $f(x)$ defined by

$$f(x) = \begin{cases} 0 & x < 0 \\ 1 - x & 0 < x < 1 \\ 0 & x > 1. \end{cases}$$

Then, for $0 < x < 1$,

$$\begin{aligned}\int_{-\infty}^{\infty} f^*(u)f(u+x)\,du &= \int_0^{1-x} (1-u)[1-(u+x)]\,du \\ &= \frac{1}{3} - \frac{x}{2} + \frac{x^3}{6}.\end{aligned}$$

The best way to determine the limits of integration is to make a graph such as the one in Fig. 3.11. Where $x > 1$, the integral is zero, and since the integral is known to be an even function of x, we have

$$\int_{-\infty}^{\infty} f^*(u)f(u+x)\,du = \begin{cases} \dfrac{1}{3} - \dfrac{|x|}{2} + \dfrac{|x|^3}{6} & -1 < x < 1 \\ 0 & |x| > 1. \end{cases}$$

The central value, obtained by putting $x = 0$, is $\frac{1}{3}$; hence

$$\gamma(x) = \begin{cases} 1 - \dfrac{3|x|}{2} + \dfrac{|x|^3}{2} & -1 < x < 1 \\ 0 & |x| > 1. \end{cases}$$

We note that $\gamma(0) = 1$, and as with the convolution integral, the area under the autocorrelation function may be checked (it is $\frac{1}{4}$) to verify that it is equal to the square of the area under $f(x)$.

A second example is furnished by

$$f(x) = \begin{cases} e^{-\alpha x} & x > 0 \\ 0 & x < 0. \end{cases}$$

Then

$$\begin{aligned}\int_{-\infty}^{\infty} f^*(u)f(u+x)\,du &= \int_0^{\infty} e^{-\alpha u}\, e^{-\alpha(x+u)}\,du \\ &= \frac{e^{-\alpha|x|}}{2\alpha},\end{aligned}$$

and

$$\gamma(x) = e^{-\alpha|x|}.$$

The autocorrelation function is often used in the study of functions representing observational data, especially observations exhibiting some degree of randomness, and ingenious computing machines have been devised to carry out the integration on data in various forms. In any case, the digital computation is straightforward in principle. In the theory of such phenomena, however, as distinct from their observation and analysis, one wishes to treat functions that run on indefinitely, which often means that the infinite integral does not exist. This may always be handled by considering a segment of the function of length X and replacing the values outside the range of this segment by zero; any difficulty associated with the fact that the original function ran on indefinitely is thus removed. The autocorrelation $\gamma(x)$ of the new function $f_X(x)$, which is zero outside the finite segment, is given, according to the definition, by

$$\gamma(x) = \frac{\int_{-\infty}^{\infty} f_X{}^*(u) f_X(u+x)\,du}{\int_{-\infty}^{\infty} f_X(u) f_X{}^*(u)\,du} = \frac{\int_{-\frac{1}{2}X}^{\frac{1}{2}X} f^*(u) f(u+x)\,du}{\int_{-\frac{1}{2}X}^{\frac{1}{2}X} f_X(u) f_X{}^*(u)\,du}.$$

Thus the infinite integrals are reduced to finite ones, and, of course, this is precisely what happens in fact when a calculation is made on a finite quantity of observational data.

Now if we are given a function $f(x)$ to which the definition of $\gamma(x)$ does not apply because of its never dying out, and we have calculated $\gamma(x)$ for finite segments, we can then make the length X of the segment as long as we wish. It may happen that as X increases, the values of $\gamma(x)$ settle down to a limit; in fact, the circumstances under which this happens are of very wide interest. This limit, when it exists, will be denoted by $C(x)$, and it is given by

$$C(x) = \lim_{X \to \infty} \frac{\int_{-\frac{1}{2}X}^{\frac{1}{2}X} f(u) f(u+x)\,du}{\int_{-\frac{1}{2}X}^{\frac{1}{2}X} [f(u)]^2\,du}.$$

As this branch of the subject is often restricted to real functions, such as signal waveforms, allowance for complex f has been dropped.

Exercise Show that when this expression is evaluated for a real function for which $\gamma(x)$ exists, then $C(x) = \gamma(x)$.

Since the limiting autocorrelation C is identical with γ in cases where γ is applicable, it would be logical to take C as defining the autocorrelation function, and this is often done. In the discussion that follows, however, we shall understand the term "autocorrelation function" to include the operation of passing to a limit only where that is necessary.

As an example of a function that does not die out and for which the infinite integral does not exist, consider a time-varying signal which is a combination of three sinusoids of arbitrary amplitudes and phases, given by

$$V(t) = A \sin(\alpha t + \phi) + B \sin(\beta t + \chi) + C \sin(\gamma t + \psi).$$

Then

$$\begin{aligned}
\int_{-\frac{1}{2}T}^{\frac{1}{2}T} V(t')V(t'+t)\,dt' &= \int_{-\frac{1}{2}T}^{\frac{1}{2}T} [A \sin(\alpha t' + \phi) A \sin(\alpha t' + \alpha t + \phi) \\
&\qquad + B \sin(\beta t' + \chi) B \sin(\beta t' + \beta t + \chi) \\
&\qquad + C \sin(\gamma t' + \psi) C \sin(\gamma t' + \gamma t + \psi) \\
&\qquad + \text{cross-product terms}]\,dt' \\
&= \int_{-\frac{1}{2}T}^{\frac{1}{2}T} \{A^2[\cos \alpha t - \cos(2\alpha t' + \alpha t + 2\phi)] \\
&\qquad + B^2[\cos \beta t - \cos(2\beta t' + \beta t + 2\chi)] \\
&\qquad + C^2[\cos \gamma t - \cos(2\gamma t' + \gamma t + 2\psi)] \\
&\qquad + \text{cross-product terms}\}\,dt' \\
&= A^2T \cos \alpha t + B^2T \cos \beta t + C^2T \cos \gamma t + F(t,T) + G(t,T),
\end{aligned}$$

where F stands for oscillatory terms and G for terms arising from cross products. As T increases, F and G become negligible with respect to the three leading terms, which increase in proportion to T. Hence

$$\frac{\int_{-\frac{1}{2}T}^{\frac{1}{2}T} V(t')V(t'+t)\,dt'}{\int_{-\frac{1}{2}T}^{\frac{1}{2}T} [V(t')]^2\,dt'} = \frac{A^2T \cos \alpha t + B^2T \cos \beta t + C^2T \cos \gamma t + F(t,T) + G(t,T)}{A^2T + B^2T + C^2T + F(0,T) + G(0,T)}$$

and

$$C(t) = \lim_{T\to\infty} \frac{\int_{-\frac{1}{2}T}^{\frac{1}{2}T} V(t')V(t'+t)\,dt'}{\int_{-\frac{1}{2}T}^{\frac{1}{2}T} [V(t')]^2\,dt'} = \frac{1}{A^2 + B^2 + C^2}(A^2 \cos \alpha t + B^2 \cos \beta t + C^2 \cos \gamma t).$$

Note that the limiting autocorrelation function $C(t)$ is a superposition of three *cosine* functions at the same frequencies as contained in the signal $V(t)$, but with different relative amplitudes, and that $C(0) = 1$.

Since the principal terms in the numerator of $\gamma(x)$ increase in proportion to X, another way to obtain a nondivergent result is to divide by X before proceeding to the limit. The expression

$$\lim_{X\to\infty} \frac{1}{X} \int_{-\frac{1}{2}X}^{\frac{1}{2}X} f(u)f(u+x)\,du$$

is not equal to unity at its origin but is generally referred to simply as the "autocorrelation function" because it becomes the same as $C(x)$ if divided by its central value (provided the central value is not zero). It may therefore be regarded as a nonnormalized form of $C(x)$.

When time is the independent variable it is customary to refer to the time average of the product $V(t)V(t + \tau)$, and write

$$\langle V(t)V(t + \tau)\rangle \equiv \lim_{T\to\infty} \frac{1}{T} \int_{-\frac{1}{2}T}^{\frac{1}{2}T} V(t)V(t + \tau)\, dt,$$

where the operation of time averaging is denoted by angular brackets according to the definition

$$\langle . \; . \; . \rangle \equiv \lim_{T\to\infty} \frac{1}{T} \int_{-\frac{1}{2}T}^{\frac{1}{2}T} . \; . \; . \; dt.$$

Using this notation for the example worked, we have

$$\langle V(t)V(t + \tau)\rangle = A^2 \cos \alpha\tau + B^2 \cos \beta\tau + C^2 \cos \gamma\tau,$$

which, of course, is a little simpler than the normalized expression. It should be noticed, however, that we do not get a useful result in cases for which γ exists.

A conspicuous feature of C in this example is the absence of any trace of the original phases. Hence, autocorrelation is not a reversible process; it is not possible to get back from the autocorrelation function to the original function from which it was derived. Autocorrelation thus involves a loss of information. In some branches of physics, such as radio interferometry and X-ray diffraction analysis of substances, it is easier to observe the autocorrelation of a desired function than to observe the function itself, and a lot of ingenuity is expended to fill in the lost information.

The character of the lost information can be seen by considering a cosinusoidal function of x. When this function is displaced relative to itself, multiplied with the unshifted function, and the result integrated, clearly the result will be the same as for a sinusoidal function of the same period and amplitude. Furthermore, the result will be the same for any harmonic function of x, with the same period and amplitude, and arbitrary origin of x. Thus the autocorrelation function does not reveal the phase of a harmonic function. Now, if a function is composed of several harmonic waves present simultaneously, then when it is displaced, multiplied, and integrated, the result can be calculated simply by considering the different periods one at a time. This is possible because the product of harmonic variations of different frequencies integrates to a negligible quantity. Consequently, each of the periodic waves may be slid along the x axis into any arbitrary phase, without affecting the autocorrelation. In particular, all the components may be shifted until they become cosine components, thus generating an even function which, among all possible functions with the same autocorrelation, possesses a certain uniqueness.

Pentagram notation for cross correlation

The cross correlation $f(x)$ of two real functions $g(x)$ and $h(x)$ is defined as

$$f(x) = g \star h = \int_{-\infty}^{\infty} g(u - x)h(u)\,du.$$

It is thus similar to the convolution integral except that the component $g(u)$ is simply displaced to $g(u - x)$, without reversal. To denote this operation of g on h we use a five-pointed star, or pentagram, instead of the asterisk that denotes convolution. While cross correlation is slightly simpler than convolution, its properties are less simple. Thus, whereas $g * h = h * g$, the cross-correlation operation of h on g is not the same as that of g on h; that is,

$$g \star h \neq h \star g.$$

By change of variables it can be seen that

$$\begin{aligned} f(x) &= g \star h = \int_{-\infty}^{\infty} g(u - x)h(u)\,du = \int_{-\infty}^{\infty} g(u)h(u + x)\,du \\ f(-x) &= h \star g = \int_{-\infty}^{\infty} h(u - x)g(u)\,du = \int_{-\infty}^{\infty} h(u)g(u + x)\,du; \\ g * h &= g(-) \star h. \end{aligned}$$

As in the case of the autocorrelation function, the cross correlation is often normalized so as to be equal to unity at the origin, and when appropriate the average $\langle g(u - x)h(u)\rangle$ is used instead of the infinite integral. When the functions are complex it is customary to define the (complex) cross-correlation function by

$$g^* \star h = \int_{-\infty}^{\infty} g^*(u - x)h(u)\,du = \int_{-\infty}^{\infty} g^*(u)h(u + x)\,du.$$

The energy spectrum

We shall refer to the squared modulus of a transform as the energy spectrum; that is, $|F(s)|^2$ is the energy spectrum of $f(x)$. The term is taken directly from the physical fields where it is used. It will be seen that there is not a one-to-one relationship between $f(x)$ and its energy spectrum, for although $f(x)$ determines $F(s)$ and hence also $|F(s)|^2$, it would be necessary to have[2] pha $F(s)$ as well as $|F(s)|$ in order to reconstitute $f(x)$.

[2] The phase angle of the complex quantity $F(s)$ is written pha $F(s)$ and is defined as follows: if there is a complex variable $z = r \exp i\theta$, then pha $z = \theta$.

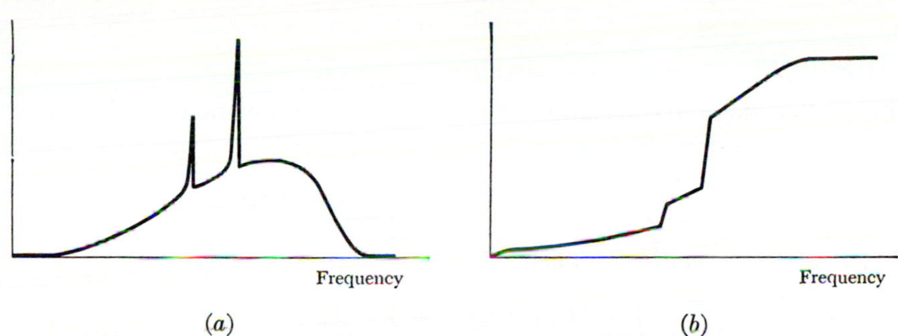

Fig. 3.12 Spectra of X radiation from molybdenum: (a) power spectrum; (b) cumulative spectrum.

Knowledge of the energy spectrum thus conveys a certain kind of information about $f(x)$ which says nothing about the phase of its Fourier components. It is the kind of information about an acoustical waveform which results from measuring the sound intensity as a function of frequency.

The information lost when only the energy spectrum can be given is of precisely the same character as that which is lost when the autocorrelation function has to do duty for the original function. The autocorrelation theorem, to be given later, expresses this equivalence.

When $f(x)$ is real, as it would be if it represented a physical waveform, the energy spectrum is an even function and is therefore fully determined by its values for $s \geqslant 0$. To stress this fact, the term "positive-frequency energy spectrum" may be used to mean the part of $|F(s)|^2$ for which $s \geqslant 0$.

Since $|F(s)|^2$ has the character of energy density measured per unit of s, it would have to become infinite if a nonzero amount of energy were associated with a discrete value of s. This is the situation with an infinitely narrow spectral line. Now we can consider a cumulative distribution function which gives the amount of energy in the range 0 to s:

$$\int_0^s |F(s)|^2 \, ds.$$

Any spectral *lines* would then appear as finite discontinuities in the cumulative energy spectrum as suggested by Fig. 3.12, and some mathematical convenience would be gained by using the cumulative spectrum in conjunction with Stieltjes integral notation. The convenience is especially marked when it is a question of using the theory of distributions for some question of rigor. However, the matter is purely one of notation, and in cases where we have to represent concentrations of energy within bands much narrower than can be resolved in the given context, we shall use the delta-symbol notation described later.

Appendix

Prove that the autocorrelation of the real nonzero function $f(x)$ is a maximum at the origin, that is,

$$\int_{-\infty}^{\infty} f(u)f(u+x)\,du \leqslant \int_{-\infty}^{\infty} [f(u)]^2\,du.$$

Let ϵ be a real number. Then

$$\int_{-\infty}^{\infty} [f(u) + \epsilon f(u+x)]^2\,du > 0$$

and $$\int_{-\infty}^{\infty} [f(u)]^2\,du + 2\epsilon \int_{-\infty}^{\infty} f(u)f(u+x)\,du + \epsilon^2 \int_{-\infty}^{\infty} [f(u+x)]^2\,du > 0;$$

that is, $$a\epsilon^2 + b\epsilon + c > 0,$$

where $$a = c = \int_{-\infty}^{\infty} [f(u)]^2\,du$$

$$b = 2\int_{-\infty}^{\infty} f(u)f(u+x)\,du.$$

Now, if the quadratic expression in ϵ may not be zero, that is, if it has no real root, then

$$b^2 - 4ac \leqslant 0.$$

Hence in this case $b/2 \leqslant a$, or

$$\frac{\int_{-\infty}^{\infty} f(u)f(u+x)\,du}{\int_{-\infty}^{\infty} [f(u)]^2\,du} \leqslant 1.$$

The equality is achieved at $x = 0$; consequently the autocorrelation function can nowhere exceed its value at the origin. The argument is the one used to establish the Schwarz inequality and readily generalizes to give the similar result for the complex autocorrelation function.

Exercise Extend the argument with a view to showing that the equality cannot be achieved for any value of x save zero.

Problems

1 Calculate the following serial products, checking the results by summation. Draw graphs to illustrate.

a. $\{6\ 9\ 17\ 20\ 10\ 1\} * \{3\ 8\ 11\}$
b. $\{1\ 1\ 1\ 1\ 1\} * \{1\ 1\ 1\ 1\}$
c. $\{1\ 4\ 2\ 3\ 5\ 3\ 3\ 4\ 5\ 7\ 6\ 9\} * \{1\ 1\}$
d. $\{1\ 4\ 2\ 3\ 5\ 3\ 3\ 4\ 5\ 7\ 6\ 9\} * \{\frac{1}{2}\ \frac{1}{2}\}$

e. {1 4 2 3 5 3 3 4 5 7 6 9} * {1 2 1}
f. {1 4 2 3 5 3 3 4 5 7 6 9} * {$\frac{1}{4}$ $\frac{1}{2}$ $\frac{1}{4}$}
g. {1 2 3 7 12 19 21 22 18 13 7 5 3 2 1} * {1 −1}
h. {1 2 3 7 12 19 21 22 21 18 13 7 5 3 2 1} * {1 1 1 1 . . . }
i. {1 1} * {1 1} * {1 1}
j. {1 1 0 0 1 0 1} * {1 1 0 0 1 0 1}
k. {1 1 0 0 1 0 1} * {1 0 1 0 0 1 1}
l. {*a* *b* *c* *d* *e*} * {*e* *d* *c* *b* *a*}
m. {1 3 1} * {1 0 0 0 0 0 ±1}
n. {1 3 1} * {1 2 2}
o. Multiply 131 by 122
p. Multiply 10,301 by 10,202
q. {1 8 1} * {1 2 2}
r. Comment on the smoothness of your results in *d* and *f* relative to the longer of the two given sequences.
s. Consider the result of *i* in conjunction with Pascal's pyramid of binomial coefficients.
t. Seek longer sequences with the same property you discovered in *k*.
u. Contemplate *j*, *k*, *l*, *m*, and *n* with a view to discerning what leads to serial products which are even.
v. Master the implication of *n*, *o*, *p*, and *q*, and design a mechanical desk computer to perform serial multiplication.

2 Derive the following results, where $H(x)$ is the Heaviside unit step function (Chapter 4):

$$\begin{aligned}
x^2H(x) * e^xH(x) &= (2e^x - x^2 - 2x - 2)H(x) \\
[\sin x\ H(x)]^{*2} &= \tfrac{1}{2}(\sin x - x\cos x)H(x) \\
[(1 - x)H(x)] * [e^xH(x)] &= xH(x) \\
H(x) * [e^xH(x)] &= (e^x - 1)H(x) \\
[e^xH(x)]^{*2} &= xe^xH(x) \\
[e^xH(x)]^{*3} &= \tfrac{1}{2}x^2e^xH(x)
\end{aligned}$$

3 Prove the commutative property of convolution, that is, that $f * g \equiv g * f$.

4 Prove the associative rule $f * (g * h) \equiv (f * g) * h$.

5 Prove the distributive rule for addition $f * (g + h) \equiv f * g + f * h$.

6 The function f is the convolution of g and h. Show that the self-convolutions of f, g, and h are related in the same way as the original functions.

7 If $f = g \star h$, show that $f \star f = (g \star g) \star (h \star h)$.

8 Show that if a is a constant, $a(f * g) = (af) * g = f * (ag)$.

9 Establish a theorem involving $f(g * h)$.

10 Prove that the autocorrelation function is hermitian, that is, that $C(-u) = C^*(u)$, and hence that when the autocorrelation function is real it is even. Note that if the autocorrelation function is imaginary it is also odd; give some thought to devising a function with an odd autocorrelation function.

11 Prove that the sum and product of two autocorrelation functions are each hermitian.

12 Alter the origin of $f(x)$ until $f * f \Big|_0$ is a maximum. Investigate the assertion that the new origin defines an axis of maximum symmetry, making any necessary modification. Investigate the merits of the parameter

$$\frac{f * f \Big|_0}{f \star f \Big|_0},$$

to be considered a measure of "degree of evenness."

13 Show that if $f(x)$ is real,

$$\int_{-\infty}^{\infty} f(x)f(-x)\,dx = \int_{-\infty}^{\infty} [E(x)]^2\,dx - \int_{-\infty}^{\infty} [O(x)]^2\,dx,$$

and note that the left-hand side is the central value of the self-convolution of $f(x)$; that is, $f * f \Big|_0$.

14 Find reciprocal sequences for $\{1\ 3\ 3\ 1\}$ and $\{1\ 4\ 6\ 4\ 1\}$.

15 Find reciprocal sequences for $\{1\ 1\}$ and $\{1\ 1\ 1\}$.

16 Establish a general procedure for finding the reciprocal of finite or semi-infinite sequences and test it on the following cases:

$$\{64\ 32\ 16\ 8\ 4\ 2\ 1\ .\ .\ .\}$$
$$\{64\ 64\ 48\ 32\ 20\ 12\ 7\ 4\ .\ .\ .\}$$
$$\left\{1\quad e^{-\frac{1}{10}}\cos\left(\frac{\pi}{10}\right)\quad e^{-\frac{2}{10}}\cos\left(\frac{2\pi}{10}\right)\quad .\ .\ .\ e^{-n/10}\cos\left(\frac{n\pi}{10}\right)\quad .\ .\ .\right\}$$

17 Find approximate numerical values for a function $f(x)$ such that

$$f(x) * E(x)$$

is zero when evaluated numerically by serial multiplication of values taken at intervals of 0.2 in x, except at the origin. Normalize $f(x)$ so that its integral is approximately unity.

18 The cross correlation $g \star h$ is to be normalized to unity at its maximum value. It is argued that

$$0 \leqslant \int [g(u) - h(u+x)]^2\,du = \int g^2\,du - 2\int g(u)h(u+x)\,du + \int h^2\,du,$$

and therefore that

$$\int g(u)h(u+x)\,du \leqslant \tfrac{1}{2}\int g^2\,du + \tfrac{1}{2}\int h^2\,du = M.$$

Consequently $(g \star h)/M$ is the desired quantity. Correct the fallacy in this argument.

Chapter 4 Notation for some useful functions

Many useful functions in Fourier analysis have to be defined piecewise because of abrupt changes. For example, we may consider the function $f(x)$ such that

$$f(x) = \begin{cases} 0 & x < 0 \\ x & 0 \leqslant x \leqslant 1 \\ 1 & x > 1. \end{cases}$$

This function, though simple in itself, is awkwardly expressed in comparison with a function such as, for example, $1 + x^2$, whose algebraic expression compactly states, over the infinite range of x, the arithmetical operations by which it is formed. For many mathematical purposes a function which is piecewise analytic is not simple to deal with, but for physical purposes a "sloping step function," to give it a name, may be at least as simple as a smoother function.

Fourier himself was concerned with the representation of functions given graphically, and according to E. W. Hobson "was the first fully to grasp the idea that a single function may consist of detached portions given arbitrarily by a graph."

To regain compactness and clarity of notation, we introduce a number of simple functions embodying various kinds of abrupt behavior. Also included here is a section dealing with sinc x, the important interpolating function, which is the transform of a discontinuous function, and some reference material on notations for the Gaussian function.

Rectangle function of unit height and base, $\Pi(x)$

The rectangle function of unit height and base, which is illustrated in Fig. 4.1, is defined by

$$\Pi(x) = \begin{cases} 0 & |x| > \frac{1}{2} \\ (\frac{1}{2} & |x| = \frac{1}{2}) \\ 1 & |x| < \frac{1}{2}. \end{cases}$$

It provides simple notation for segments of functions which have simple expressions, for example, $f(x) = \Pi(x) \cos \pi x$ is compact notation for

$$f(x) = \begin{cases} 0 & x < -\frac{1}{2} \\ \cos \pi x & -\frac{1}{2} < x < \frac{1}{2} \\ 0 & \frac{1}{2} < x \end{cases}$$

(see Fig. 4.2). We may note that $h\Pi[(x - c)/b]$ is a displaced rectangle function of height h and base b, centered on $x = c$ (see Fig. 4.3). Hence, purely by multiplication by a suitably displaced rectangle function, we

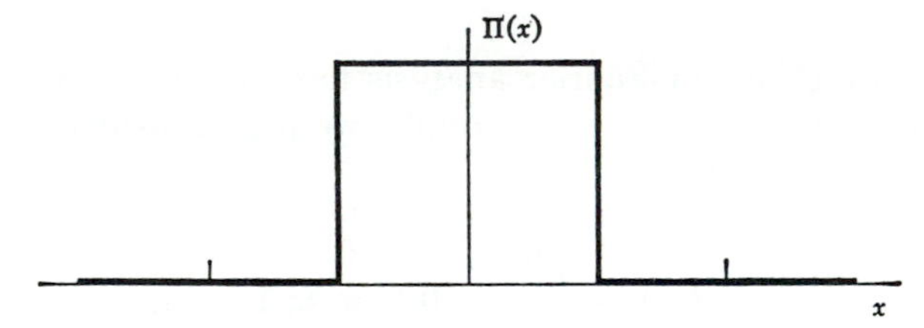

Fig. 4.1 The rectangle function of unit height and base, $\Pi(x)$.

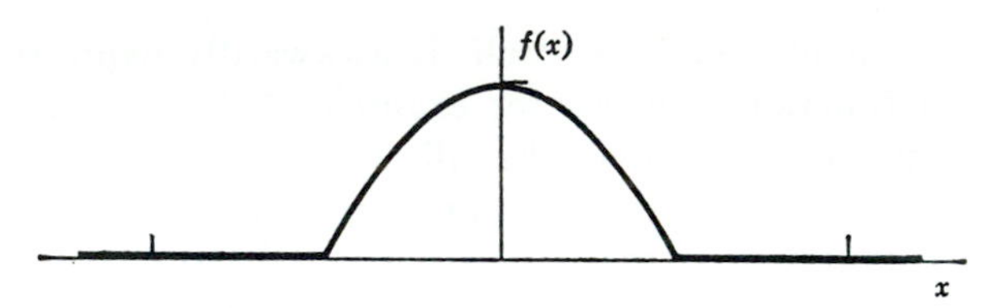

Fig. 4.2 A segmented function expressed by $\Pi(x) \cos \pi x$.

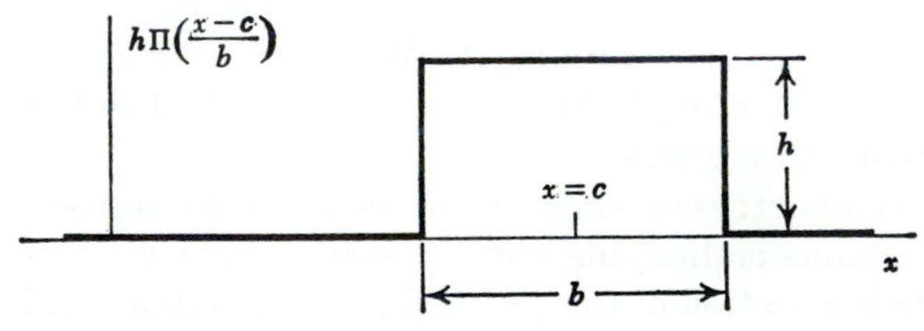

Fig. 4.3 A displaced rectangle function of arbitrary height and base expressed in terms of $\Pi(x)$.

can select, or gate, any segment of a given function, with any amplitude, and reduce the rest to zero.

The rectangle function also enters through convolution into expressions for running means, and, of course, in the frequency domain multiplication by the rectangle function is an expression of ideal low-pass filtering. It is important in the theory of convergence of Fourier series, where it is generally known as Dirichlet's discontinuous factor. The notation rect x, which has been suggested for this function, does not lend itself readily to notation such as $\overline{\Pi}$, $\Pi * \Pi$, and Πf.

For reasons explained below, it is almost never important to specify the values at $x = \pm\frac{1}{2}$, that is, at the points of discontinuity, and we shall normally omit mention of those values. Likewise, it is not necessary or desirable to emphasize the values $\Pi(\pm\frac{1}{2}) = \frac{1}{2}$ in graphs; it is preferable to show graphs which are reminiscent of high-quality oscillograms (which, of course, would never show extra brightening halfway up the discontinuity).

The triangle function of unit height and area, $\Lambda(x)$

By definition,

$$\Lambda(x) = \begin{cases} 0 & |x| > 1 \\ 1 - |x| & |x| < 1. \end{cases}$$

This function, which is illustrated in Fig. 4.4, gains its importance largely from being the self-convolution of $\Pi(x)$, but it has other uses—for example, in giving compact notation for polygonal functions (continuous functions consisting of linear segments).

Note that $h\Lambda(x/\frac{1}{2}b)$ is a triangle function of height h, base b, and area $\frac{1}{2}hb$.

Various exponentials and Gaussian and Rayleigh curves

Figure 4.5 shows, from left to right, a rising exponential, a falling exponential, a truncated falling exponential, and a double-sided falling exponential.

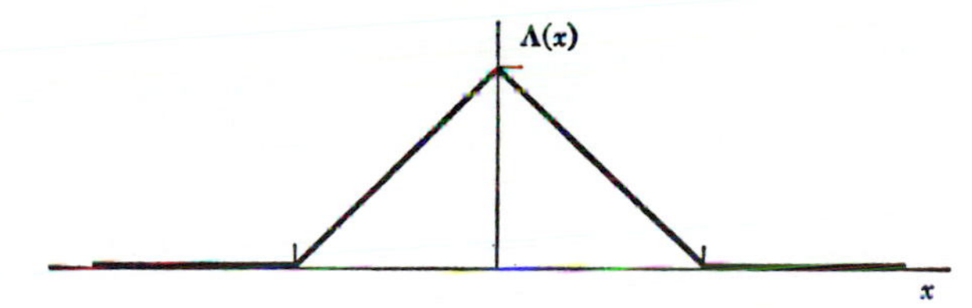

Fig. 4.4 The triangle function of unit height and area, $\Lambda(x)$.

Figure 4.6 shows the Gaussian function exp $(-\pi x^2)$. Of the various ways of normalizing this function, we have chosen one in which both the central ordinate and the area under the curve are unity, and certain advantages follow from this choice. The Fourier transform of the Gaussian function is also Gaussian, and proves to be normalized in precisely the same way under our choice. In statistics the Gaussian distribution is referred to as the "normal (error) distribution with zero mean" and is normalized so that the area and the standard deviation are unity. We may use the term "probability ordinate" to distinguish the form

$$\frac{1}{(2\pi)^{\frac{1}{2}}} e^{-\frac{1}{2}x^2}.$$

When the standard deviation is σ, the probability ordinate is

$$\frac{1}{\sigma(2\pi)^{\frac{1}{2}}} e^{-x^2/2\sigma^2},$$

and the area under the curve remains unity. The central ordinate is equal to $0.3989/\sigma$. Prior to the strict standardization now prevailing

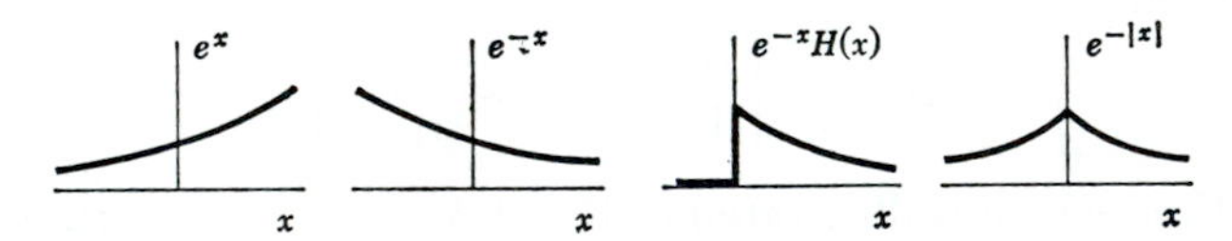

Fig. 4.5 Various exponential functions.

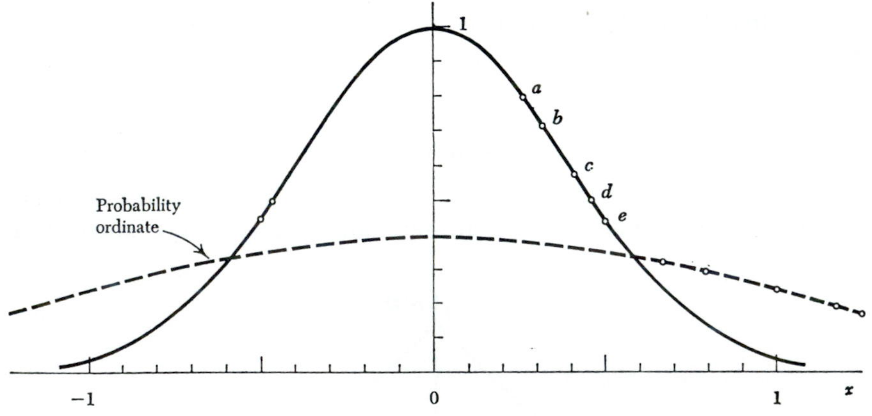

Fig. 4.6 The Gaussian function exp $(-\pi x^2)$ and the probability ordinate $(2\pi)^{-\frac{1}{2}}$ exp $(-\frac{1}{2}x^2)$.

and that it dies away more rapidly than any power of x; that is,

$$\lim_{x \to \infty} x^n e^{-x^2} = 0$$

for all n.

Figure 4.7 shows these two important sequences; later we emphasize that corresponding members are Fourier transform pairs.

Heaviside's unit step function, $H(x)$

An indispensable aid in the representation of simple discontinuities, the unit step function is defined by

$$H(x) = \begin{cases} 0 & x < 0 \\ (\frac{1}{2} & x = 0) \\ 1 & x > 0, \end{cases}$$

and is illustrated in Fig. 4.8. It represents voltages which are suddenly switched on or forces which begin to act at a definite time and are constant thereafter. Furthermore, any function with a jump can be decomposed into a continuous function plus a step function suitably displaced. As a simple example of additive use, consider the rectangle function $\Pi(x)$, which has two unit discontinuities, one positive and one negative. If these are removed, nothing remains. Hence $\Pi(x)$ is expressible entirely in terms of step functions as follows (see Fig. 4.9):

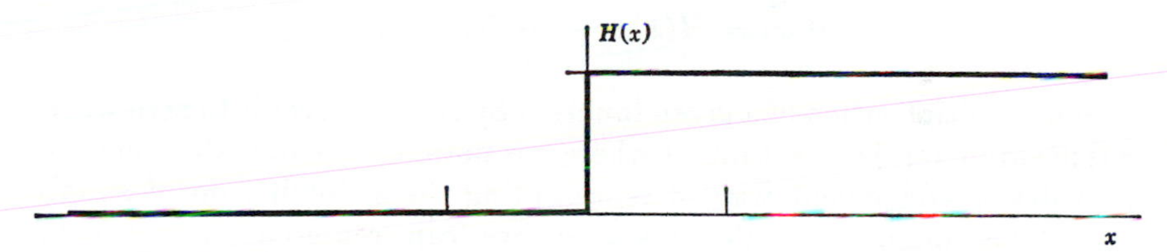

Fig. 4.8 The unit step function.

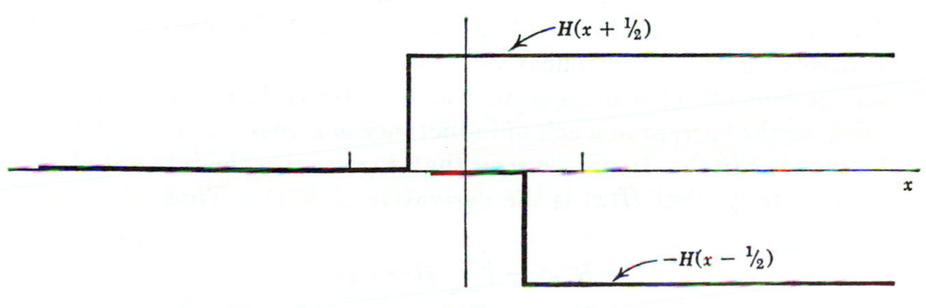

Fig. 4.9 Two functions whose sum is $\Pi(x)$.

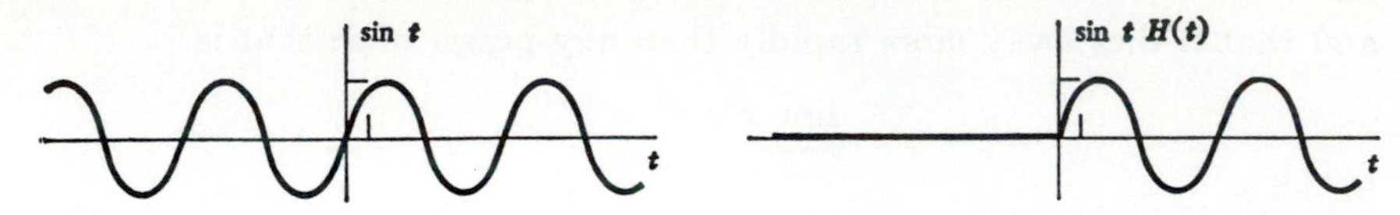

Fig. 4.10 Graphs of sin t and sin t $H(t)$ which exhibit a principal use of the unit step function.

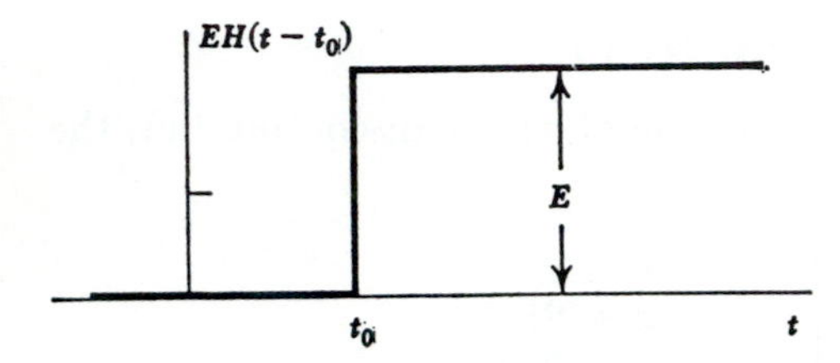

Fig. 4.11 A voltage E which appears at $t = t_0$ represented in step-function notation by $EH(t - t_0)$.

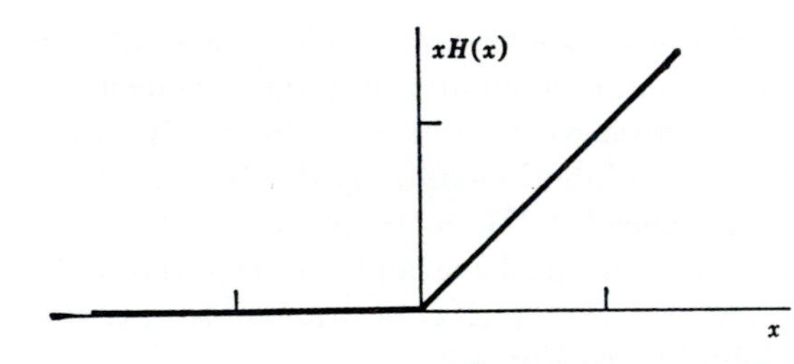

Fig. 4.12 The ramp function $xH(x)$.

$$\Pi(x) = H(x + \tfrac{1}{2}) - H(x - \tfrac{1}{2}).$$

Since multiplication of a given function by $H(x)$ reduces it to zero where x is negative but leaves it intact where x is positive, the unit step function provides a convenient way of representing the switching on of simply expressible quantities. For instance, we can represent a sinusoidal quantity which switches on at $t = 0$ by $\sin t\, H(t)$ (see Fig. 4.10). A voltage which has been zero until $t = t_0$ and then jumps to a steady value E is represented by $EH(t - t_0)$ (see Fig. 4.11).

The ramp function $R(x) = xH(x)$ furnishes a further example of notation involving $H(x)$ multiplicatively (see Fig. 4.12). $(F/m)R(t)$ represents the velocity of a mass m to which a steady force $FH(t)$ has been applied, or the current in a coil of inductance m across which the potential difference is $FH(t)$. It will be seen that $R(x)$ is also the integral of $H(x)$ and, conversely, that $H(x)$ is the derivative of $R(x)$. Thus

$$R(x) = \int_{-\infty}^{x} H(x')\, dx'$$

and

$$R'(x) = H(x).$$

Step-function notation plays a role in simplifying integrals with variable limits of integration by reducing the integrand to zero in the range beyond the original limits. Then constant limits such as $-\infty$ to ∞ or 0 to ∞ can be used. For example, the function $H(x - x')$ is zero where $x' > x$, and therefore $\int_{-\infty}^{x} f(x')\,dx'$ can always be written $\int_{-\infty}^{\infty} f(x')H(x - x')\,dx'$. Thus in the example appearing above we can write

$$R(x) = \int_{-\infty}^{x} H(x')\,dx' = \int_{-\infty}^{\infty} H(x')H(x - x')\,dx'.$$

This last integral reveals the character of $R(x)$ as a convolution integral,

$$R(x) = H(x) * H(x),$$

and in general we may now note that convolution with $H(x)$ means integration (see Fig. 4.13):

$$H(x) * f(x) = \int_{-\infty}^{\infty} f(x')H(x - x')\,dx' = \int_{-\infty}^{x} f(x')\,dx'$$

or

$$f(x) = \frac{d}{dx}[H(x) * f(x)].$$

Usually, it is not important to define $H(0)$, but for the sake of compatibility with the theory of single-valued functions it is desirable to assign a value, usually $\frac{1}{2}$. Certain internal consistencies are then likely to be observed. For instance, in the equation $R'(x) = H(x)$ it is clear that $R'(0+) = H(0+) = 1$ and that $R'(0-) = H(0-) = 0$, and if we deem $R'(0)$ to mean $\lim_{\Delta x \to 0} [R(x + \frac{1}{2}\Delta x) - R(x - \frac{1}{2}\Delta x)]/\Delta x$, we find $R'(0) = \frac{1}{2}$. Furthermore, the Fourier integral, when it converges at a point of discontinuity, gives the midvalue.

However, there is no obligation to take $H(0) = \frac{1}{2}$; it is not uncommonly taken as zero. This is a natural consequence of a point of view according to which (see Fig. 4.14)

$$\hat{H}(x) = \lim_{\tau \to 0} [(1 - e^{-x/\tau})H(x)].$$

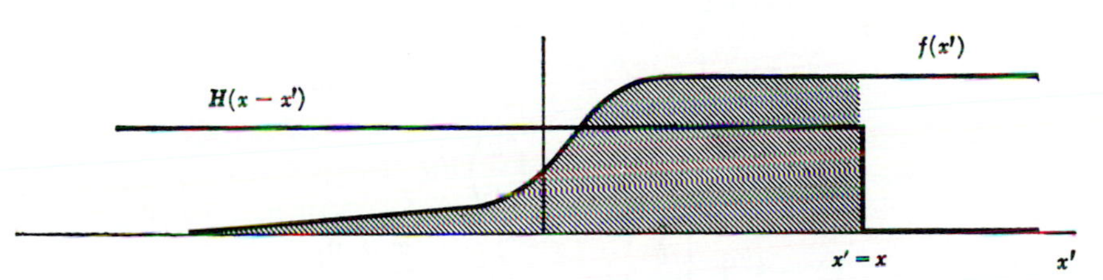

Fig. 4.13 The shaded area is $\int_{-\infty}^{x} f(x')\,dx'$, *or a value of the convolution of* $f(x)$ *with* $H(x)$.

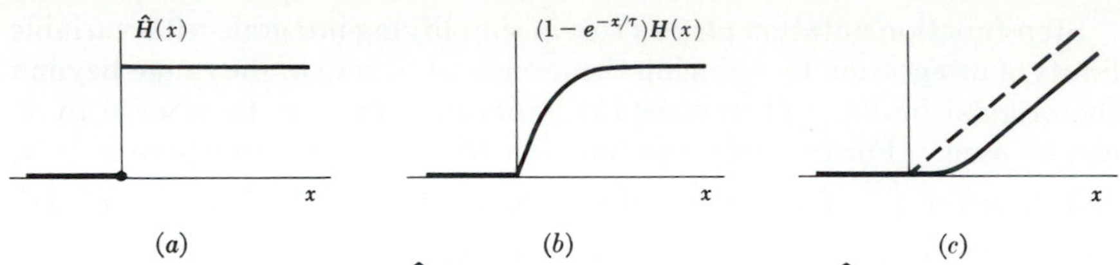

(a) (b) (c)

Fig. 4.14 (a) *The function* $\hat{H}(x)$; (b) *an approximation to* $\hat{H}(x)$; (c) *the integral of* (b).

Then $\hat{H}(0) = 0$. The circumflex indicates that there is a slight difference from the definition already decided on.

Thus
$$\hat{H}(x) = \begin{cases} 1 & x > 0 \\ 0 & x \leqslant 0 \end{cases}$$
and
$$H(x) = \hat{H}(x) + \tfrac{1}{2}\delta^0(x)$$
where
$$\delta^0(x) = \begin{cases} 0 & x \neq 0 \\ 1 & x = 0. \end{cases}$$

The function $\delta^0(x)$ is one of a class of *null functions* referred to below. No discrepancy need arise from the relation $R'(x) = \hat{H}(x)$, since the integral of $\hat{H}(x)$, regarded as

$$\lim_{\tau \to 0} \int_{-\infty}^{x} (1 - e^{-u/\tau})H(u)\, du,$$

is certainly $R(x)$; and $R'(0)$, regarded as the limiting slope at $x = 0$ of approximations such as those shown in Fig. 4.14*c*, is certainly equal to $\hat{H}(0)$, namely, 0.

It is sometimes useful to have continuous approximations to $H(x)$. The following examples all approach $H(x)$ as a limit for all x as $\tau \to 0$.

$$\tfrac{1}{2} + \frac{1}{\pi} \arctan \frac{x}{\tau}$$

$$\tfrac{1}{2} \operatorname{erfc}\left(-\frac{x}{\tau}\right) = \frac{1}{\pi^{\frac{1}{2}}} \int_{-x}^{\infty} \tau^{-1} e^{-u^2/\tau^2}\, du$$

$$\tfrac{1}{2} + \frac{1}{\pi} \operatorname{Si} \frac{\pi x}{\tau} = \int_{-\infty}^{x} \tau^{-1} \operatorname{sinc} \frac{u}{\tau}\, du$$

$$\int_{-\infty}^{x} \tau^{-1} \Pi\left(\frac{u}{\tau}\right) du$$

$$\tfrac{1}{2} + \begin{cases} \tfrac{1}{2}(1 - e^{-x/\tau}) & x > 0. \\ -\tfrac{1}{2}(1 - e^{x/\tau}) & x < 0. \end{cases}$$

An example which approaches $H(x)$ as $\tau \to 0$ for all x except $x = 0$ is

$$f(x,\tau) = \begin{cases} 0 & x < -\tau \\ 1 - e^{-(x+\tau)/\tau} & x > -\tau. \end{cases}$$

In this case

$$\lim_{\tau \to 0} f(x,\tau) = H(x) + (\tfrac{1}{2} - e^{-1})\, \delta^0(x),$$

since $f(0,\tau) = 1 - e^{-1}$ for all τ. A further example which approaches $H(x)$ as $\tau \to 0$ is

$$\int_{-\infty}^{x} \tau^{-1}\Lambda\left(\frac{u - \frac{1}{2}\tau}{\tau}\right) du.$$

The difference between $H(x)$ and any version such that $H(0) \neq \frac{1}{2}$ is a null function whose integral is always zero. If it were necessary to make physical observations of a quantity varying as $H(x)$ or $\hat{H}(x)$, with the finite resolving power to which physical observations are limited, it would not be possible to distinguish between the mathematically distinct alternatives, since the weighted means over nonzero intervals, which are the only quantities measurable, would be unaffected by the presence or absence of null functions. For physical applications of $H(x)$ it is therefore perhaps more graceful not to mention $H(0)$.

The sign function, sgn x

The function sgn x (pronounced signum x) is equal to $+1$ or -1, according to the sign of x (see Fig. 4.15). Thus

$$\operatorname{sgn} x = \begin{cases} -1 & x < 0 \\ 1 & x > 0. \end{cases}$$

Clearly it differs little from the step function $H(x)$ and has most of its properties. It has a positive discontinuity of 2. The relation to $H(x)$ is

$$\operatorname{sgn} x = 2H(x) - 1.$$

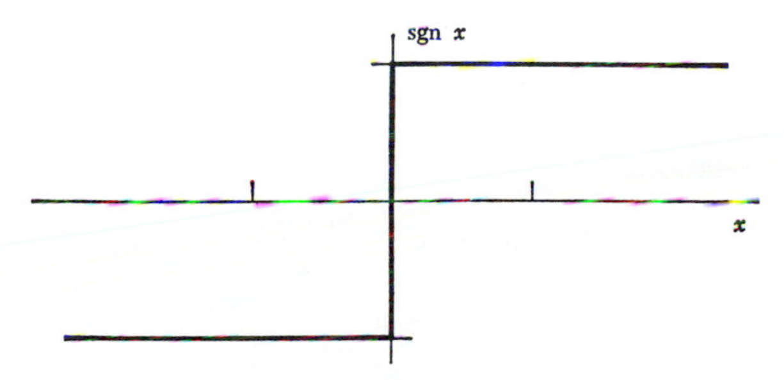

Fig. 4.15 The odd function sgn x.

However, $\lim_{A\to\infty} \int_{-A}^{A} \operatorname{sgn} x \, dx = 0$, whereas $\int_{-\infty}^{\infty} H(x)$ does not exist. Furthermore, we may note that sgn x, unlike $H(x)$, is an odd function, and this property sometimes makes it more useful in symbolic expressions.

The filtering or interpolating function, sinc x

We define

$$\operatorname{sinc} x = \frac{\sin \pi x}{\pi x},$$

a function with the properties that

$$\begin{aligned} \operatorname{sinc} 0 &= 1 \\ \operatorname{sinc} n &= 0 \qquad n = \text{nonzero integer} \\ \int_{-\infty}^{\infty} \operatorname{sinc} x \, dx &= 1. \end{aligned}$$

This function will appear so frequently, in many different connections, that it is convenient to have a special symbol for it, especially in some agreed normalized form. In the form chosen here, the central ordinate is unity and the total area under the curve is unity (see Fig. 4.16*a*); the word "sinc" appears in Woodward's book[1] and has achieved some currency. A table of the sinc function appears on page 368.

The unique properties of sinc x go back to its spectral character: it contains components of all frequencies up to a certain limit and none beyond. Furthermore, the spectrum is flat up to the cutoff frequency. By our choice of notation, sinc x and $\Pi(s)$ are a Fourier transform pair; the cutoff frequency of sinc x is thus 0.5 (cycles per unit of x).

When sinc x enters into convolution it performs ideal low-pass filtering; that is, it removes all components above its cutoff and leaves all below unaltered, and under certain special circumstances discussed later under the sampling theorem, it performs an important kind of interpolation.

In terms of the widely tabulated sine integral Si x (shown in Fig. 4.16*c*), where

$$\operatorname{Si} x = \int_0^x \frac{\sin u}{u} \, du,$$

we have the relations

$$\begin{aligned} \int_0^x \operatorname{sinc} u \, du &= \frac{\operatorname{Si}(\pi x)}{\pi} \\ \operatorname{sinc} x &= \frac{d}{dx} \frac{\operatorname{Si}(\pi x)}{\pi}, \end{aligned}$$

[1] P. M. Woodward, "Probability and Information Theory with Applications to Radar," McGraw-Hill Book Company, New York, 1953.

and for the integral of sinc x (see Fig. 4.16*b*) we have

$$H(x) * \operatorname{sinc} x = \int_{-\infty}^{x} \operatorname{sinc} u \, du = \frac{1}{2} + \frac{\operatorname{Si}(\pi x)}{\pi}.$$

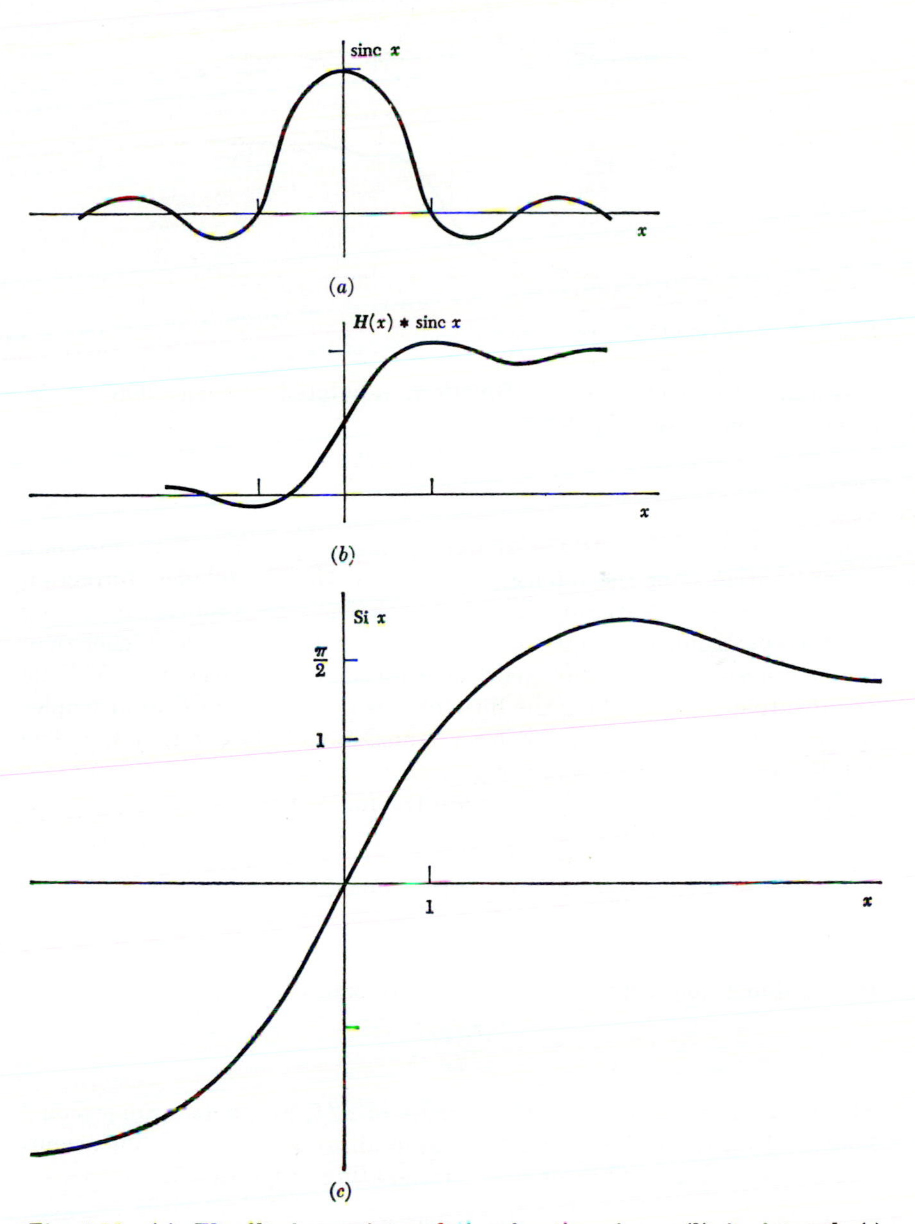

Fig. 4.16 (a) The filtering or interpolating function sinc x; (b) its integral; (c) the sine integral Si x.

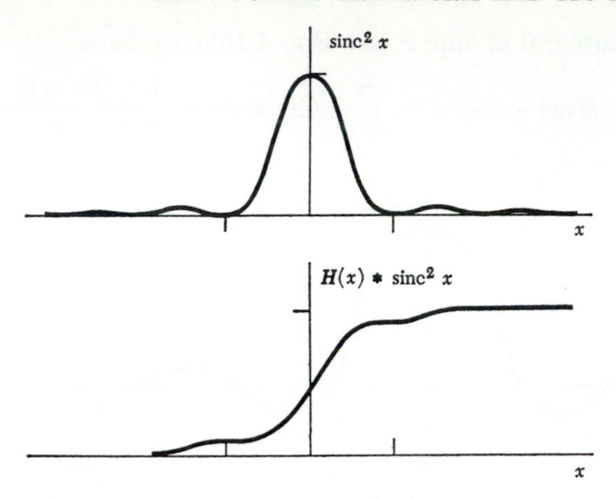

Fig. 4.17 The square of sinc x and its integral.

Another frequently needed function, tabulated on page 368, is the square of sinc x (see Fig. 4.17):

$$\operatorname{sinc}^2 x = \left(\frac{\sin \pi x}{\pi x}\right)^2.$$

This function represents the power radiation pattern of a uniformly excited antenna, or the intensity of light in the Fraunhofer diffraction pattern of a slit. Naturally, it shares with sinc x the property of having a cutoff spectrum, since squaring cannot generate frequencies higher than the sum-frequency of any pair of sinusoidal constituents. A little quantitative thought along the lines of this appeal to physical principles would soon reveal that the Fourier transform of $\operatorname{sinc}^2 x$ is $\Lambda(s)$. The cutoff frequency is one cycle per unit of x.

Among the properties of $\operatorname{sinc}^2 x$ are the following:

$$\operatorname{sinc}^2 0 = 1$$

$$\operatorname{sinc}^2 n = 0 \qquad n = \text{nonzero integer}$$

$$\int_{-\infty}^{\infty} \operatorname{sinc}^2 x = 1.$$

In two dimensions a function analogous to sinc x is

$$\frac{J_1(\pi r)}{2r},$$

which has unit volume, a central value of $\pi/4$, and a two-dimensional Fourier transform $\Pi(s)$. Another generalization to two dimensions, which has analogous filtering and interpolating properties, is

$$\operatorname{sinc} x \operatorname{sinc} y.$$

The two-dimensional Fourier transform of this function is $\Pi(u)\Pi(v)$.

Pictorial representation

Certain conventions in graphical representations will be adopted for the purpose of clarity. For example, theorems relating to Fourier transforms will be illustrated, where possible, by examples which are real. In these diagrams the points where the abscissa and ordinate are equal to unity are marked if it is appropriate to do so.

In representing purely imaginary quantities a dashed line is always

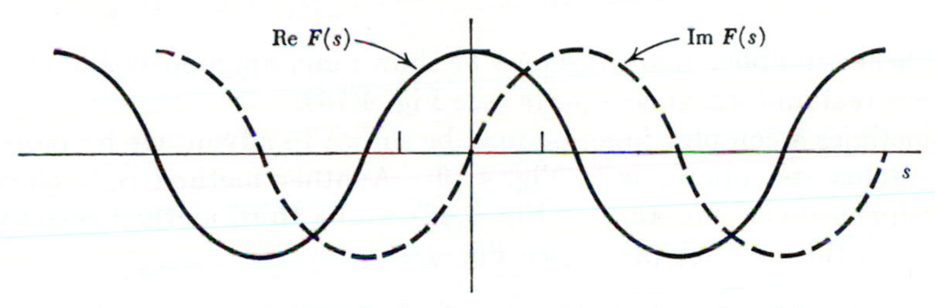

Fig. 4.18 Representation of a complex function $F(s)$ by its real and imaginary parts.

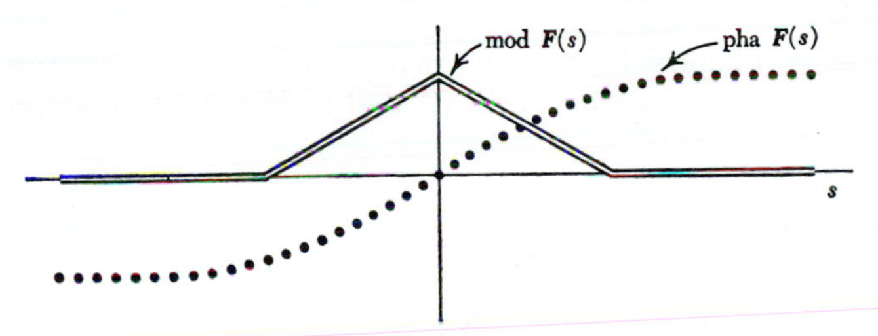

Fig. 4.19 Complex function shown in modulus and phase.

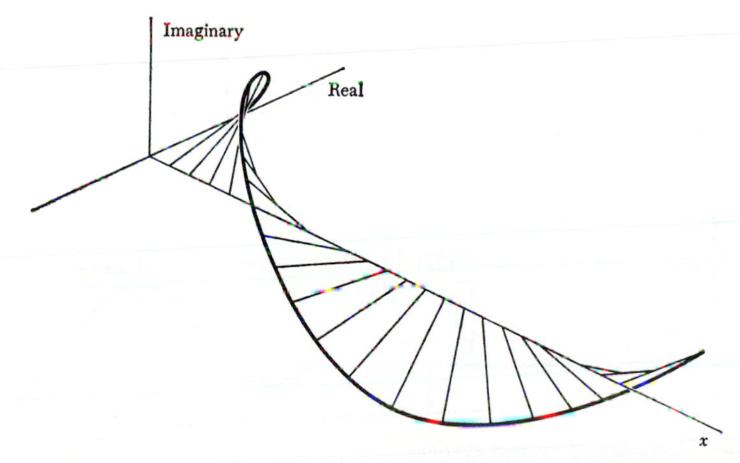

Fig. 4.20 Complex function in three dimensions.

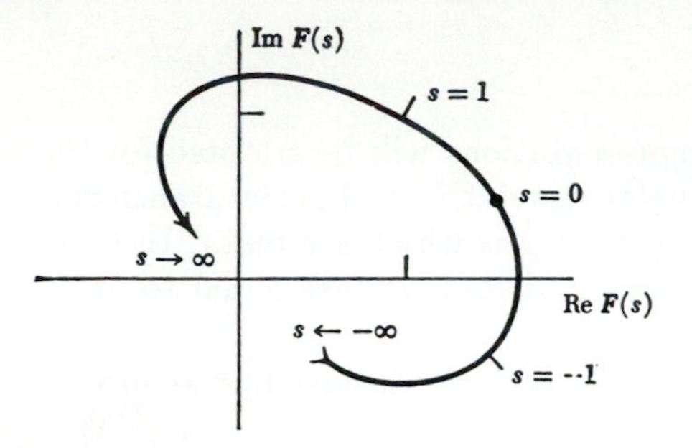

Fig. 4.21 Values of $F(s)$ on the complex plane of $F(s)$.

used; hence complex quantities may be shown unambiguously and clearly by their real and imaginary parts (see Fig. 4.18).

Sometimes a complex function may be shown to advantage by plotting its modulus and phase, as in Fig. 4.19. Another method is to show a three-dimensional diagram (see Fig. 4.20), and a third method is to show a locus on the complex plane (see Fig. 4.21).

Summary of special symbols

A small number of special symbols which are used extensively throughout this work are summarized in Table 4.1 and in Fig. 4.22 for reference.

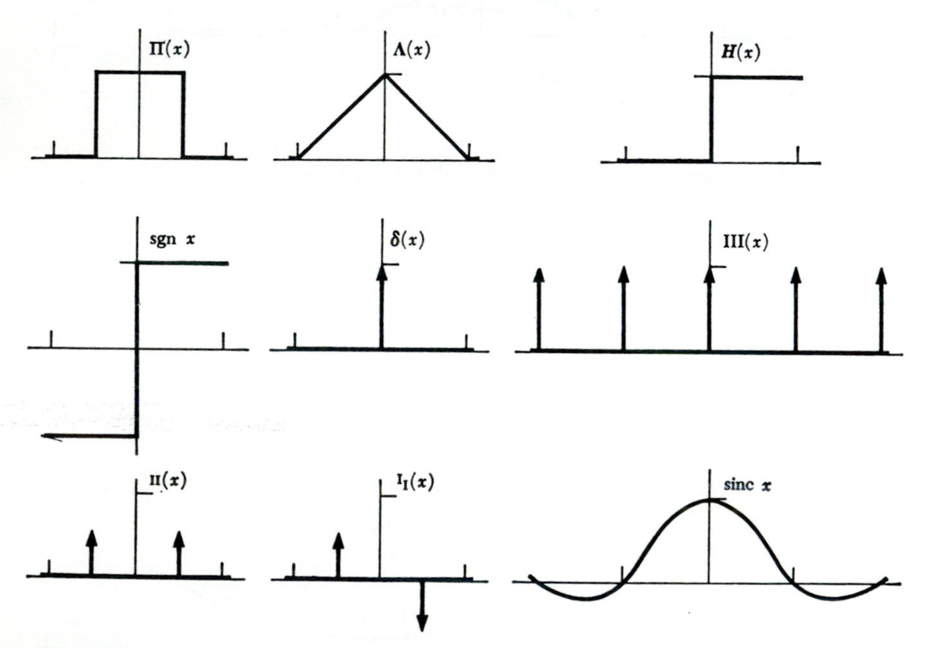

Fig. 4.22 Summary of special symbols.

Table 4.1 Special symbols

Function	*Notation*		
Rectangle	$\Pi(x) = \begin{cases} 1 \\ 0 \end{cases}$	$\begin{matrix} \|x\| < \frac{1}{2} \\ \|x\| > \frac{1}{2} \end{matrix}$	
Triangle	$\Lambda(x) = \begin{cases} 1 - \|x\| \\ 0 \end{cases}$	$\begin{matrix} \|x\| < 1 \\ \|x\| > 1 \end{matrix}$	
Heaviside unit step	$H(x) = \begin{cases} 1 \\ 0 \end{cases}$	$\begin{matrix} x > 0 \\ x < 0 \end{matrix}$	
Sign (signum)	$\operatorname{sgn} x = \begin{cases} 1 \\ -1 \end{cases}$	$\begin{matrix} x > 0 \\ x < 0 \end{matrix}$	
Impulse symbol*	$\delta(x)$		
Sampling or replicating symbol*	$\mathrm{III}(x) = \sum_{-\infty}^{\infty} \delta(x - n)$		
Even impulse pair	$\mathrm{II}(x) = \frac{1}{2}\delta(x + \frac{1}{2}) + \frac{1}{2}\delta(x - \frac{1}{2})$		
Odd impulse pair	${}^{\mathrm{I}}{}_{\mathrm{I}}(x) = \frac{1}{2}\delta(x + \frac{1}{2}) - \frac{1}{2}\delta(x - \frac{1}{2})$		
Filtering or interpolating	$\operatorname{sinc} x = \frac{\sin \pi x}{\pi x}$		
Asterisk notation for convolution	$f(x) * g(x) \triangleq \int_{-\infty}^{\infty} f(u)g(x - u)\, du$		(p. 25)
Asterisk notation for serial products	$\{f_i\} * \{g_i\} \triangleq \left\{\sum_j f_j g_{i-j}\right\}$		(p. 32)
Pentagram notation	$f(x) \star g(x) \triangleq \int_{-\infty}^{\infty} f(u)g(x + u)\, du$		(p. 40)
Various two-dimensional functions	${}^2\Pi(x,y) = \Pi(x)\Pi(y)$ ${}^2\delta(x,y) = \delta(x)\,\delta(y)$ ${}^2\mathrm{III}(x,y) = \mathrm{III}(x)\mathrm{III}(y)$ ${}^2\operatorname{sinc}(x,y) = \operatorname{sinc} x \operatorname{sinc} y$		

* See Chapter 5.

Problems

1 Show that

$$H(ax + b) = \begin{cases} H\left(x + \dfrac{b}{a}\right) & a > 0 \\ H\left(-x - \dfrac{b}{a}\right) & a < 0, \end{cases}$$

and hence that

$$H(ax + b) = H\left(x + \frac{b}{a}\right) H(a) + H\left(-x - \frac{b}{a}\right) H(-a).$$

2 Discuss the function $\frac{1}{2}[1 + x/(x^2)^{\frac{1}{2}}]$ used by Cauchy.

3 Show that the operation $H(x)\,*$ is an integrating operation in the sense that

$$H(x) * [f(x)H(x)] = \int_0^x f(x)\,dx.$$

4 Calculate $(d/dx)\,[\Pi(x) * H(x)]$ and prove that $(d/dx)[f(x) * H(x)] = f(x)$.

5 By evaluating the integral, prove that $\text{sinc } x * \text{sinc } x = \text{sinc } x$.

6 Prove that $\text{sinc } x * J_0(\pi x) = J_0(\pi x)$.

7 Prove that $4 \text{ sinc } 4x * \sin x = \sin x$.

8 Show that

$$\begin{aligned}\Pi(x) &= H(x + \tfrac{1}{2}) - H(x - \tfrac{1}{2})\\ &= H(\tfrac{1}{2} + x) + H(\tfrac{1}{2} - x) - 1\\ &= H(\tfrac{1}{4} - x^2)\\ &= \tfrac{1}{2}[\text{sgn } (x + \tfrac{1}{2}) - \text{sgn } (x - \tfrac{1}{2})]\end{aligned}$$

and that $\Pi(x^2) = \Pi(x/2^{\frac{1}{2}})$.

9 Show that

$$\begin{aligned}\Lambda(x) &= \Pi(x) * \Pi(x)\\ &= \Pi(x) * H(x + \tfrac{1}{2}) - \Pi(x) * H(x - \tfrac{1}{2}).\end{aligned}$$

10 Experiment with the equation $f[f(x)] = f(x)$ and note that $f(x) = \text{sgn } x$ is a solution. Find other solutions and attempt to write down the general solution compactly with the aid of step-function notation.

11 Show that $\text{erf } x = 2\Phi(2^{\frac{1}{2}}\sigma x) - 1$, where

$$\Phi(x) = \int_{-\infty}^{x} \frac{1}{\sigma(2\pi)^{\frac{1}{2}}}\, e^{-u^2/2\sigma^2}\,du.$$

12 Show that the first derivative of $\Lambda(x)$ is given by

$$\Lambda'(x) = -\Pi\left(\frac{x}{2}\right) \text{sgn } x$$

and calculate the second derivative.

13 In abbreviated notation the relation of $\Lambda(x)$ to $\Pi(x)$ could be written $\Lambda = \Pi * \Pi$ or $\Lambda = \Pi^{*2}$. Show that

$$\Pi^{*3} = \Pi * \Lambda = \tfrac{1}{2}(x + 1\tfrac{1}{2})^2\Pi(x + 1) + (\tfrac{3}{4} - x^2)\Pi(x) + \tfrac{1}{2}(x - 1\tfrac{1}{2})^2\Pi(x - 1).$$

Show also that

$$\begin{aligned}\Pi^{*3} = \tfrac{1}{2}(x + 1\tfrac{1}{2})^2H(x + 1\tfrac{1}{2}) - \tfrac{3}{2}(x + \tfrac{1}{2})^2H(x + \tfrac{1}{2}) + \tfrac{3}{2}(x - \tfrac{1}{2})^2H(x - \tfrac{1}{2})\\ - \tfrac{1}{2}(x - 1\tfrac{1}{2})^2H(x - 1\tfrac{1}{2}).\end{aligned}$$

14 Examine the derivatives of Π^{*3} at $x = \frac{1}{2}\pm$ and $x = 1\frac{1}{2}\pm$ and reach some conclusion about the continuity of slope and curvature.

15 Show that $(d/dx)|x| = \text{sgn } x$ and that $(d/dx) \text{ sgn } x = 2\delta(x)$. Comment on the fact that

$$\frac{d^2|x|}{dx^2} = \frac{d^2}{dx^2}\,[2xH(x)] = 2\delta(x).$$

Chapter 5 The impulse symbol

It is convenient to have notation for intense unit-area pulses so brief that measuring equipment of a given resolving power is unable to distinguish between them and even briefer pulses. This concept is covered in mechanics by the term "impulse." The important attribute of an impulse is its integral; the precise details of its form are unimportant. The idea has been current for a century[1] or more in mathematical circles and was extensively employed by Heaviside, for example, who used the symbol $p1$.[2] The notation $\delta(x)$, which was subsequently introduced into quantum mechanics by Dirac,[3] is now in general use. The underlying concept permeates physics. Point masses, point charges, point sources, concentrated forces, line sources, surface charges, and the like are familiar and accepted entities in physics. Of course, these things do not exist. Their conceptual value stems from the fact that the impulse response—the effect associated with the impulse (point mass, point charge, and the like)—may be indistinguishable, given measuring equipment of specified resolving power, from the response due to a physically realizable pulse. It is then a convenience to have a name for pulses which are so brief and intense that making them any briefer and more intense does not matter.

We have in mind an infinitely brief or concentrated, infinitely strong

[1] For historical examples from the writings of Hermite, Cauchy, Poisson, Kirchhoff, Helmholtz, Kelvin, and Heaviside, see B. van der Pol and H. Bremmer, "Operational Calculus Based on the Two-sided Laplace Integral," Cambridge University Press, Cambridge, England, 1955.

[2] This symbol means the derivative of the unit step function.

[3] P. A. M. Dirac, "The Principles of Quantum Mechanics," 3d ed., Oxford University Press, Oxford, 1947.

unit-area impulse and therefore wish to write

$$\delta(x) = 0 \qquad x \neq 0$$

and

$$\int_{-\infty}^{\infty} \delta(x)\, dx = 1.$$

However, the impulse symbol[4] $\delta(x)$ does not represent a function in the sense in which the word is used in analysis (to stress this fact Dirac coined the term "improper function"), and the above integral is not a meaningful quantity until some *convention* for interpreting it is declared. Here we use it to mean

$$\lim_{\tau \to 0} \int_{-\infty}^{\infty} \tau^{-1}\Pi\left(\frac{x}{\tau}\right) dx.$$

The function $\tau^{-1}\Pi(x/\tau)$ is a rectangle function of height τ^{-1} and base τ and has unit area; as τ tends to zero a sequence of unit-area pulses of ever-increasing height is generated. The limit of the integral is, of course, equal to unity. In other words, to interpret expressions containing the impulse symbol, we fall back on certain sequences of finite unit-area pulses of brief but nonzero duration, and of some particular shape. We perform the operations indicated, such as integration, differentiation, multiplication, and then discuss limits as the duration approaches zero. There is some convenience mathematically in taking the pulse shape always Gaussian; obviously, one has to prepare for possible awkwardness in retaining $\Pi(x)$ as a choice, where differentiation is involved. However, the essence of the physics is that the pulse shape should not matter, and we therefore proceed under the expectation that the choice of pulse shape will remain at our disposal.

We adapt our approach to each case as it arises. Later we give a systematic presentation of the theory of generalized functions, a recently developed exposition of the sequence idea in tidied-up mathematical form.

The need to broaden Fourier transform theory was mentioned earlier in connection with functions, such as periodic functions, which do not possess Fourier transforms. The term "transform pair in the limit" was introduced to describe cases where one or both members of the transform pair are generalized functions. All these cases can be expressed with the aid of the impulse symbol and its derivatives, $\delta(x)$, $\delta'(x)$, and so on, which thus furnish the notation for the broadened theory.

The convenience of the impulse symbol lies in its reserve over detail. As a specific example of the relevance of this feature to physical systems, consider an electrical network, say a low-pass filter. An applied pulse of voltage produces a certain transient response, and it is readily observa-

[4] We use the word "symbol" systematically to signpost the entities that are not functions; we may also use the term "generalized function," introduced in 1953 by Temple.

ble, as the applied pulses are made briefer and briefer, that the response settles down to a definite form. It is also observable that the form of the response is then independent of the input pulse shape, be it rectangular, triangular, or even a pair of pulses. This happens because the high-frequency components, which distinguish the different applied pulses, produce negligible response. The network is thus characterized by a certain definite, readily observable form of response, which can be elicited by a multiplicity of applied waveforms, the details of which are irrelevant; it is necessary only that they be brief enough. Since the response may be scrutinized with an oscilloscope of the highest precision and time resolution, we must, of course, be prepared to keep the applied pulse duration shorter than the minimum set by the quality of the measuring instrument. The impulse symbol enables us to make abbreviated statements about arbitrarily shaped indefinitely brief pulses.

An intimate relationship between the impulse symbol and the unit step function follows from the property that $\int_{-\infty}^{x} \delta(x')\,dx'$ is unity if x is positive but zero if x is negative. Hence

$$\int_{-\infty}^{x} \delta(x')\,dx' = H(x).$$

This equation furnishes an opportunity to illustrate the interpretation of an expression containing the impulse symbol. First we replace $\delta(x)$ by the pulse sequence $\tau^{-1}\Pi(x/\tau)$ and contemplate the sequence of integrals

$$\int_{-\infty}^{x} \tau^{-1}\Pi\left(\frac{x'}{\tau}\right) dx'.$$

As long as τ is not zero or infinite, each such integral is a function of x that may be described as a ramp-step function, as illustrated in Fig. 5.1. Now fix x, and consider the limit of the sequence of values of the integral generated as τ approaches zero. We see that if we have fixed on a positive (negative) x, then the limit of the integral will be unity (zero). Therefore, in accordance with the definition of $H(x)$ we can write

$$\lim_{\tau \to 0} \int_{-\infty}^{x} \tau^{-1}\Pi\left(\frac{x'}{\tau}\right) dx' = H(x).$$

The equation

$$\int_{-\infty}^{x} \delta(x')\,dx' = H(x)$$

is shorthand for this.

Since under ordinary circumstances

$$\int_{-\infty}^{x} f(x')\,dx' = g(x)$$

implies that

$$f(x) = g'(x),$$

one writes by analogy

$$\delta(x) = \frac{d}{dx} H(x)$$

and states that "the derivative of the unit step function is the impulse symbol." Since the unit step function does not in fact possess a derivative at the origin, this statement must be interpreted as shorthand for "the derivatives of a sequence of differentiable functions that approach $H(x)$ as a limit constitute a suitable defining sequence for $\delta(x)$." The ramp-step functions of Fig. 5.1 are differentiable and approach $H(x)$. Since the amount of the step is in all cases unity, the area under each derivative function is unity, thus qualifying the sequence of derivative functions as a suitable sequence to define a unit impulse.

It is not *necessary* to deal in terms of sequences of rectangle functions to discuss impulses. Representations of $\delta(x)$ in terms of various pulse shapes include the following sequences, generated as τ approaches zero (through positive values).

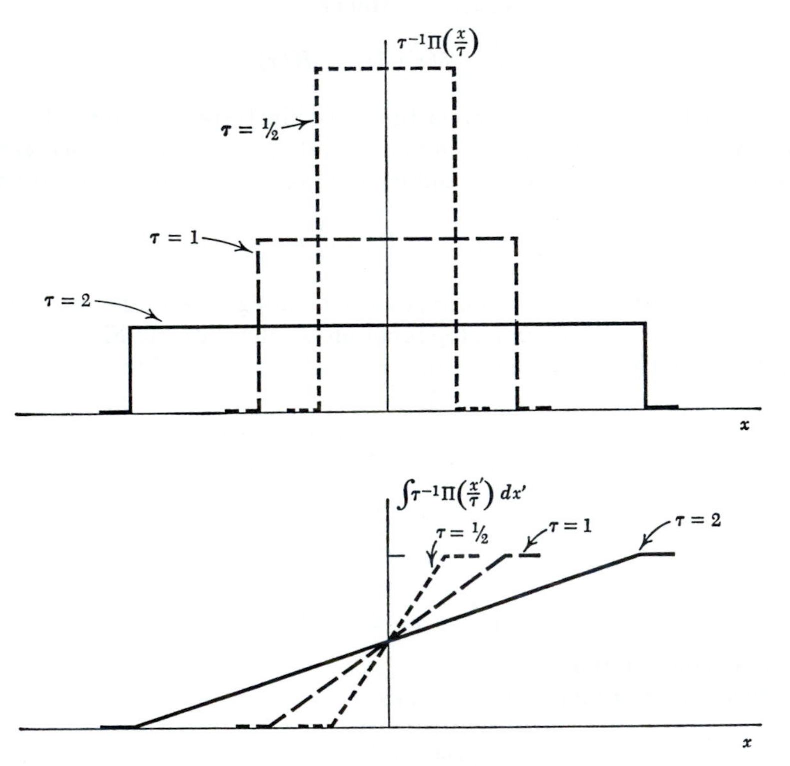

Fig. 5.1 A rectangular pulse sequence and the sequence of ramp-step functions obtained by integration.

The sequence

$$\tau^{-1}\Pi\left(\frac{x - \frac{1}{2}\tau}{\tau}\right)$$

is composed of rectangle functions, all having their left-hand edge at the origin; it is often considered in the analysis of circuits where transients cannot exist prior to $t = 0$, the instant at which some switch is thrown. If one uses this sequence instead of the centered sequence, the results will not necessarily be the same. This is discussed below in connection with the sifting property. The sequence

$$\tau^{-1} e^{-\pi x^2/\tau^2}$$

of Gaussian profiles, which has been mentioned previously, has the convenient property that derivatives of all orders exist. On the other hand, these profiles lack the convenience of being nonzero over only a finite range of x. The sequence

$$\tau^{-1}\Lambda\left(\frac{x}{\tau}\right)$$

of triangle functions is useful for discussing situations where first derivatives are needed, since the profiles are continuous. In addition, they have an advantage in being zero outside the interval in which $|x| < \tau$. The sequence

$$\tau^{-1} \operatorname{sinc} \frac{x}{\tau}$$

has the curious property of not dying out to zero where $x \neq 0$; at any value of x not equal to zero the value oscillates without diminishing as $\tau \rightarrow 0$. The sequence serves perfectly well to define $\delta(x)$ for a reason that is given below in connection with the sifting property. The resonance profiles

$$\frac{\tau}{\pi(x^2 + \tau^2)}$$

decay rather slowly with increasing x. A product with an arbitrary function may well have infinite area. To eliminate this possibility completely, one is led to contemplate sequences that are zero outside a finite range, thus obtaining freedom to accept products with functions having any kind of asymptotic behavior. In addition, one would like to have derivatives of all orders exist.

$$\tau^{-1} \exp\left[\frac{-1}{1 - (x/\tau)^2}\right] \Pi\left(\frac{x}{2\tau}\right)$$

is a specific example of a sequence of profiles, each of which is zero outside the interval in which $|x| < \tau$, and each of which possesses derivatives of all orders. To see that it is possible for a function to descend to zero with zero slope, zero curvature, and all higher derivatives zero, differentiate the function

$$e^{-1/x}H(x),$$

and evaluate the derivatives at $x = 0$. There may seem to be a clash with the Maclaurin formula, according to which

$$f(x) = f(0) + xf'(0) + \frac{x^2}{2}f''(0) + \ . \ . \ . + \frac{x^n}{n!}f^{(n)}(0) + \text{remainder.}$$

In the case of many functions familiar from analysis, the remainder term can be shown to vanish as $n \to \infty$. In the case we have chosen here, however, the first $n + 1$ terms vanish and the remainder term contains the whole value of the function.

The sifting property

Following our rule for interpreting expressions containing the impulse symbol, we may try to assign a meaning to

$$\int_{-\infty}^{\infty} \delta(x)f(x)\,dx.$$

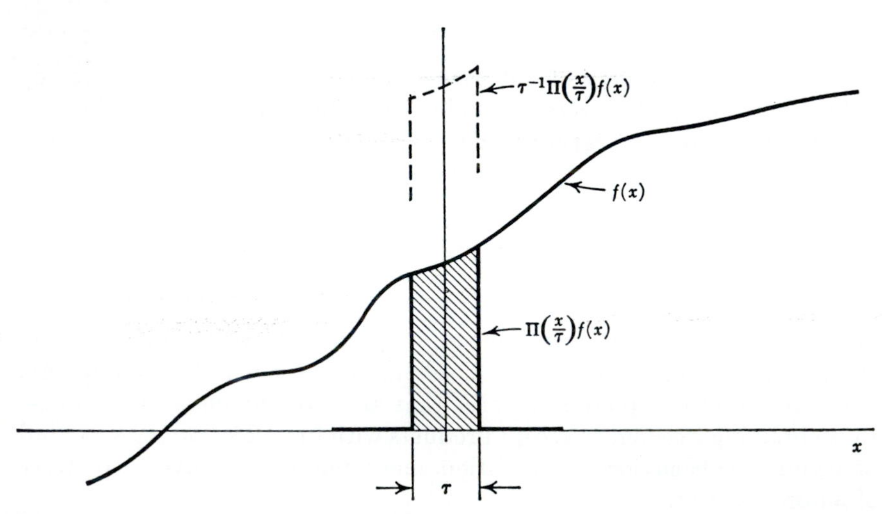

Fig. 5.2 Explaining the sifting property. The shaded area is approximately $\tau f(0)$.

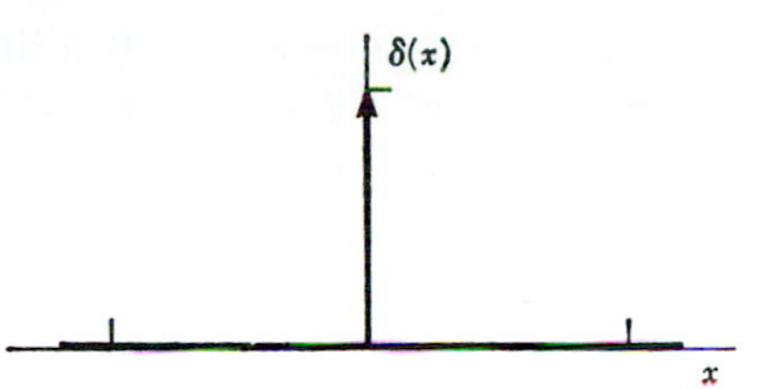

Fig. 5.3 Graphical representation of the impulse symbol $\delta(x)$ as a spike of unit height.

Thus we substitute the sequence $\tau^{-1}\Pi(x/\tau)$ for $\delta(x)$, perform the multiplication and integration, and finally take the limit of the integral as $\tau \rightarrow 0$:

$$\lim_{\tau\rightarrow 0} \int_{-\infty}^{\infty} \tau^{-1}\Pi\left(\frac{x}{\tau}\right) f(x)\, dx.$$

In Fig. 5.2 the integrand is indicated in broken outline. Its area is τ^{-1} times the shaded area. The shaded area, whose width is τ and whose average height is approximately $f(0)$, amounts to approximately $\tau f(0)$. Hence the area under the integrand approaches $f(0)$ as τ approaches zero. Thus we write

$$\int_{-\infty}^{\infty} \delta(x) f(x)\, dx = f(0)$$

and refer to this statement as the sifting property of the impulse symbol, since the operation on $f(x)$ indicated on the left-hand side sifts out a single value of $f(x)$. It will be seen that it is immaterial what sort of pulse is incorporated in the integrand, and this fact is the essence of the utility of $\delta(x)$. It just stands for a unit pulse whose duration is much smaller than any interval of interest, and consequently whose pulse *shape* means nothing to us; only its integral counts. It will be represented on graphs (see Fig. 5.3) as a spike of unit height, and impulses in general will be shown as spikes of height equal to their integral.

It is clear that we can also write

$$\int_{-\infty}^{\infty} \delta(x-a) f(x)\, dx = f(a)$$

and

$$\int_{-\infty}^{\infty} \delta(x) f(x-a)\, dx = f(-a).$$

The resemblance to the convolution integral can be emphasized by writing

$$\int_{-\infty}^{\infty} \delta(x') f(x-x')\, dx' = \int_{-\infty}^{\infty} \delta(x-x') f(x')\, dx' = f(x)$$

or, in asterisk notation,

$$\delta(x) * f(x) = f(x) * \delta(x) = f(x).$$

If $f(x)$ has a jump at $x = 0$, a little thought devoted to a diagram such as the one Fig. 5.2 will show that the sifting integral will have a limiting value of $\frac{1}{2}[f(0+) + f(0-)]$. Consequently, it is more general to write

$$\delta(x) * f(x) = \frac{f(x+) + f(x-)}{2}.$$

The expression on the right-hand side differs from $f(x)$ only by a null function, and hence the refinement is ordinarily not important. This does not alter the fact that $\frac{1}{2}[f(x+) + f(x-)]$ can be different in value from $f(x)$.

The asymmetrical sequence $\tau^{-1}\Pi[(x - \frac{1}{2}\tau)/\tau]$ mentioned above will be seen to have the property of sifting out $f(x+)$. At points of discontinuity of $f(x)$, the use of this asymmetrical sequence therefore gives a different result. In transient analysis, where discontinuities at the switching instant $t = 0$ are particularly common, the choice of sequence can thus appear to give different answers. The difference, however, can only be instantaneous.

The impulse symbol has many fascinating properties, most of which can be proved easily. An important one which must be watched carefully in algebraic manipulation is

$$\delta(ax) = \frac{1}{|a|}\,\delta(x);$$

that is, if the scale of x is compressed by a factor a, thus reducing the area of the pulses which previously had unit area, then the strength of the impulse is reduced by the factor $|a|$. The modulus sign allows for the property

$$\delta(-x) = \delta(x).$$

From this it would seem that the impulse symbol has the property of evenness; however, we gave an equation earlier involving a sequence of displaced rectangle functions which were not themselves even (Prob. 19).

It can easily be shown by considering sequences of pulses that we may write, if $f(x)$is continuous at $x = 0$,

$$f(x)\,\delta(x) = f(0)\,\delta(x).$$

From the sifting property, putting $f(x) = x$, we have

$$\int_{-\infty}^{\infty} x\,\delta(x)\,dx = 0.$$

One generally writes

$$x\,\delta(x) \equiv 0,$$

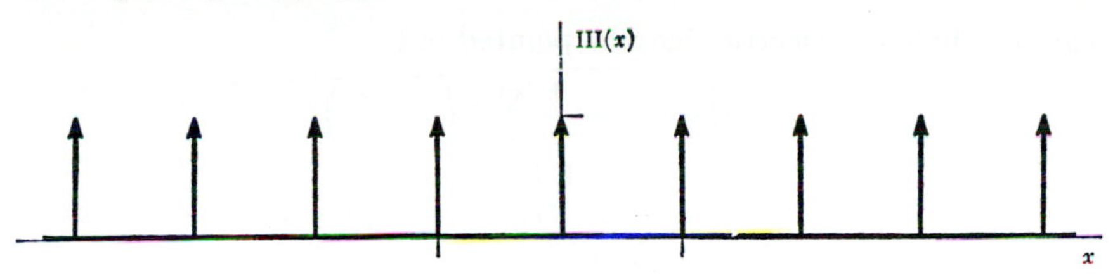

Fig. 5.4 The shah symbol III(x).

although if the prelimit graphs are contemplated, it will be seen that this equation conceals a nonvanishing component reminiscent of the Gibbs phenomenon in Fourier series. Thus it is true that

$$\lim_{\tau \to 0} \left[x\tau^{-1}\Pi\left(\frac{x}{\tau}\right) \right] = 0 \qquad \text{for all } x;$$

nevertheless,

$$\lim_{\tau \to 0} \left[x\tau^{-1}\Pi\left(\frac{x}{\tau}\right) \right]_{\max} = \tfrac{1}{2},$$

and, moreover, the limit of the minimum value is $-\frac{1}{2}$. Consequently, among those functions which are identically zero, $x\delta(x)$ is rather curious, and one has the feeling that if it could be applied to the deflecting electrodes of an oscilloscope, one would see spikes.

The sampling or replicating symbol III(x)

Consider an infinite sequence of unit impulses spaced at unit interval as shown in Fig. 5.4. Any reservations that apply to the impulse symbol $\delta(x)$ apply equally in this case; indeed, even more may be needed because we have to deal with an infinite number of infinite discontinuities and a nonconvergent infinite integral. For example, *all* the conditions for existence of a Fourier transform are violated. The conception of an infinite sequence of impulses proves, however, to be extremely useful—and easy to manipulate algebraically.

To describe this conception we introduce the *shah*[5] symbol III(x) and write

$$\text{III}(x) = \sum_{n=-\infty}^{\infty} \delta(x - n).$$

[5] The symbol III is pronounced *shah* after the Cyrillic character III, which is said to have been modeled on the Hebrew letter ש (*shin*), which in turn may derive from the Egyptian , a hieroglyph depicting papyrus plants along the Nile.

Various obvious properties may be pointed out:

$$\begin{aligned}
\mathrm{III}(ax) &= \frac{1}{|a|}\sum \delta\left(x - \frac{n}{a}\right)\\
\mathrm{III}(-x) &= \mathrm{III}(x)\\
\mathrm{III}(x+n) &= \mathrm{III}(x) \qquad n \text{ integral}\\
\mathrm{III}(x-\tfrac{1}{2}) &= \mathrm{III}(x+\tfrac{1}{2})\\
\int_{n-\frac{1}{2}}^{n+\frac{1}{2}} \mathrm{III}(x)\,dx &= 1\\
\mathrm{III}(x) &= 0 \qquad x \neq n.
\end{aligned}$$

Evidently, $\mathrm{III}(x)$ is periodic with unit period.

A periodic *sampling* property follows as a generalization of the sifting integral already discussed in connection with the impulse symbol. Thus multiplication of a function $f(x)$ by $\mathrm{III}(x)$ effectively samples it at unit intervals:

$$\mathrm{III}(x)f(x) = \sum_{n=-\infty}^{\infty} f(n)\,\delta(x-n).$$

The information about $f(x)$ in the intervals between integers where $\mathrm{III}(x) = 0$ is not contained in the product; however, the values of $f(x)$ at integral values of x are preserved (see Fig. 5.5).

The sampling property makes $\mathrm{III}(x)$ a valued tool in the study of a wide variety of subjects (for example, the radiation patterns of antenna arrays, the diffraction patterns of gratings, raster scanning in television and radar, pulse modulation, data sampling, Fourier series, and computing at discrete tabular intervals).

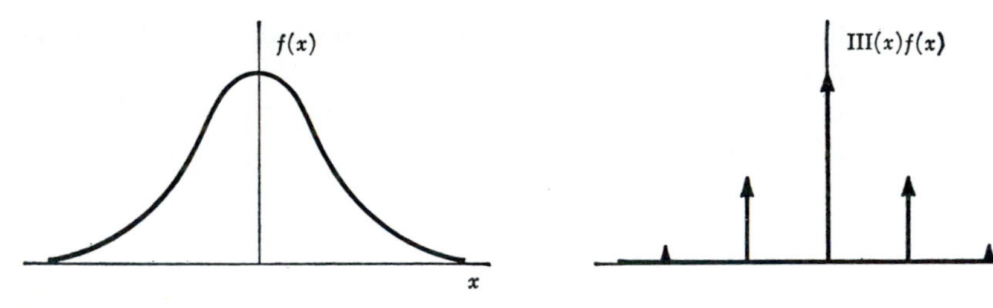

Fig. 5.5 The sampling property of $\mathrm{III}(x)$.

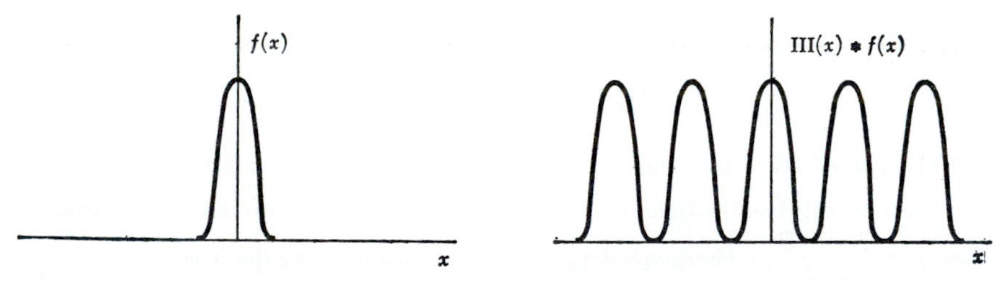

Fig. 5.6 The replicating property of $\mathrm{III}(x)$.

Just as important as the sampling property under multiplication is a *replicating* property exhibited when $\mathrm{III}(x)$ enters into convolution with a function $f(x)$. Thus

$$\mathrm{III}(x) * f(x) = \sum_{n=-\infty}^{\infty} f(x-n);$$

as shown in Fig. 5.6, the function $f(x)$ appears in replica at unit intervals of x *ad infinitum* in both directions. Of course, if $f(x)$ spreads over a base more than one unit wide, there is overlapping.

The III symbol is thus also applicable wherever there are periodic structures. This twofold character is not accidental, but is connected with the fact that III is its own Fourier transform (in the limit), which of course makes it twice as useful as it otherwise would have been.

The self-reciprocal property under the Fourier transformation is derived later.

The even and odd impulse pairs $\mathrm{II}(x)$ *and* $\mathrm{I_I}(x)$

Figure 5.7 shows the often-needed impulse-pair symbols defined by

$$\mathrm{II}(x) = \tfrac{1}{2}\delta(x+\tfrac{1}{2}) + \tfrac{1}{2}\delta(x-\tfrac{1}{2}),$$
$$\mathrm{I_I}(x) = \tfrac{1}{2}\delta(x+\tfrac{1}{2}) - \tfrac{1}{2}\delta(x-\tfrac{1}{2}).$$

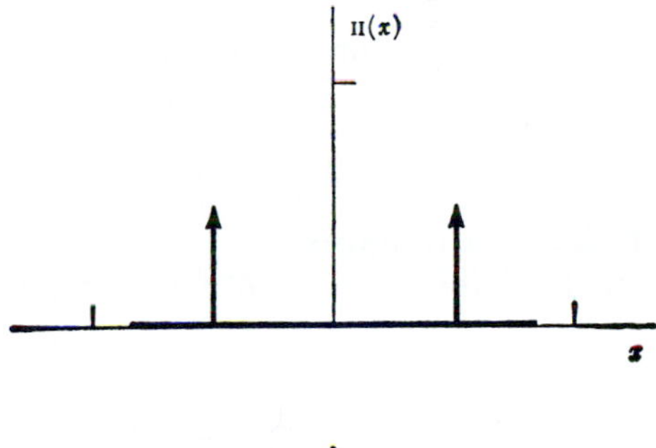

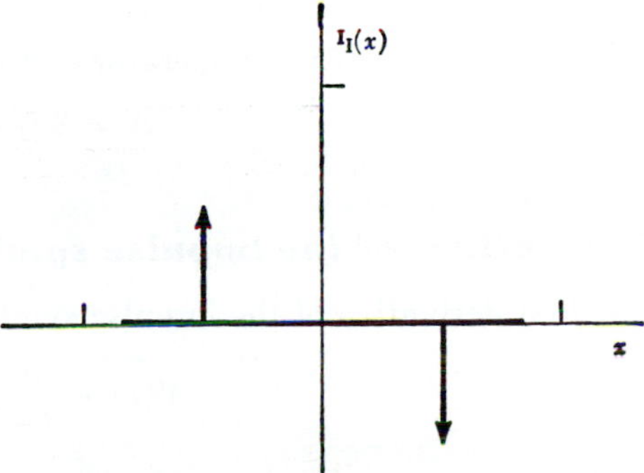

Fig. 5.7 The even and odd impulse pairs, $\mathrm{II}(x)$ *and* $\mathrm{I_I}(x)$.

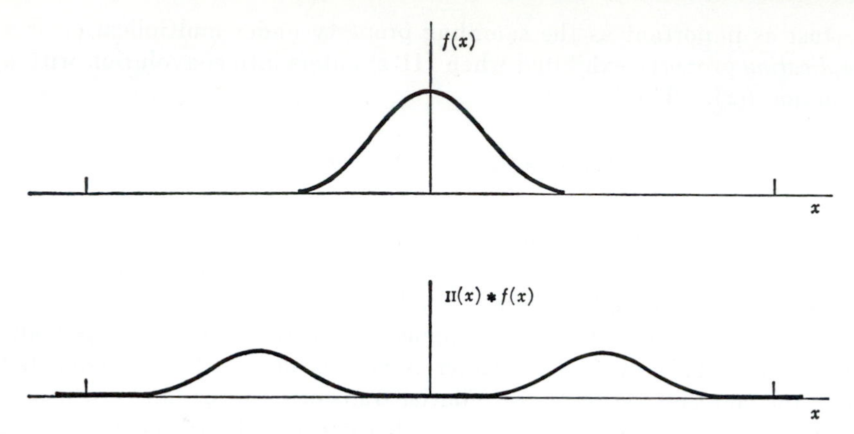

Fig. 5.8 The convolution of $\mathrm{II}(x)$ *with* $f(x)$.

The impulse pairs derive importance from their transform relationship to the cosine and sine functions. Thus

$$\mathrm{II}(x) \supset \cos \pi s \qquad \cos \pi x \supset \mathrm{II}(s)$$
$$\mathrm{I}_{\mathrm{I}}(x) \supset i \sin \pi s \qquad \sin \pi x \supset i\,\mathrm{I}_{\mathrm{I}}(s).$$

When convolved with a function $f(x)$, the even impulse pair $\mathrm{II}(x)$ has a duplicating property. Thus, as illustrated in Fig. 5.8,

$$\mathrm{II}(x) * f(x) = \tfrac{1}{2}f(x + \tfrac{1}{2}) + \tfrac{1}{2}f(x - \tfrac{1}{2}).$$

There are occasions when $\mathrm{II}(x)$ might better consist of two unit impulses, but as defined it is normalized to unit area; that is,

$$\int_{-\infty}^{\infty} \mathrm{II}(x)\, dx = 1,$$

which has advantages.

If the finite difference of $f(x)$ is defined by

$$\Delta f(x) = f(x + \tfrac{1}{2}) - f(x - \tfrac{1}{2}),$$

then

$$\Delta f(x) = 2\, \mathrm{I}_{\mathrm{I}}(x) * f(x).$$

Thus the finite difference operator can be expressed as

$$\Delta \equiv 2\, \mathrm{I}_{\mathrm{I}} * .$$

Derivatives of the impulse symbol

The first derivative of the impulse symbol is defined symbolically by

$$\delta'(x) = \frac{d}{dx}\, \delta(x).$$

The mental picture that accompanies this conception is the same as that involved in the conception of an infinitesimal dipole in electrostatics, and it is well known that the idea of an infinitesimal dipole is convenient and easy to think with physically. The $\delta'(x)$ notation carries this facility over into mathematical form, but, of course, there are difficulties because we cannot ask a function to go positively infinite just to the left of the origin and negatively infinite just to the right, and to be zero where $|x| > 0$. To cap this we would wish to write $\delta'(0) = 0$.

For rigorous interpretation of statements involving $\delta'(x)$ we may fall back on sequences of pulses such as were invoked in connection with $\delta(x)$, and consider their derivatives (two examples are given in Fig. 5.9). Then statements such as

$$\int_{-\infty}^{\infty} \delta'(x)\, dx = 0$$

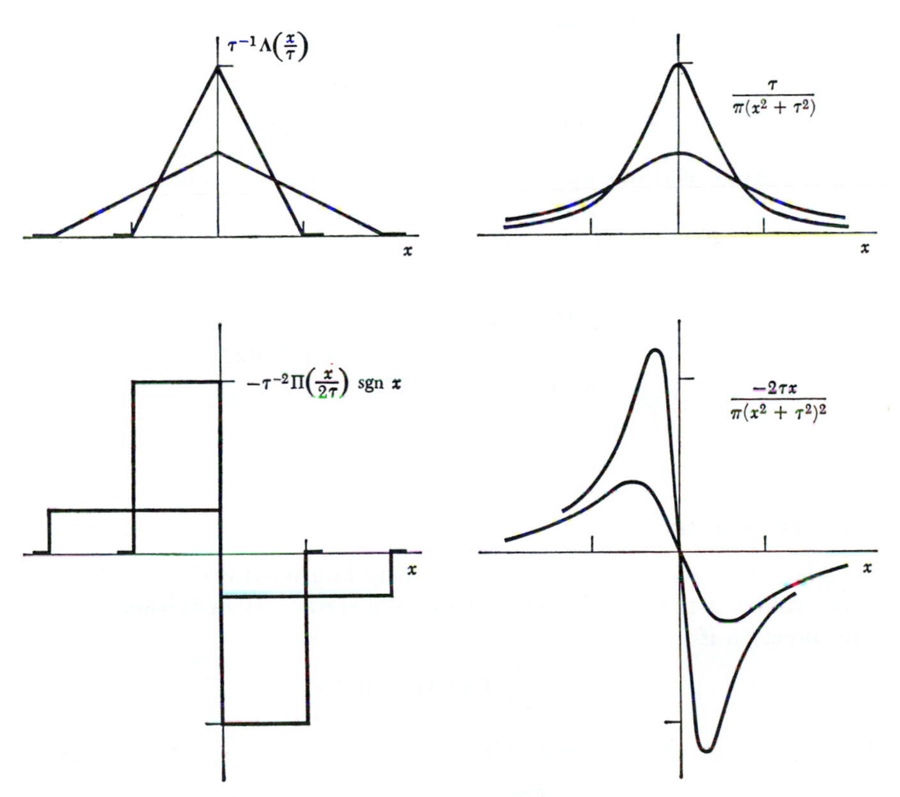

Fig. 5.9 Pulse sequences (above) and their derivatives (below) which, as $\tau \rightarrow 0$*, are used for contemplating the meaning of* $\delta(x)$ *and* $\delta'(x)$*.*

are deemed to be shorthand for statements such as

$$\lim_{\tau \to 0} \int_{-\infty}^{\infty} \left[\frac{-2\tau x}{\pi(x^2 + \tau^2)^2} \right] dx = 0,$$

where the quantity in brackets is the derivative of one of the pulse shapes considered previously in discussing $\delta(x)$. The precise form of pulse adopted is unimportant—even a rectangular one will do—but in later work the possibility that a differentiable pulse shape may offer an advantage should be considered.

A derivative-sifting property

$$\delta' * f \equiv \int_{-\infty}^{\infty} \delta'(x - x')f(x')\,dx' = f'(x)$$

may be established in this way. Further properties are

$$\begin{aligned}
\int_{-\infty}^{\infty} x\,\delta'(x)\,dx &= -1 \\
\int_{-\infty}^{\infty} |\delta'(x)|\,dx &= \infty \\
x^2\,\delta'(x) &= 0 \\
\delta'(-x) &= -\delta'(x) \qquad x\,\delta'(x) = -\delta(x) \\
f(x)\,\delta'(x) &= f(0)\,\delta'(x) - f'(0)\,\delta(x).
\end{aligned}$$

The following relations apply to derivatives of higher order.

$$\begin{aligned}
\int_{-\infty}^{\infty} \delta''(x)\,dx &= 0 \\
\delta''(x) * f(x) &= f''(x) \\
\int_{-\infty}^{\infty} x^2\,\delta''(x)\,dx &= 2 \\
\delta^{(n)}(x) &= (-1)^n n! x^{-n}\,\delta(x) \\
\delta^{(n)}(x) * f(x) &= f^{(n)}(x) \\
\int_{-\infty}^{\infty} \delta^{(n)}(x) f(x)\,dx &= (-1)^n f^{(n)}(0)
\end{aligned}$$

Null functions

Null functions are known chiefly for having Fourier transforms which are zero, while not themselves being identically zero. By definition, $f(x)$ is a null function if

$$\int_a^b f(x)\,dx = 0$$

for all a and b. An alternative statement is

$$\int_{-\infty}^{\infty} |f(x)|\,dx = 0,$$

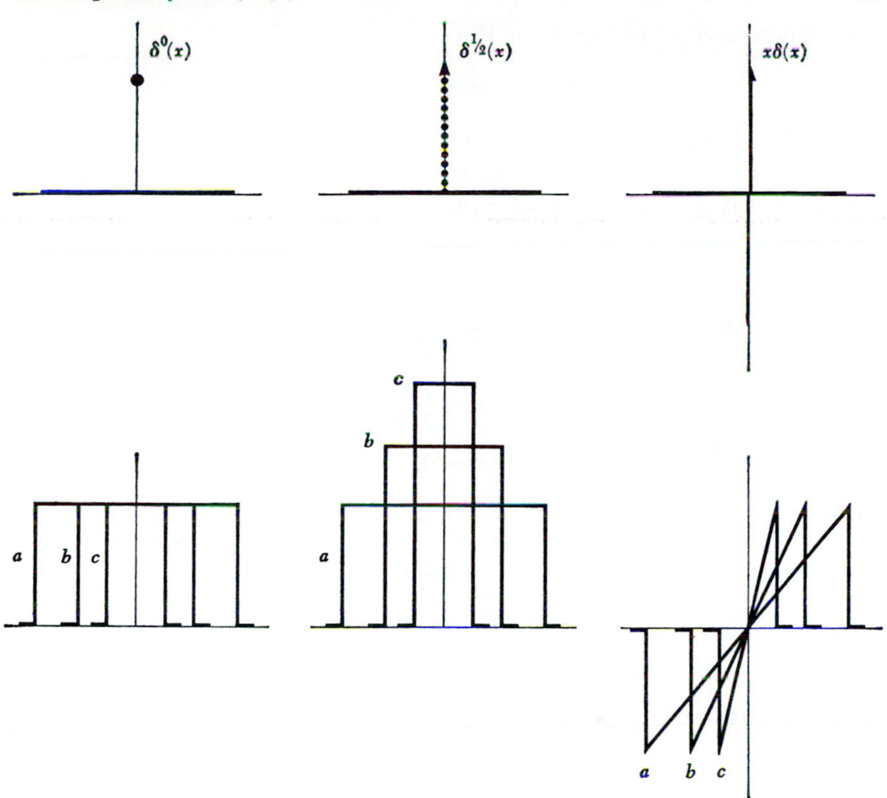

Fig. 5.10 Sequences (below) defining $\delta^0(x)$, $\delta^{\frac{1}{2}}(x)$, and $x\,\delta(x)$ (above).

Null functions arise in connection with the one-to-one relationship between a function and its transform, a relationship defined by Lerch's theorem, which states that if two functions $f(x)$ and $g(x)$ have the same transform, then $f(x) - g(x)$ is a null function.

An example of a null function (see Fig. 5.10) is $\delta^0(x)$, an ordinary single-valued function defined by

$$\delta^0(x) = \begin{cases} 0 & x \neq 0 \\ 1 & x = 0, \end{cases}$$

which thus has a discontinuity at $x = 0$ similar in a way to that possessed by $H(x)$ [defined so that $H(0) = \frac{1}{2}$]. Under the ordinary rules of integration, the integral of $\delta^0(x)$ is certainly zero. However, it aptly describes the current taken by a series combination of a resistance and a capacitance from a battery, in the limit as the capacitance approaches zero.

We can now more succinctly state the relation between $H(x)$ and the step function $\hat{H}(x)$ of Fig. 4.14; thus

$$H(x) = \hat{H}(x) + \tfrac{1}{2}\delta^0(x),$$

the difference between the two being a null function.

The symbol $\delta^{\frac{1}{2}}(x)$ has to be considered in terms of sequences of pulses in the same way as $\delta(x)$. Consider the sequence

$$\tau^{-\frac{1}{2}}\Pi\left(\frac{x}{\tau}\right)$$

as $\tau \to 0$. Then we can attach meaning to the statements

$$\delta^{\frac{1}{2}}(x) = \begin{cases} 0 & x \neq 0 \\ \infty & x = 0, \end{cases}$$

$$\int_{-\infty}^{\infty} \delta^{\frac{1}{2}}(x)\,dx = 0,$$

and

$$\int_{-\infty}^{\infty} [\delta^{\frac{1}{2}}(x)]^2\,dx = 1.$$

We could describe $\delta^{\frac{1}{2}}(x)$ as a null symbol.

Exercise Would we wish to call $\delta'(x)$ a null symbol?

Some functions in two and more dimensions

One encounters the two- and three-dimensional impulse symbols ${}^2\delta(x,y)$, ${}^3\delta(x,y,z)$, as natural generalizations of $\delta(x)$. For example, ${}^2\delta(x,y)$ describes the pressure distribution over the xy plane when a concentrated unit force is applied at the origin; ${}^3\delta(x,y,z)$ describes the charge density in a volume containing a unit charge at the point (0,0,0). In establishing properties of ${}^2\delta(x,y)$ one considers a sequence, as $\tau \to 0$, of functions such as $\tau^{-2}\Pi(x/\tau)\Pi(y/\tau)$ or $(4/\pi)\tau^{-2}\Pi[(x^2 + y^2)^{\frac{1}{2}}/\tau]$, which have unit volume

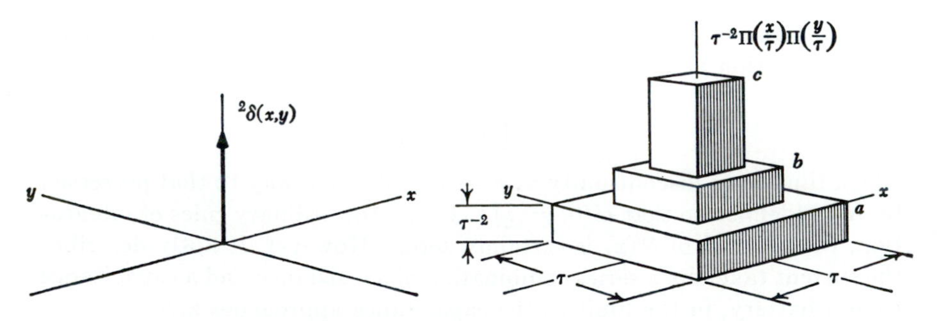

Fig. 5.11 The two-dimensional impulse symbol ${}^2\delta(x,y)$ and a defining sequence of functions a,b,c.

(Fig. 5.11). Then we have

$$^2\delta(x,y) = \begin{cases} 0 & x^2 + y^2 \neq 0 \\ \infty & x^2 + y^2 = 0, \end{cases}$$

$$\int_{-\infty}^{\infty}\int_{-\infty}^{\infty} {}^2\delta(x,y)\,dx\,dy = 1,$$

$$^2\delta(ax,by) = \frac{1}{|ab|}\,{}^2\delta(x,y),$$

and the very interesting relation

$$^2\delta(x,y) = \delta(x)\,\delta(y).$$

Introducing the radial coordinate r such that $r^2 = x^2 + y^2$, we can express $^2\delta(x,y)$ in terms of $\delta(r)$:

$$^2\delta(x,y) = \frac{\delta(r)}{\pi|r|}.$$

In three dimensions,

$$^3\delta(x,y,z) = \begin{cases} 0 & x^2 + y^2 + z^2 \neq 0 \\ \infty & x^2 + y^2 + z^2 = 0, \end{cases}$$

$$\int_{-\infty}^{\infty}\int_{-\infty}^{\infty}\int_{-\infty}^{\infty} {}^3\delta(x,y,z)\,dx\,dy\,dz = 1,$$

and $$^3\delta(x,y,z) = \delta(x)\,\delta(y)\,\delta(z) = {}^2\delta(x,y)\,\delta(z).$$

In cylindrical coordinates $r^2 = x^2 + y^2$,

$$^3\delta(x,y,z) = \frac{\delta(r)\,\delta(z)}{\pi|r|},$$

and with $\rho^2 = x^2 + y^2 + z^2$,

$$^3\delta(x,y,z) = \frac{\delta(\rho)}{2\pi\rho^2}.$$

For describing arrays in two dimensions we have the bed-of-nails symbol $^2\mathrm{III}(x,y)$, illustrated in Fig. 5.12 and defined by

$$^2\mathrm{III}(x,y) = \sum_{m=-\infty}^{\infty}\sum_{n=-\infty}^{\infty} {}^2\delta(x-m,\,y-n).$$

Figure 5.13 shows an approach to the discussion of its properties. It has the property

$$^2\mathrm{III}(x,y) = \mathrm{III}(x)\mathrm{III}(y)$$

and various extensions of the integral properties of $^2\delta(x,y)$ and $\mathrm{III}(x)$, for example,

$$\int_{-\infty}^{\infty}\int_{-\infty}^{\infty} f(x,y)\,{}^2\mathrm{III}(x,y)\,dx\,dy = \sum_m\sum_n f(m,n).$$

It is doubly periodic,

$$^2\text{III}(x + m, y + n) = {}^2\text{III}(x,y) \qquad m,n \text{ integral,}$$

and

$$\frac{1}{|XY|}\, {}^2\text{III}\left(\frac{x}{X}, \frac{y}{Y}\right)$$

represents a doubly periodic array of two-dimensional unit impulses with period X in the x direction and Y in the y direction.

Tabulation at discrete intervals of two independent variables (two-dimensionally sampled data), and the coefficients of double Fourier series are handled through the relation

$$f(x,y)\, {}^2\text{III}(x,y) = \sum_m \sum_n f(m,n)\, {}^2\delta(x - m, y - n).$$

Convolution with $^2\text{III}(x,y)$ describes replication in two dimensions, such as one has in a two-dimensional array of identical antennas, and, as with III in one dimension, ^2III has the distinction of being its own two-dimensional Fourier transform (in the limit).

The scheme of multidimensional notation introduced here permits various self-explanatory extensions which are occasionally useful for compactness. Thus

$$^3\text{III}(x,y,z) = \text{III}(x)\text{III}(y)\text{III}(z)$$

$$^2\Pi(x,y) = \Pi(x)\Pi(y)$$

$$^2\text{sinc}\,(x,y) = \frac{\sin \pi x \sin \pi y}{\pi^2 xy}$$

$$^2\text{II}(x,y) = \text{II}(x)\text{II}(y).$$

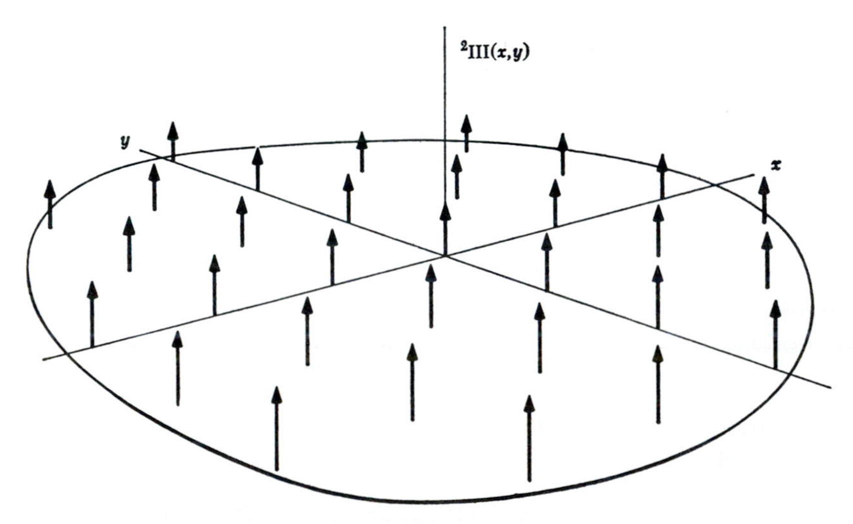

Fig. 5.12 The bed-of-nails symbol, $^2\text{III}(x,y)$.

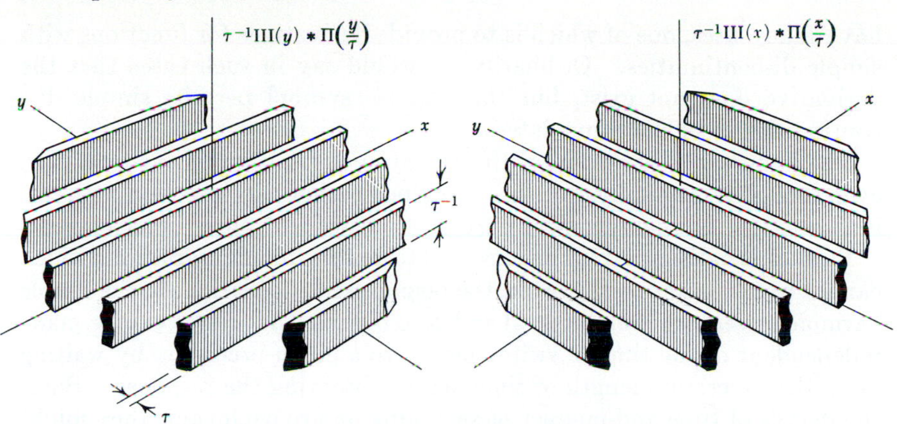

Fig. 5.13 Two functions whose product is suitable for discussing $^2\text{III}(x,y)$.

Two other important two-dimensional distributions do not need new symbols. The row of spikes (see Fig. 5.14) is adequately expressed by $\text{III}(x)\ \delta(y)$ and the grating by $\text{III}(x)$. These two distributions form a two-dimensional Fourier transform pair and are suitable for discussing phenomena such as the diffraction of light by a row of pinholes or by a diffraction grating.

The concept of generalized function

As has been seen, a good deal of convenience attends the use of the impulse symbol $\delta(x)$ and other combinations of impulses such as $\text{III}(x)$ and $\text{II}(x)$. The word "symbol" has been used to call attention to the fact that these entities are not functions, but despite their apparent lack of status they

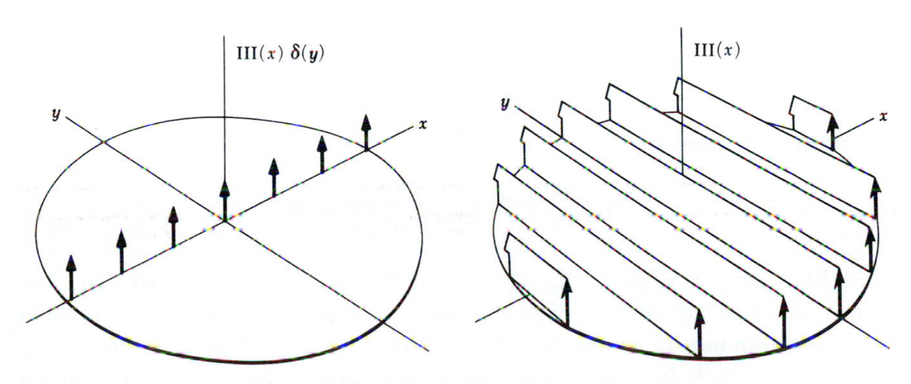

Fig. 5.14 The row of spikes (left) and grating (right).

have many uses, one of which is to provide derivatives for functions with simple discontinuities. Ordinarily we would say in such cases that the derivative does not exist, but the impulse symbol permits simple discontinuities to be accommodated.

The importance of dealing with discontinuous and impulsive behavior, even though it is nonphysical, was explained earlier in connection with the indispensable but nonphysical pure alternating and pure direct current.

Pure alternating current means an eternal harmonic variation, which cannot be generated. However, the response to a variation that is simple harmonic over a certain interval and zero outside that range can be made independent of the time of switching on, to a given precision, by waiting more than a certain length of time before observing the response. Since the details of time and manner of switching on are irrelevant, they might as well be relegated to the infinitely remote past, thus making it convenient to refer to the observable response as the response to pure alternating current.

Impulses are likewise impossible to generate physically, but the responses to different sufficiently brief but finite pulses can be made indistinguishable to an instrument of given finite temporal resolving power.

Since precision of measurement is known to be limited by the temporal and spectral resolution of the measuring instrument, the physical limitations referred to in the preceding paragraphs can be characterized as "finite resolution." The feature of finite resolving power invoked in the explanation contains the key to the mathematical interpretation of impulse-symbol notation: integrals containing the impulse symbol are to be interpreted as limits of a sequence of integrals in which the impulse is replaced by a sequence of unit-area rectangular pulses $\tau^{-1}\Pi(x/\tau)$. The limit of the sequence of integrals may exist, even though the rectangular pulses grow *without* limit.

Thus the statement

$$\int_{-\infty}^{\infty} \delta(x) f(x)\, dx = f(0)$$

is deemed to mean

$$\lim_{\tau \to 0} \int_{-\infty}^{\infty} \tau^{-1}\Pi\left(\frac{x}{\tau}\right) f(x)\, dx = f(0).$$

In this interpretation of the statement the integrals can exist, and the limit of the integrals as $\tau \to 0$ can exist. The physical situation to which this corresponds is a sequence of ever more compact stimuli producing responses which become indistinguishable under observation to a given precision, no matter how high that precision is.

A satisfactory mathematical formulation of the theory of impulses has been evolved along these lines and is expounded in the books of Lighthill[1] and Friedman.[2] Lighthill credits Temple with simplifying the mathematical presentation; Temple[3] in turn credits the Polish mathematician Mikusiński[4] with introducing the presentation in terms of sequences in 1948. Schwartz's two volumes[5] on the theory of distributions unify "in one systematic theory a number of partial and special techniques proposed for the analytical interpretation of 'improper' or 'ideal' functions and symbolic methods."[6]

The idea of sequences was current in physical circles before 1948, however.[7]

The introduction of rectangular pulse sequences was not meant to imply that other pulse shapes are not equally valid. In fact the essence of the approach is that the detailed pulse shape is unimportant. The advantage of rectangular pulses is the purely practical one of facilitating integration. However, rectangular pulses do not lend themselves to discussing the derivative of an impulse. For that we need something smoother which does not itself have an impulsive derivative. Now for a general theory in which we wish to discuss derivatives of any order it is advantageous to have a pulse sequence such that derivatives of all orders exist. Schwartz and Temple introduce pulse shapes which have all derivatives and furthermore are zero outside a finite range; an example mentioned earlier in this chapter is[8]

$$\begin{cases} \tau^{-1}e^{-\tau^2/(\tau^2-x^2)} & |x| < \tau \\ 0 & |x| \geqslant \tau. \end{cases}$$

In actual fact one never inserts such a function explicitly into an integral; when it becomes necessary to integrate, a pulse shape with a *sufficient* number of derivatives is chosen. Often a rectangular pulse suffices.

Particularly well-behaved functions The term "generalized function" may be defined as follows. First we consider the class S of functions which possess derivatives of all orders at all points and which,

[1] M. J. Lighthill, "An Introduction to Fourier Analysis and Generalised Functions," Cambridge University Press, Cambridge, England, 1958.

[2] B. Friedman, "Principles and Techniques of Applied Mathematics," John Wiley & Sons, New York, 1956.

[3] G. Temple, Theories and Applications of Generalised Functions, *J. Lond. Math. Soc.*, vol. 28, p. 181, 1953.

[4] J. G.-Mikusiński, Sur la méthode de généralisation de Laurent Schwartz et sur la convergence faible, *Fundamenta Mathematicae*, vol. 35, p. 235, 1948.

[5] L. Schwartz, "Théorie des distributions," vols. 1 and 2, Herman & Cie, Paris, 1950 and 1951.

[6] Temple, *op. cit.*, p. 175.

[7] B. van der Pol, Discontinuous Phenomena in Radio Communication, *J. Inst. Elec. Engrs.*, vol. 81, p. 381, 1937.

[8] Schwartz, *op. cit.*, p. 22.

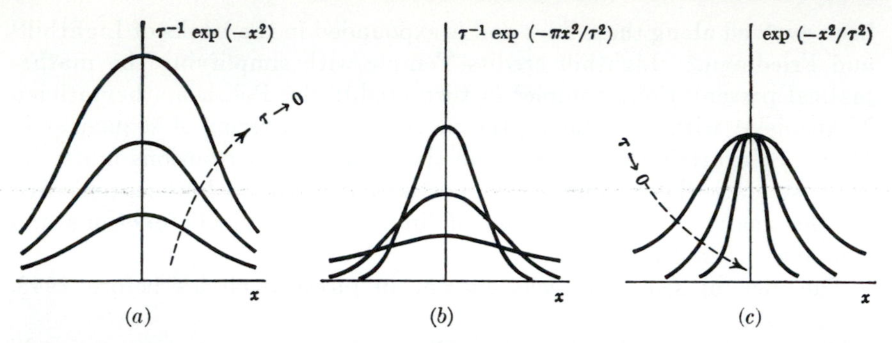

Fig. 5.15 Sequences of particularly well-behaved functions: (a) not regular; (b) regular; (c) exercise.

together with all the derivatives, die off at least as rapidly as $|x|^{-N}$ as $|x| \to \infty$, no matter how large N may be. We shall refer to members of the class S as particularly well-behaved functions. We note that the derivative and the Fourier transform of a particularly well-behaved function are also particularly well behaved. To prove the second of these statements let

$$\bar{F}(s) = \int_{-\infty}^{\infty} F(x)e^{-i2\pi sx}\,dx.$$

The conditions met by the particularly well-behaved function more than suffice to ensure that the Fourier integral exists. Differentiating p times, we have

$$\bar{F}^{(p)}(s) = \int_{-\infty}^{\infty} [(-i2\pi x)^p F(x)]e^{-i2\pi sx}\,dx,$$

and integrating by parts N times, we have

$$\begin{aligned}|\bar{F}^{(p)}(s)| &= \left| \frac{1}{(i2\pi s)^N} \int_{-\infty}^{\infty} \frac{d^N}{dx^N} [(-i2\pi x)^p F(x)]e^{-i2\pi sx}\,dx \right| \\ &\leqslant \frac{(2\pi)^{p-N}}{|s|^N} \int_{-\infty}^{\infty} \left| \frac{d^N}{dx^N} [x^p F(x)] \right| dx \\ &= O(|s|^{-N});\end{aligned}$$

hence $\bar{F}(s)$ belongs to S.

Regular sequences Among sequences of particularly well-behaved functions we distinguish sequences $p_\tau(x)$ which lead to limits when multiplied by any other particularly well-behaved function $F(x)$ and integrated. Thus if

$$\lim_{\tau \to 0} \int_{-\infty}^{\infty} p_\tau(x)F(x)\,dx$$

exists, then we call $p_\tau(x)$ a *regular* sequence of particularly well-behaved functions.

A sequence of particularly well-behaved functions which is not regular (see Fig. 5.15) is

$$\tau^{-1}e^{-x^2}.$$

An example of a regular sequence is

$$\tau^{-1}e^{-\pi x^2/\tau^2}.$$

In this example each member of the sequence has unit area, but that is not essential; for instance, consider the regular sequence $(1 + \tau^{-1}) \exp(-x^2/\tau^2)$.

Generalized functions A generalized function $p(x)$ is then taken to be defined by a regular sequence $p_\tau(x)$ of particularly well-behaved functions. In fact the generalized function *is* the regular sequence, and since the limit to which a regular sequence leads can be the same for more than one sequence, a generalized function is finally defined as the class of all regular sequences of particularly well-behaved functions equivalent to a given regular sequence. The symbol $p(x)$ thus represents an entity rather different from an ordinary function. It stands for a class of functions, and is itself not a function. Therefore, when we write it in a context where ordinary functions are customary, the meaning to be assigned must be stated. For example, we shall deem that

$$\int_{-\infty}^{\infty} p(x)F(x)\,dx,$$

where $p(x)$ is a generalized function and $F(x)$ is any particularly well-behaved function, shall mean

$$\lim_{\tau\to 0} \int_{-\infty}^{\infty} p_\tau(x)F(x)\,dx,$$

where $p_\tau(x)$ is any regular sequence of particularly well-behaved functions defining $p(x)$.

Two points may be noticed. First, the sequences $\tau^{-1}\Pi(x/\tau)$ and $\tau^{-1}\Lambda(x/\tau)$ do not define a generalized function in the present sense, for the members of these sequences do not possess derivatives of all orders at all points. Second, the function $F(x)$ on which the limiting process is tested has to be particularly well-behaved.

These highly restrictive conditions enable one to make a logical development at the cost of appearing to exclude the simple sequences and simple functions we ordinarily handle. We must bear in mind, however, that where differentiability is not in question we may fall back on rectangular pulses, and that where only first or second derivatives are required, Π^{*2} and Π^{*3} suffice. The requirement on asymptotic behavior is met by the rectangular pulse. On the other hand, the sequence τ^{-1} sinc (x/τ), which satisfies the requirement on differentiability, does not die away quickly enough to be a regular sequence in the strict sense; nevertheless, it is usable when it enters into a product with a function which is zero outside a finite interval.

The advantage of the Gaussian pulse as a special but simple case of a particularly well-behaved function has been exploited systematically by Lighthill,[9] whose line of development is followed.

The sequence $\exp(-\tau^2x^2)$ defines a generalized function $I(x)$, for

$$\lim_{\tau\to 0}\int_{-\infty}^{\infty} e^{-\tau^2x^2}F(x)\,dx = \int_{-\infty}^{\infty} F(x)\,dx,$$

which integral exists. Hence we can make the following statement about $I(x)$:

$$\int_{-\infty}^{\infty} I(x)F(x)\,dx = \int_{-\infty}^{\infty} F(x)\,dx,$$

where $F(x)$ is any particularly well-behaved function.

The sequence $\tau^{-1}\exp(-\pi x^2/\tau^2)$ defines a generalized function, for

$$\lim_{\tau\to 0}\int_{-\infty}^{\infty} \tau^{-1}e^{-\pi x^2/\tau^2}F(x)\,dx$$

exists and is equal to $F(0)$, where $F(x)$ is any particularly well-behaved function. To prove this note that

$$\begin{aligned}\left|\int_{-\infty}^{\infty} \tau^{-1}e^{-\pi x^2/\tau^2}F(x)\,dx - F(0)\right| &= \left|\int_{-\infty}^{\infty} \tau^{-1}e^{-\pi x^2/\tau^2}[F(x) - F(0)]\,dx\right| \\ &\leqslant \max|F'(x)|\int_{-\infty}^{\infty} \tau^{-1}e^{-\pi x^2/\tau^2}|x|\,dx \\ &= \frac{\tau}{\pi}\max|F'(x)|,\end{aligned}$$

which approaches zero as $\tau \to 0$. The generalized function defined by this and equivalent sequences we call $\delta(x)$, and we can state immediately that

$$\int_{-\infty}^{\infty} \delta(x)F(x)\,dx = F(0),$$

where $F(x)$ is any particularly well-behaved function.

Algebra of generalized functions We have introduced one rule for handling the symbol standing for a generalized function, namely, that giving the meaning of

$$\int_{-\infty}^{\infty} p(x)F(x)\,dx.$$

Further rules are needed for handling the symbols for generalized functions where they appear in other algebraic situations.

Let $p(x)$ and $q(x)$ be two generalized functions, defined by the regular

[9] Lighthill, *op. cit.*

sequences $p_\tau(x)$ and $q_\tau(x)$, respectively. Now consider the sequence $p_\tau(x) + q_\tau(x)$. First, we note that it is a sequence of particularly well-behaved functions. Next, we see whether the sequence is a regular one, that is, whether

$$\lim_{\tau \to 0} \int_{-\infty}^{\infty} [p_\tau(x) + q_\tau(x)]F(x)\,dx$$

exists, where $F(x)$ belongs to S. The integral splits into two terms, each of which has a limit, since $p_\tau(x)$ and $p_\tau(x)$ are by definition regular sequences. The sum of the two limits is the limit whose existence thus establishes that $p_\tau(x) + q_\tau(x)$ is a regular sequence that consequently defines a generalized function. This generalized function we would wish to assign as the denotation of

$$p(x) + q(x);$$

it remains only to verify that the result is the same irrespective of the choice of the defining sequences $p_\tau(x)$ and $q_\tau(x)$, and indeed we see that the defining sequences $p_\tau(x) + q_\tau(x)$ are equivalent, since the sum of the two limits is independent of the choice of $p_\tau(x)$ and $q_\tau(x)$.

We now have a meaning for the addition of generalized functions.

Let $p(x)$ be a generalized function defined by a sequence $p_\tau(x)$. From the formula for integration by parts,

$$\int_{-\infty}^{\infty} p_\tau'(x)F(x)\,dx = -\int_{-\infty}^{\infty} p_\tau(x)F'(x)\,dx,$$

where $F(x)$ is any particularly well-behaved function and so therefore is $F'(x)$. Since $F'(x)$ is a particularly well-behaved function, and since $p_\tau(x)$ is by definition a regular sequence, it follows that

$$-\lim_{\tau \to 0} \int_{-\infty}^{\infty} p_\tau(x)F'(x)\,dx$$

exists, hence

$$\lim_{\tau \to 0} \int_{-\infty}^{\infty} p_\tau'(x)F(x)\,dx$$

exists. Thus $p_\tau'(x)$ is a regular sequence of particularly well-behaved functions, and all such sequences are equivalent. To the generalized function so defined we assign the notation

$$p'(x).$$

This gives us a meaning for the derivative of a generalized function. Here is an example of a statement which can be made about the derivative $p'(x)$ of a generalized function $p(x)$:

$$\int_{-\infty}^{\infty} p'(x)F(x)\,dx = -\int_{-\infty}^{\infty} p(x)F'(x)\,dx.$$

Similarly, $\int_{-\infty}^{\infty} p^{(n)}(x)F(x)\,dx = (-1)^n \int_{-\infty}^{\infty} p(x)F^{(n)}(x)\,dx.$

Since by definition $F^{(n)}(x)$ exists, however large n may be, it follows that we have an interpretation for the nth derivative of a generalized function, for any n.

Differentiation of ordinary functions Generalized functions possess derivatives of all orders, and if an ordinary function could be regarded as a generalized function, then there would be a satisfactory basis for formulas such as

$$\frac{d}{dx}[H(x)] = \delta(x).$$

If $f(x)$ is an ordinary function and we form a sequence $f_\tau(x)$ such that

$$\lim_{\tau\to 0} \int_{-\infty}^{\infty} f_\tau(x)F(x)\,dx = \int_{-\infty}^{\infty} f(x)F(x)\,dx,$$

where $F(x)$ is any particularly well-behaved function, then the sequence defines a *generalized* function, which we may denote by the same symbol $f(x)$. The symbol $f(x)$ then has two meanings. We shall limit attention to functions $f(x)$ which as $|x| \to \infty$ behave as $|x|^{-N}$ for some value of N.

A suitable sequence $f_\tau(x)$ is given by

$$[\tau^{-1}e^{-\pi x^2/\tau^2}] * [f(x)e^{-\tau^2 x^2}].$$

With this enlargement of the notion of generalized functions we can embrace the unit step function $H(x)$ as a generalized function and assign meaning to its derivative $H'(x)$. Thus

$$\begin{aligned}\int_{-\infty}^{\infty} H'(x)F(x)\,dx &= -\int_{-\infty}^{\infty} H(x)F'(x)\,dx\\ &= -\int_{0}^{\infty} F'(x)\,dx\\ &= \int_{\infty}^{0} F'(x)\,dx\\ &= F(0),\end{aligned}$$

but
$$\int_{-\infty}^{\infty} \delta(x)F(x)\,dx = F(0),$$

hence
$$H'(x) = \delta(x).$$

The generalized function $\delta(x)$ is thus the derivative of the generalized function $H(x)$, and this is how we interpret formulas such as $H'(x) = \delta(x)$; we take the symbol for an ordinary function such as $H(x)$ to stand for the corresponding generalized function.

Problems

1 What is the even part of

$$\delta(x+3)+\delta(x+2)-\delta(x+1)+\tfrac{1}{2}\delta(x)+\delta(x-1)-\delta(x-2)-\delta(x-3)?$$

2 Attempting to clarify the meaning of $\delta(xy)$, a student gave the following explanation. "Where u is zero, $\delta(u)$ is infinite. Now xy is zero where $x = 0$ and where $y = 0$, therefore $\delta(xy)$ is infinite along the x and y axes. Hence $\delta(xy) = \delta(x) + \delta(y)$." Explain the fallacy in this argument, and show that

$$\delta(xy)=\frac{\delta(x)+\delta(y)}{(x^2+y^2)^{\frac{1}{2}}}.$$

3 Show that

$$\Pi(x)=\delta(2x^2-\tfrac{1}{2})$$

and that

$$\delta(x^2-a^2)=\tfrac{1}{2}|a|^{-1}\{\delta(x-a)+\delta(x+a)\}.$$

4 Show that

$$\int_{-\infty}^{\infty}e^{-i2\pi xs}\,ds=\delta(x)$$

and that

$$\int_{-\infty}^{\infty}\delta(x)e^{i2\pi sx}\,dx=1.$$

5 Show that

$$\delta(ax+b)=\frac{1}{|a|}\,\delta\left(x+\frac{a}{b}\right),\qquad a\neq 0.$$

6 If $f(x) = 0$ has roots x_n, show that

$$\delta[f(x)]=\sum_n\frac{\delta\,(x-x_n)}{|f'(x_n)|}$$

wherever $f'(x_n)$ exists and is not zero. Consider the ideas suggested by $\delta(x^3)$ and $\delta(\operatorname{sgn} x)$.

7 Show that

$$\pi\delta(\sin\pi x)=\text{III}(x)$$

and

$$\delta(\sin x)=\pi^{-1}\,\text{III}\left(\frac{x}{\pi}\right).$$

8 Show that

$$\text{III}(x)+\text{III}(x-\tfrac{1}{2})=2\,\text{III}(2x)=\text{III}(x)*4\Pi(2x-\tfrac{1}{2}).$$

9 Show that

$$\text{III}(x)\Pi\left(\frac{x}{8}\right)=\text{III}(x)\Pi\left(\frac{x}{7}\right)+\frac{\Pi(x/8)}{8}$$

and also that

$$\text{III}(x)\Pi\left(\frac{x}{6\frac{1}{2}}\right)=\text{III}(x)\Pi\left(\frac{x}{7\frac{1}{2}}\right).$$

10 Can the following equation be correct?

$$x\,\delta(x-y) = y\,\delta(x-y).$$

11 Show that $\Lambda(x) * \sum_{-\infty}^{\infty} a_n\,\delta(x-n)$ is the polygon through the points (n,a_n).

12 Prove that

$$\delta'(-x) = -\delta'(x)$$
$$x\,\delta'(x) = -\delta(x).$$

Show also that

$$f(x)\,\delta'(x) = f(0)\,\delta'(x) - f'(0)\,\delta(x),$$

for example, by differentiating $f(x)\,\delta(x)$.

13 In attempting to show that $\delta'(x) = -\delta(x)/x$ a student presented the following argument. "A suitable sequence, as τ approaches zero, for defining $\delta(x)$ is $\tau/\pi(x^2+\tau^2)$. Therefore a suitable sequence for $\delta'(x)$ is the derivative

$$\frac{d}{dx}\frac{\tau}{\pi(x^2+\tau^2)} = \frac{-2\tau x}{\pi(x^2+\tau^2)^2}$$
$$= \frac{-2x}{x^2+\tau^2}\,\frac{\tau}{\pi(x^2+\tau^2)}.$$

The second factor is the sequence for $\delta(x)$, and the first factor goes to $-2/x$ in the limit as τ approaches zero. Therefore $\delta'(x) = -2\delta(x)/x$." Explain the fallacy in this argument.

14 Show that

$$x^n\delta^{(n)}(x) = (-1)^n n!\delta(x)$$

and hence that $\quad x^2\,\delta''(x) = 2\delta(x)$

and $\quad x^3\,\delta''(x) = 0.$

15 The function $[x]$ is here defined as the mean of the greatest integer less than x and the greatest integer less than or equal to x. Show that

$$[x]' = \mathrm{III}(x)$$

and also that $\quad \frac{d}{dx}\{[x]H(x)\} = \mathrm{III}(x)H(x) - \tfrac{1}{2}\delta(x).$

(The common definition of $[x]$ as the greatest integer less than x is not fully suitable for the needs of this exercise; the two definitions differ by the null function which is equal to $\frac{1}{2}$ for integral values of x and is zero elsewhere.)

16 The sawtooth function $Sa(x)$ is defined by $Sa(x) = [x] - x + \frac{1}{2}$. Show that

$$Sa'(x) = \mathrm{III}(x) - 1$$

and that $\quad \frac{d}{dx}[Sa(x)H(x)] = [\mathrm{III}(x) - 1]H(x).$

17 Show that $\operatorname{sgn}^2 x = 1 - \delta^0(x)$.

18 The Kronecker delta is defined by

$$\delta_{ij} = \begin{cases} 1 & i = j \\ 0 & i \neq j. \end{cases}$$

Show that it may be expressed as a null function of $i - j$ as follows:

$$\delta_{ij} = \delta^0(i - j).$$

19 We wish to consider the suitability of a sequence of asymmetrical profiles, such as $\tau^{-1}\{\Lambda(x/\tau) + \frac{1}{2}\Lambda[(x - \tau)/\tau]\}$, for representing the impulse symbol. Discuss the sifting property that leads to a result of the form

$$\delta_a * f = \mu\, \delta_+ * f + \nu\, \delta_- * f,$$

where δ_a is a symbol based on the asymmetrical sequence, δ_+ is based on the sequence $\tau^{-1}\Pi[(x - \frac{1}{2}\tau)/\tau]$, and δ_- is based on the sequence $\tau^{-1}\Pi[(x + \frac{1}{2}\tau)/\tau]$ (τ positive).

20 Prove the relation ${}^2\delta(x,y) = \delta(r)/\pi|r|$.

21 Illustrate on an isometric projection the meaning you would assign to $\text{III}[(x^2 + y^2)^{\frac{1}{2}}]$. How would you express something which on this diagram would have the appearance of equally spaced concentric rings of equal height?

22 The function $f_r(x)$ is formed from $f(x)$ by reversing it; that is, $f_r(x) = f(-x)$. Show that the operation of forming f_r from f can be expressed with the aid of the impulse symbol by

$$f \star \delta$$

and hence that

$$(f \star \delta) \star \delta = f.$$

23 Under what conditions could we say that $(f \star \delta) \star \delta = f \star (\delta \star \delta)$?

24 All the sequences $f(x,\tau)$ given on page 73 have the property that $f(0,\tau)$ increases without limit as $\tau \to 0$. Show that $\frac{1}{2}\tau^{-1}\Lambda[(x/\tau) - 1] + \frac{1}{2}\tau^{-1}\Lambda[(x/\tau) + 1]$ is an equivalent sequence which, however, possesses a limit of zero, as $\tau \to 0$, for all x. Show that $f(0,\tau)$, far from needing to approach ∞ as $\tau \to 0$, may indeed approach $-\infty$.

25 Show that

$$f(x)\,\delta''(x) = f(0)\,\delta''(x) - 2f'(0)\,\delta'(x) + f''(0)\,\delta(x)$$

and that in general

$$f(x)\,\delta^{(n)}(x) = f(0)\,\delta^{(n)}(x) - \binom{n}{1} f'(0)\,\delta^{(n-1)}(x) + \cdots - \binom{n}{n-1} f^{(n-1)}(0)\,\delta'(x) + f^{(n)}(0)\,\delta(x).$$

Chapter 6 The basic theorems

A small number of theorems play a basic role in thinking with Fourier transforms. Most of them are familiar in one form or another, but here we collect them as simple mathematical properties of the Fourier transformation. Most of their derivations are quite simple, and their applicability to impulsive functions can readily be verified by consideration of sequences of rectangular or other suitable pulses. As a matter of interest, proofs based on the algebra of generalized functions as given in Chapter 5 are gathered for illustration at the end of this chapter.

The emphasis in this chapter, however, is on illustrating the *meaning* of the theorems and gaining familiarity with them. For this purpose a stock-in-trade of particular transform pairs is first provided so that the meaning of each theorem may be shown as it is encountered.

A few transforms for illustration

Six transform pairs for reference are listed below. They are all well known, and the integrals are evaluated in Chapter 7; we content ourselves at this point with asserting that the following integrals may be verified.

$$\int_{-\infty}^{\infty} e^{-\pi x^2} e^{-i2\pi xs}\,dx = e^{-\pi s^2} \quad \text{and} \quad \int_{-\infty}^{\infty} e^{-\pi s^2} e^{+i2\pi sx}\,ds = e^{-\pi x^2}$$

$$\int_{-\infty}^{\infty} \operatorname{sinc} x\, e^{-i2\pi xs}\,dx = \Pi(s) \quad \text{and} \quad \int_{-\infty}^{\infty} \Pi(s) e^{+i2\pi sx}\,ds = \operatorname{sinc} x$$

$$\int_{-\infty}^{\infty} \operatorname{sinc}^2 x\, e^{-i2\pi xs}\,dx = \Lambda(s) \quad \text{and} \quad \int_{-\infty}^{\infty} \Lambda(s) e^{+i2\pi sx}\,ds = \operatorname{sinc}^2 x$$

Thus the transform of the Gaussian function is the same Gaussian function, the transform of the sinc function is the unit rectangle function, and the transform of the sinc² function is the triangle function of unit height and area.

These formulas are illustrated as the first three transform pairs in Fig. 6.1. Note that item (2) of the figure, which says that $\Pi(s)$ is the transform of sinc x, could be supplemented by a second figure, with left and right graphs interchanged, which would say that sinc s is the transform of $\Pi(x)$. A consequence of the reciprocal property of the Fourier transformation, this extra figure would appear redundant. However, the statement

$$\Pi(s) = \int_{-\infty}^{\infty} \operatorname{sinc} x \, e^{-i2\pi xs} \, dx$$

has quite a different character from

$$\operatorname{sinc} s = \int_{-\infty}^{\infty} \Pi(x) e^{-i2\pi xs} \, dx.$$

The first statement tells us that the integral of the product of certain rather ordinary functions is equal to unity for absolute values of the constant s less than $\frac{1}{2}$. Whether s is equal to say 0.3 or 0.35, the value of the integral is unchanged. However, if $|s|$ exceeds $\frac{1}{2}$, the situation changes abruptly, because the integral now comes to nothing and continues to do so, regardless of the precise value of s. Thus

$$\int_{-\infty}^{\infty} \frac{\sin \pi x}{\pi x} e^{-i2\pi xs} \, dx = \begin{cases} 1 & |s| < \frac{1}{2} \\ 0 & |s| > \frac{1}{2}. \end{cases}$$

This rather curious behavior is typical of many situations where the Fourier integral connects ordinary, continuous, and differentiable functions, on the one hand, with awkward, abrupt functions requiring piecewise definition, on the other. The second statement may be rewritten

$$\frac{\sin \pi s}{\pi s} = \int_{-\frac{1}{2}}^{\frac{1}{2}} e^{-i2\pi xs} \, dx.$$

Here an elementary definite integral of the exponential function is equal to an ordinary function of the parameter s. Thus the direct and inverse transforms express different things. Two separate and distinct physical meanings will later be seen to be associated with each transform pair.

Three further transforms required for illustrating the basic theorems are transform pairs in the limiting sense discussed earlier.

Taking the result for the Gaussian function, and making a simple substitution of variables, we have[1]

$$\int_{-\infty}^{\infty} e^{-\pi(ax)^2} e^{-i2\pi sx} \, dx = |a|^{-1} e^{-\pi(s/a)^2}.$$

[1] In this formula the absolute value of a is used in order to counteract the sign reversal associated with the interchange of the limits of integration when a is negative.

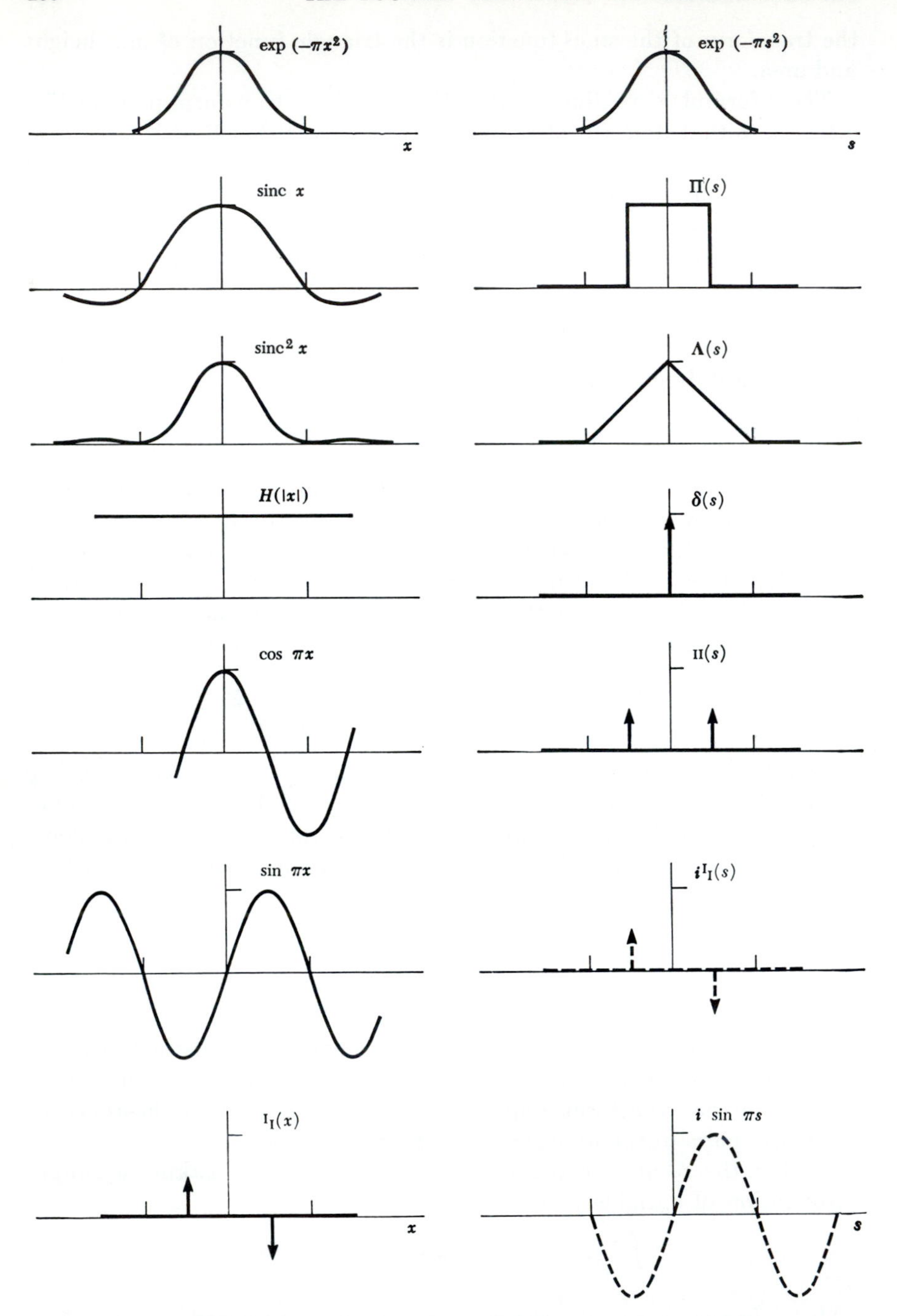

Fig. 6.1 Some Fourier transform pairs for reference.

As $a \to 0$, the right-hand side represents a defining sequence for $\delta(s)$; the left-hand side is the Fourier transform of what in the limit is unity. It follows that

$$1 \text{ is the Fourier transform in the limit of } \delta(s).$$

The remaining two examples come from the verifiable relation

$$\int_{-\infty}^{\infty} e^{i\pi x} e^{-(ax)^2} e^{-i2\pi sx}\, dx = |a|^{-1} e^{-[(s-\frac{1}{2})/a]^2},$$

whence

$$e^{i\pi x} \text{ is the Fourier transform in the limit of } \delta(s - \tfrac{1}{2});$$

or, splitting the left-hand side into real and imaginary parts and the right-hand side into even and odd parts, $\cos \pi x$ is the Fourier transform in the limit of

$$\tfrac{1}{2}\delta(s + \tfrac{1}{2}) + \tfrac{1}{2}\delta(s - \tfrac{1}{2}) = \text{II}(s)$$

and $i \sin \pi x$ is the minus-i Fourier transform in the limit of

$$-\tfrac{1}{2}\delta(s + \tfrac{1}{2}) + \tfrac{1}{2}\delta(s - \tfrac{1}{2}) = -{}^{\text{I}}\text{I}(s).$$

Summarizing the examples,

$$\begin{aligned}
e^{-\pi x^2} &\supset e^{-\pi s^2} \\
\operatorname{sinc} x &\supset \Pi(s) \\
\operatorname{sinc}^2 x &\supset \Lambda(s) \\
1 &\supset \delta(s) \\
\cos \pi x &\supset \text{II}(s) = \tfrac{1}{2}\delta(s + \tfrac{1}{2}) + \tfrac{1}{2}\delta(s - \tfrac{1}{2}) \\
\sin \pi x &\supset i\,{}^{\text{I}}\text{I}(s) = \tfrac{1}{2}i\delta(s + \tfrac{1}{2}) - \tfrac{1}{2}i\delta(s - \tfrac{1}{2}) \\
{}^{\text{I}}\text{I}(x) &\supset i \sin \pi s.
\end{aligned}$$

All the transform pairs chosen for illustration have physical interpretations, which will be brought out later. Many properties appear among the transform pairs chosen for reference, including discontinuity, impulsiveness, limited extent, nonnegativeness, and oddness. The only examples exhibiting complex or nonsymmetrical properties are

$$e^{i\pi x} \supset \delta(s - \tfrac{1}{2})$$

and

$$\delta(x - \tfrac{1}{2}) \supset e^{-i\pi s}.$$

Similarity theorem

If $f(x)$ has the Fourier transform $F(s)$, then $f(ax)$ has the Fourier transform $|a|^{-1}F(s/a)$.

Derivation:

$$\int_{-\infty}^{\infty} f(ax)e^{-i2\pi xs}\,dx = \frac{1}{|a|}\int_{-\infty}^{\infty} f(ax)e^{-i2\pi(ax)(s/a)}\,d(ax)$$
$$= \frac{1}{|a|}F\left(\frac{s}{a}\right).$$

This theorem is well known in its application to waveforms and spectra, where compression of the time scale corresponds to expansion of the frequency scale. However, as one member of the transform pair expands

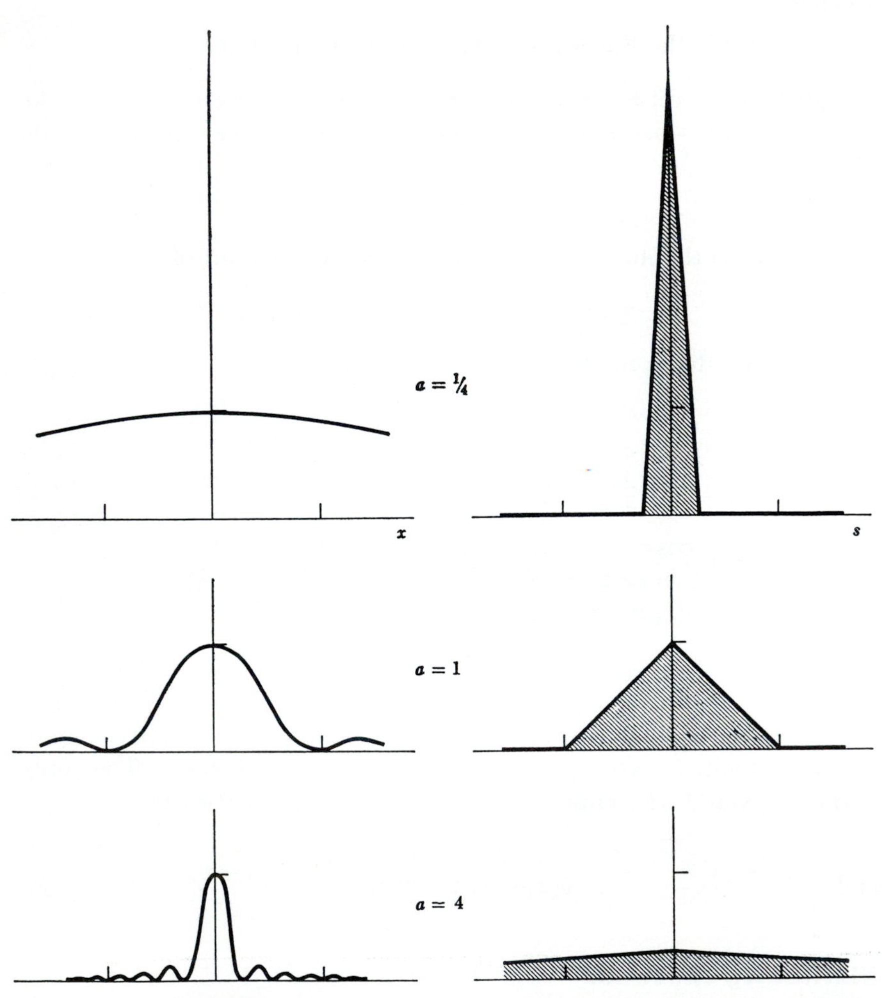

Fig. 6.2 The effect of changes in the scale of abscissas as described by the similarity theorem. The shaded area remains constant.

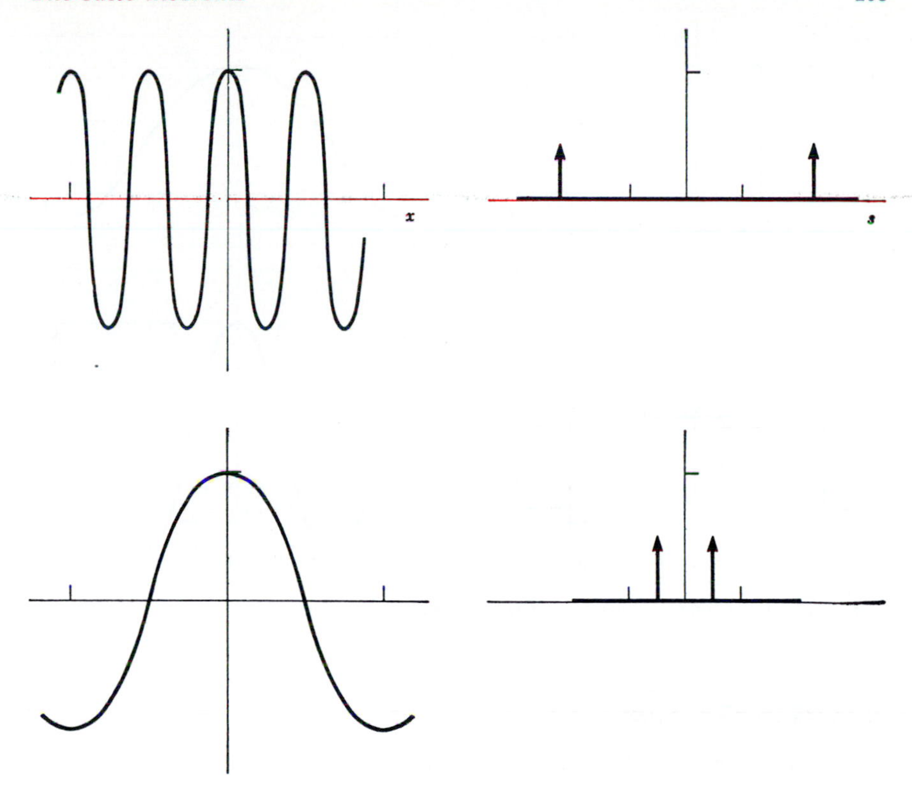

Fig. 6.3 Expansion of a cosinusoid and corresponding shifts in its spectrum.

horizontally, the other not only contracts horizontally but also grows vertically in such a way as to keep constant the area beneath it, as shown in Fig. 6.2.

A special case of interest arises with periodic functions and impulses. As Fig. 6.3 shows, expansion of a cosinusoid leads simply to shifts of the impulses constituting the transform. This is not simply a compression of the scale of s, for that would entail a reduction in strength of the impulses.

In a more symmetrical version of this theorem,

If $f(x)$ has the Fourier transform $F(s)$ then $|a|^{\frac{1}{2}}f(ax)$ has the Fourier transform $|b|^{\frac{1}{2}}F(bs)$, where $b = a^{-1}$.

Then, as each function expands or contracts it also shrinks or grows vertically (see Fig. 6.4) to compensate (in such a way that the integral of its square is maintained constant, as will be seen later from the power theorem).

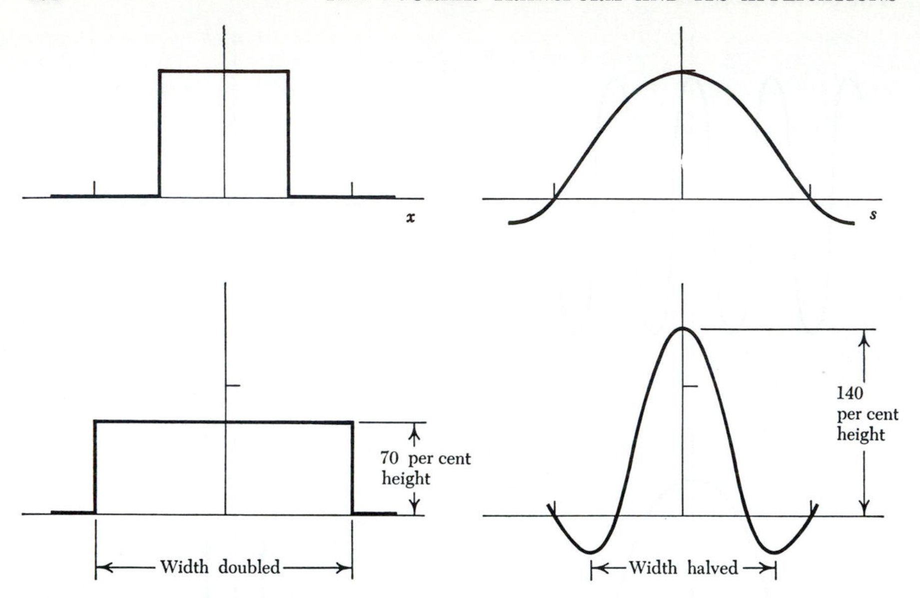

Fig. 6.4 A symmetrical version of the similarity theorem.

Addition theorem

If $f(x)$ and $g(x)$ have the Fourier transforms $F(s)$ and $G(s)$, respectively, then $f(x) + g(x)$ has the Fourier transform $F(s) + G(s)$.

Derivation:

$$\begin{aligned}\int_{-\infty}^{\infty} [f(x) + g(x)]e^{-i2\pi xs}\,dx &= \int_{-\infty}^{\infty} f(x)e^{-i2\pi xs}\,dx + \int_{-\infty}^{\infty} g(x)e^{-i2\pi xs}\,dx \\ &= F(s) + G(s).\end{aligned}$$

This theorem, which is illustrated by an example in Fig. 6.5, reflects the suitability of the Fourier transform for dealing with linear problems. A corollary is that $af(x)$ has the transform $aF(s)$, where a is a constant.

Shift theorem

If $f(x)$ has the Fourier transform $F(s)$, then $f(x - a)$ has the Fourier transform $e^{-2\pi ias}F(s)$.

Derivation:

$$\begin{aligned}\int_{-\infty}^{\infty} f(x - a)e^{-i2\pi xs}\,dx &= \int_{-\infty}^{\infty} f(x - a)e^{-i2\pi(x-a)s}e^{-i2\pi as}d(x - a) \\ &= e^{-i2\pi as}F(s).\end{aligned}$$

If a given function is shifted in the positive direction by an amount a, no Fourier component changes in amplitude; it is therefore to be expected that the changes in its Fourier transform will be confined to phase changes. According to the theorem, each component is delayed in phase by an amount proportional to s; that is, the higher the frequency, the greater the change in phase angle. This occurs because the absolute shift a occupies a greater fraction of the period s^{-1} of a harmonic component in proportion to its frequency. Hence the phase delay is a/s^{-1} cycles or $2\pi as$ radians. The constant of proportionality describing the linear change of phase with s is $2\pi a$, the rate of change of phase with frequency being greater as the shift a is greater.

The shift theorem is one of those which are self-evident in a chosen physical embodiment. Consider parallel light falling normally on an aperture. To shift the diffracted beam through a small angle, one changes the angle of incidence by that amount. But this is simply a way of causing the phase of the illumination to change linearly across the aper-

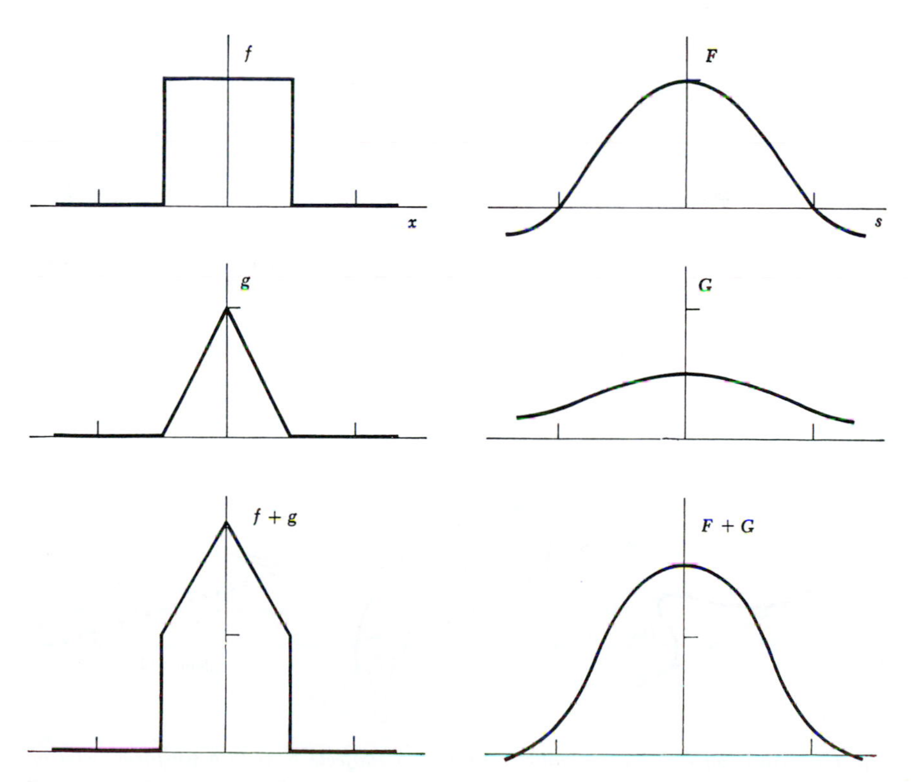

Fig. 6.5 The addition theorem $f + g \supset F + G$.

ture; another way is to insert a prism. These well-understood procedures for shifting the direction of a light beam are shown in Chapter 13 to exemplify the shift theorem.

In the example of Fig. 6.6, a function $f(x)$ is shown whose transform $F(s)$ is real. A shifted function $f(x - \frac{1}{4})$ has a transform which is derivable by subjecting $F(s)$ to a uniform twist of $\pi/2$ per unit of s. The figure attempts to show that the plane containing $F(s)$ has been deformed into a helicoid. The practical difficulties of representing a complex function of s in a three-dimensional plot are overcome by showing the modulus and phase of $F(s)$ separately; however, the three-dimensional diagram often gives a better insight.

The second example (see Fig. 6.7) shows familiar results for the cosine and sine functions and for the intermediate cases which arise as the cosine slides along the axis of x. In this case the helicoidal surface is not shown. An alternative representation in terms of real and imaginary parts is given, incorporating the convention introduced earlier of showing the imaginary part by a broken line. A small shift evidently leaves the real

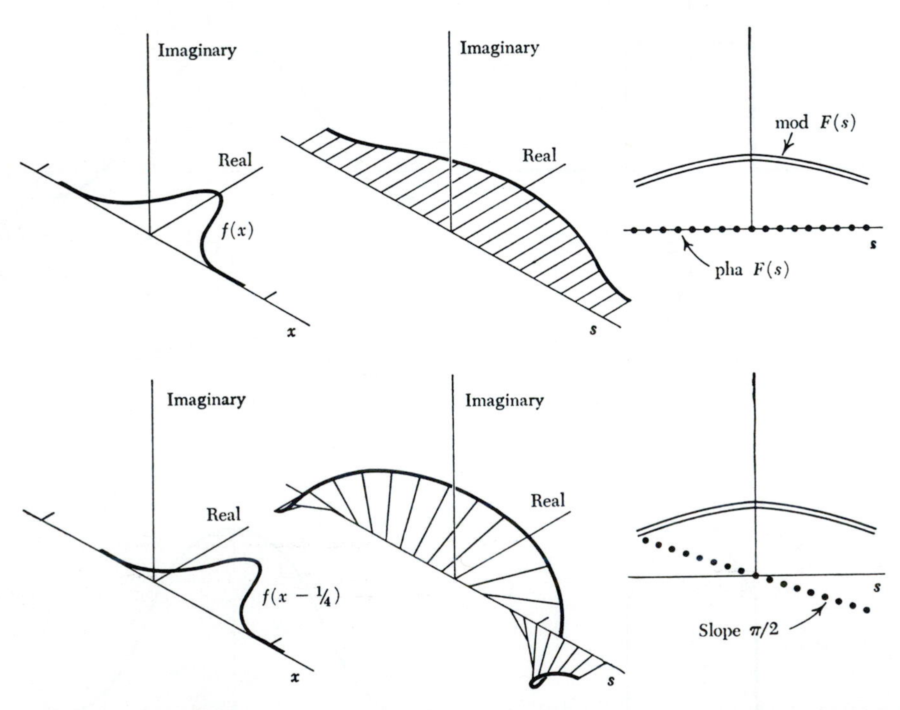

Fig. 6.6 Shifting $f(x)$ by one quarter unit of x subjects $F(s)$ to a uniform twist of 90 deg per unit of s.

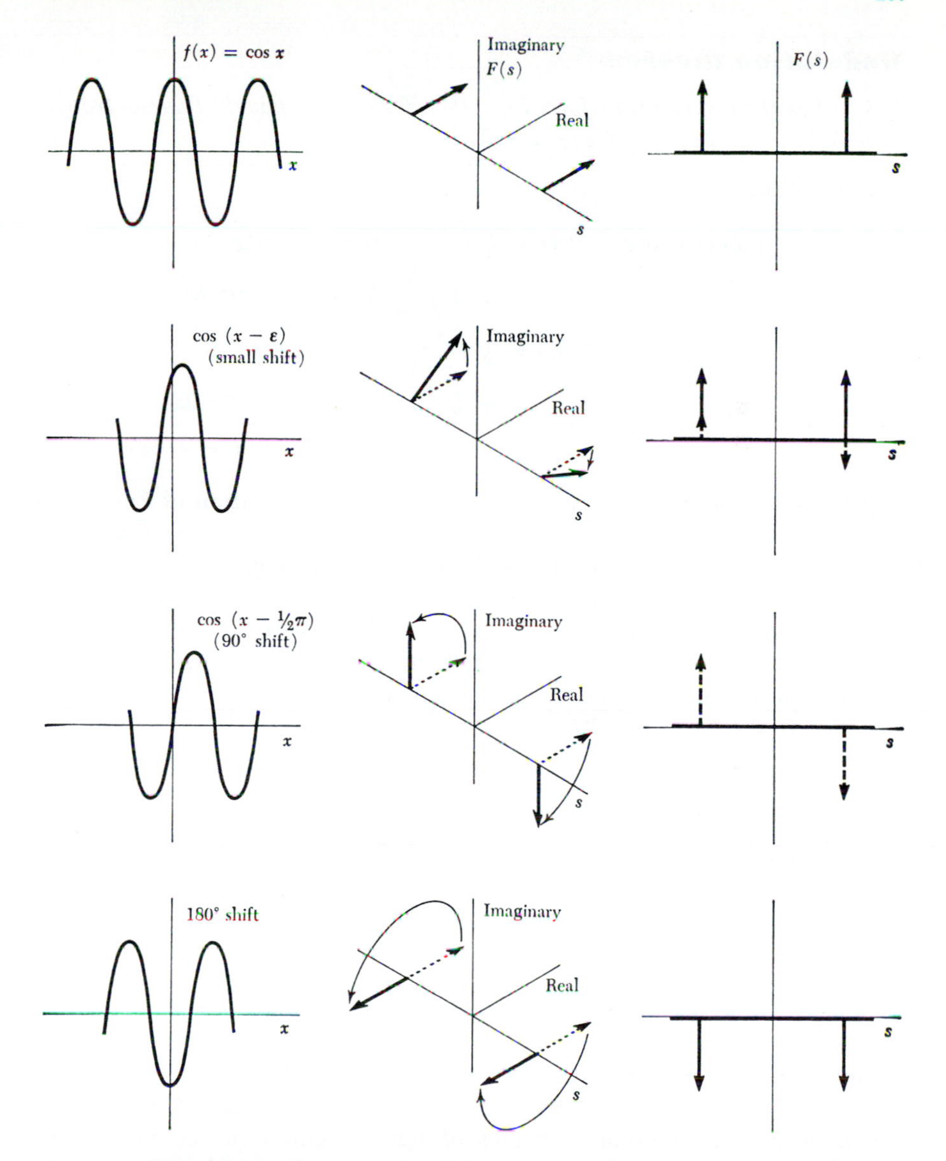

Fig. 6.7 The effect of shifting a cosinusoid.

part of the transform almost intact but introduces an odd imaginary part. With further shift the imaginary part increases until at a shift of $\pi/2$ there is no real part left. Then the real part reappears with opposite sign until at a shift of π both components have undergone a full reversal of phase.

Modulation theorem

If $f(x)$ has the Fourier transform $F(s)$, then $f(x)$ cos ωx has the Fourier transform $\frac{1}{2}F(s - \omega/2\pi) + \frac{1}{2}F(s + \omega/2\pi)$.

Derivation:

$$\begin{aligned}\int_{-\infty}^{\infty} f(x) \cos \omega x \, e^{-i2\pi xs} \, dx &= \tfrac{1}{2} \int_{-\infty}^{\infty} f(x) e^{i\omega x} e^{-i2\pi xs} \, dx \\ &\quad + \tfrac{1}{2} \int_{-\infty}^{\infty} f(x) e^{-i\omega x} e^{-i2\pi xs} \, dx \\ &= \tfrac{1}{2} \int_{-\infty}^{\infty} f(x) e^{-i2\pi(s-\omega/2\pi)x} \, dx \\ &\quad + \tfrac{1}{2} \int_{-\infty}^{\infty} f(x) e^{-i2\pi(s+\omega/2\pi)x} \, dx \\ &= \tfrac{1}{2}F(s - \omega/2\pi) + \tfrac{1}{2}F(s + \omega/2\pi).\end{aligned}$$

The new transform will be recognized as the convolution of $F(s)$ with $\frac{1}{2}\delta(s + \omega/2\pi) + \frac{1}{2}\delta(s - \omega/2\pi) = (\pi/\omega)\text{II}(\pi s/\omega)$. This is a special case of the convolution theorem, but it is important enough to merit special mention. It is well known in radio and television, where a harmonic carrier wave is modulated by an envelope. The spectrum of the envelope is separated into two parts, each of half the original strength. These two replicas of the original are then shifted along the s axis by amounts $\pm\omega/2\pi$, as shown in Fig. 6.8.

Convolution theorem

As stated earlier, the convolution of two functions f and g is another function h defined by the integral

$$h(x) = \int_{-\infty}^{\infty} f(u) g(x - u) \, du.$$

A great deal is implied by this expression. For instance, $h(x)$ is a linear functional of $f(x)$; that is, $h(x_1)$ is a linear sum of values of $f(x)$, duly weighted as described by $g(x)$. However, it is not the most general linear functional; it is the particular kind for which any other value $h(x_2)$ is given by a linear combination of values of $f(x)$ weighted in the *same* way. Another way of conveying this special property of convolution is to say that a shift of $f(x)$ along the x axis results simply in an equal shift of $h(x)$; that is, if $h(x) = f(x) * g(x)$, then

$$f(x - a) * g(x) = h(x - a).$$

Suppose that a train is slowly crossing a bridge. The load at the point x is $f(x)$, and the deflection at x is $h(x)$. Since the structural members are

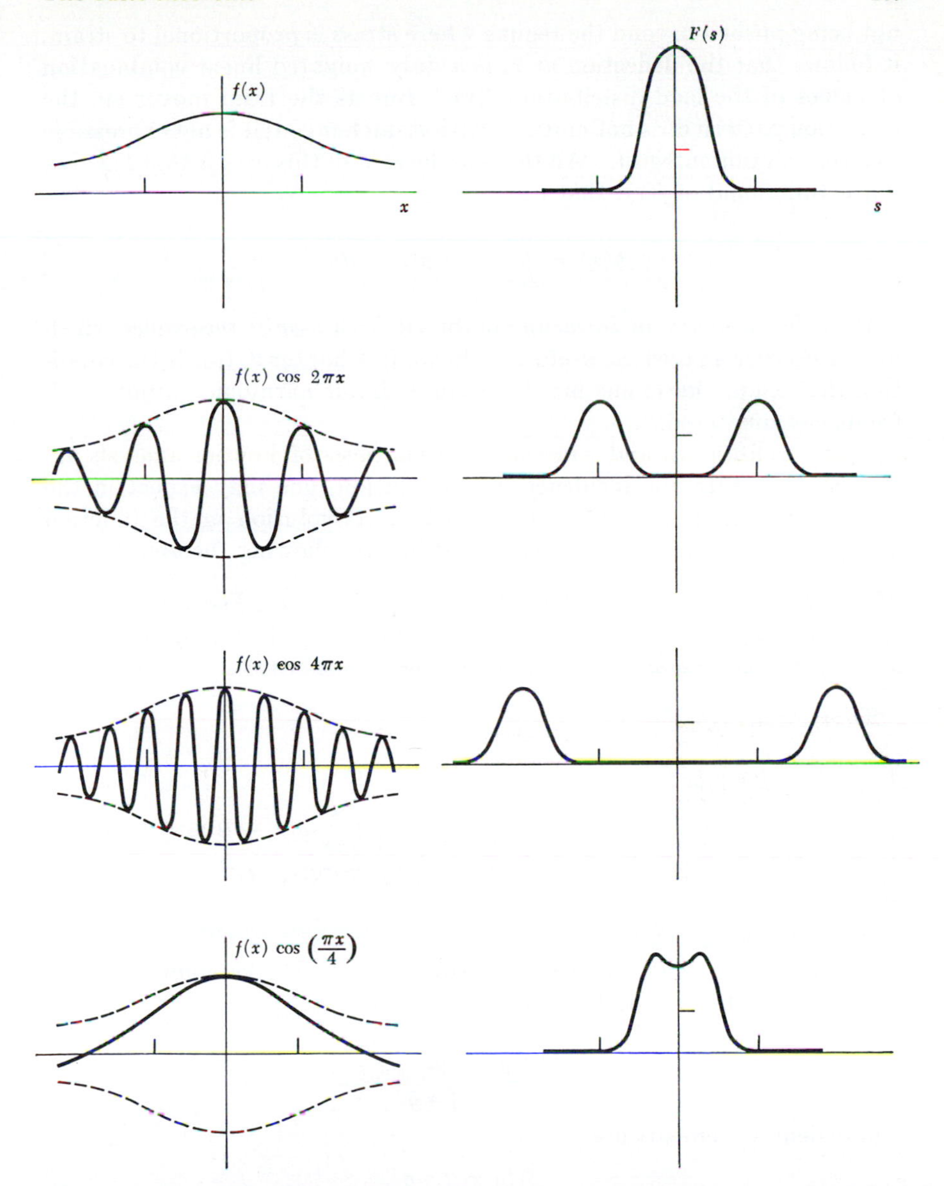

Fig. 6.8 An envelope function $f(x)$ multiplied by cosinusoids of various frequencies, with the corresponding spectra.

not being pushed beyond the regime where stress is proportional to strain, it follows that the deflection at x_1 is a duly weighted linear combination of values of the load distribution $f(x)$. But as the train moves on, the deflection pattern does not move on with it unchanged; it is not expressible as a convolution integral. All that can be said in this case is that $h(x)$ is a linear functional of $f(x)$; that is,

$$h(x) = \int_{-\infty}^{\infty} f(u)g(x,u)\,du.$$

It is the property of *linearity* combined with *x-shift invariance* which makes Fourier analysis so useful; as shown in Chapter 9, this is the condition that simple harmonic inputs produce simple harmonic outputs with frequency unaltered.

If the well-known and widespread advantages of Fourier analysis are concomitant with the incidence of convolution, one may expect in the transform domain a simple counterpart of convolution in the function domain. This counterpart is expressed in the following theorem.

If $f(x)$ has the Fourier transform $F(s)$ and $g(x)$ has the Fourier transform $G(s)$, then $f(x) * g(x)$ has the Fourier transform $F(s)G(s)$; that is, convolution of two functions means multiplication of their transforms.

Derivation:

$$\begin{aligned}\int_{-\infty}^{\infty}\left[\int_{-\infty}^{\infty} f(x')g(x-x')\,dx'\right] e^{-i2\pi xs}\,dx \\ &= \int_{-\infty}^{\infty} f(x')\left[\int_{-\infty}^{\infty} g(x-x')e^{-i2\pi xs}\,dx\right] dx' \\ &= \int_{-\infty}^{\infty} f(x')e^{-i2\pi x's}G(s)\,dx' \\ &= F(s)G(s).\end{aligned}$$

Using bars to denote Fourier transforms, we can give compact statements of the theorem and its converse. Thus

$$\overline{f * g} = \bar{f}\bar{g},$$
$$\overline{fg} = \bar{f} * \bar{g}.$$

Equivalent statements are

$$\bar{\bar{f\bar{g}}}\; \overline{\bar{f}\bar{g}} = f * g$$
$$\overline{\bar{f} * \bar{g}} = fg.$$

We have stated earlier that

$$\begin{aligned} f * g &= g * f && \text{(commutative)} \\ f * (g * h) &= (f * g) * h && \text{(associative)} \\ f * (g + h) &= f * g + f * h && \text{(distributive).} \end{aligned}$$

expressed as the square of one variable alone. The theorem does not have a distinctive name of its own; some authors refer to it as Parseval's theorem, which is the well-established name of a theorem in the theory of Fourier series (Chapter 10).

Autocorrelation theorem

If $f(x)$ has the Fourier transform $F(s)$, then its autocorrelation function $\int_{-\infty}^{\infty} f^*(u)\,f(u+x)\,du$ has the Fourier transform $|F(s)|^2$.

Derivation:

$$\begin{aligned}\int_{-\infty}^{\infty} |F(s)|^2 e^{i2\pi xs}\,ds &= \int_{-\infty}^{\infty} F(s)F^*(s)e^{i2\pi xs}\,ds\\ &= f(x) * f^*(-x)\\ &= \int_{-\infty}^{\infty} f(u)f^*(u-x)\,du\\ &= \int_{-\infty}^{\infty} f^*(u)f(u+x)\,du.\end{aligned}$$

A special case of the convolution theorem, the autocorrelation theorem is familiar in communications in the form that the autocorrelation function of a signal is the Fourier transform of its power spectrum.[3] It is illustrated in Fig. 6.12. The unique feature of this theorem, as contrasted with a theorem that could be stated for the self-convolution, is that information about the phase of $F(s)$ is entirely missing from $|F(s)|^2$. The autocorrelation function correspondingly contains no information about the phase of the Fourier components of $f(x)$, being unchanged if phases are allowed to alter, as was shown on p. 45.

Exercise Show that the normalized autocorrelation function $\gamma(x)$, for which $\gamma(0) = 1$ (see p. 41), has as its Fourier transform the normalized power spectrum $|\Phi(s)|^2$ whose infinite integral is unity, and which is defined by

$$|\Phi(s)|^2 = \frac{|F(s)|^2}{\int_{-\infty}^{\infty} |F(s)|^2\,ds}.$$

A statement may also be added about the function $C(x)$, which was defined in Chapter 3 by the sequence of autocorrelation functions $\gamma_X(x)$ generated from the functions $f(x)\Pi(x/X)$ as $X \to \infty$. If $\gamma_X(x)$ approached a limit, then the limit was called $C(x)$. Corresponding to the sequence of normalized autocorrelation functions, $\gamma_X(x)$ is the sequence of normalized power spectra $|\Phi_X(s)|^2$. If $\gamma_X(x)$ approaches a limit as the segment length X increases, then the normalized power

[3] The corresponding theorem for signals that do not tend to zero as time advances is sometimes referred to as Wiener's theorem (see N. Wiener, "Extrapolation, Interpolation, and Smoothing of Stationary Time Series," John Wiley and Sons, New York, 1949).

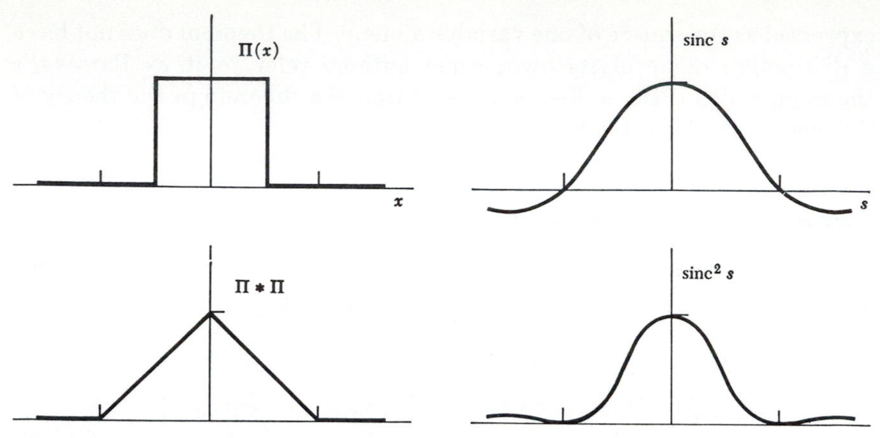

Fig. 6.12 The autocorrelation theorem: autocorrelating a function corresponds to squaring (the modulus of) its transform.

spectrum settles down to a limiting form $|\Phi_\infty(s)|^2$. In these circumstances the autocorrelation theorem takes the form

$$C(x) \supset |\Phi_\infty(s)|^2$$

and one generally says, as before, that the autocorrelation is the Fourier transform of the power spectrum, suiting the definitions to the needs of the case.

Clearly it may happen that the sequence of transforms of $\gamma_X(x)$ does not approach limits for all s but is of a character describable with impulse symbols $\delta(s)$. Therefore situations may be entertained where the transform of $C(x)$ is a generalized function. For example, an ideal line spectrum such as is possessed by a signal carrying finite power at a single frequency is such a case. We know that if $f(x) = \cos \alpha x$, then $C(x) = \cos \alpha x$. The Fourier transform of $C(x)$ is thus a generalized function $\frac{1}{2}\delta(s + \alpha/2\pi) + \frac{1}{2}\delta(s - \alpha/2\pi)$, and if necessary we could work out the sequence of transforms of $\gamma_X(x)$ that define it. The interesting point here, however, is that the power spectrum as a generalized function is not deducible from the autocorrelation theorem, for no interpretation has been given for products such as $[\delta(x)]^2$.

Exercise Give an interpretation for $[\delta(x)]^2$ by attempting to apply the autocorrelation theorem to $f(x) = \cos \alpha x$ and test it on some other simple example such as $f(x) = 1$.

Exercise Show that the situation cannot arise where the sequence $\gamma_X(x)$ calls for the use of $\delta(x)$ in representing $C(x)$.

Exercise We wish to discuss the ideal situation of a power spectrum which is flat and extends to infinite frequency. Determine $C(x)$ and its transform. Show that the nonnormalized autocorrelation function lends itself to this requirement, and that a good version of the autocorrelation theorem can be devised in which the power spectrum is normalized so as to be equal to unity at its origin. What does this form of the theorem say when $f(x) = \cos \alpha x$?

Derivative theorem

If $f(x)$ has the Fourier transform $F(s)$ then $f'(x)$ has the Fourier transform $i2\pi sF(s)$.

Derivation:

$$\begin{aligned}\int_{-\infty}^{\infty} f'(x)e^{-i2\pi xs}\,dx &= \int_{-\infty}^{\infty} \lim \frac{f(x+\Delta x) - f(x)}{\Delta x}\, e^{-i2\pi xs}\,dx \\ &= \lim \int_{-\infty}^{\infty} \frac{f(x+\Delta x)}{\Delta x}\, e^{-i2\pi xs}\,dx - \lim \int_{-\infty}^{\infty} \frac{f(x)}{\Delta x}\, e^{-i2\pi xs}\,dx \\ &= \lim \frac{e^{i2\pi\,\Delta xs}F(s) - F(s)}{\Delta x} \\ &= i2\pi sF(s).\end{aligned}$$

Since taking the derivative of a function multiplies its transform by $i2\pi s$, we can say that differentiation enhances the higher frequencies, attenuates the lower frequencies, and suppresses any zero-frequency component. Examples are given in Figs. 6.13 and 6.14.

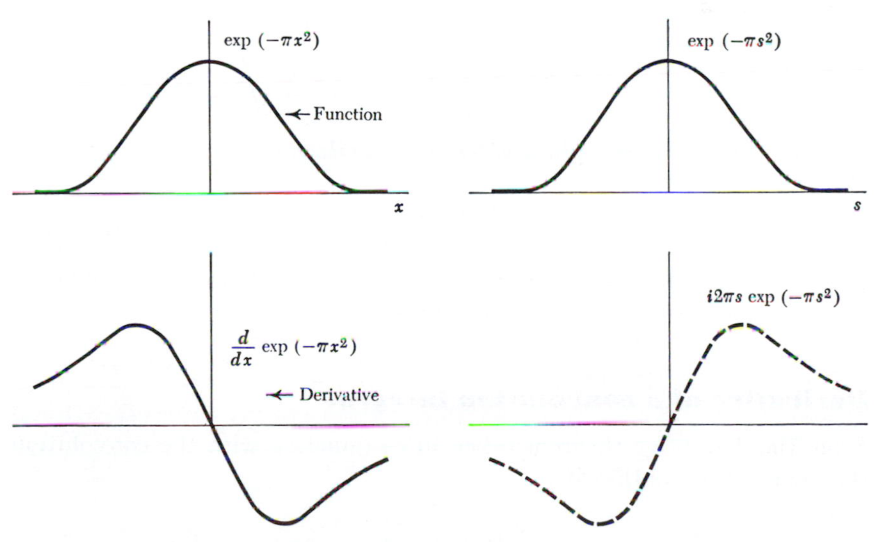

Fig. 6.13 Differentiation of a function incurs multiplication of the transform by $i2\pi s$.

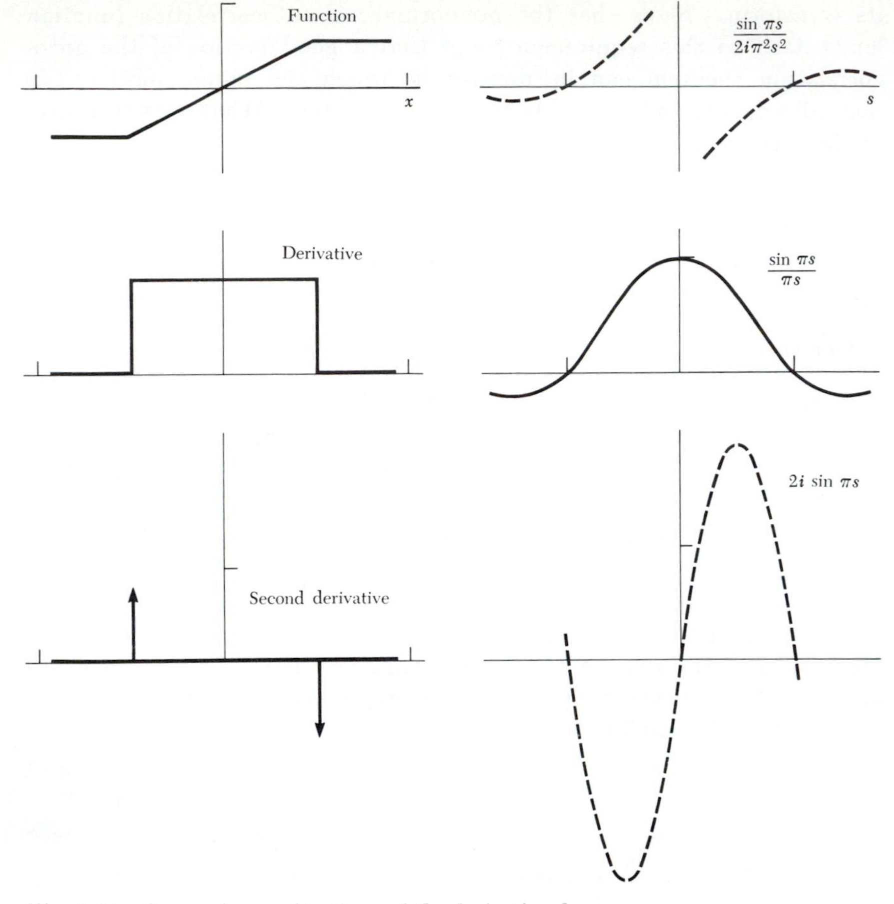

Fig. 6.14 Successive applications of the derivative theorem.

It happens quite frequently that the multiplication by $i2\pi s$ causes the integral of $|i2\pi sF(s)|$ to diverge. Correspondingly, the derivative $f'(x)$ will exhibit infinite discontinuities. Such situations are accommodated by the impulse symbol and its derivatives.

Derivative of a convolution integral

From the derivative theorem taken in conjunction with the convolution theorem, it follows that if

$$h = f * g,$$

then

$$h' = f' * g,$$

and also

$$h' = f * g'.$$

Derivation:

$$\frac{d}{dx}[f(x) * g(x)] \supset i2\pi s[F(s)G(s)]$$
$$f'(x) * g(x) \supset [i2\pi sF(s)]G(s)$$
$$f(x) * g'(x) \supset F(s)[i2\pi sG(s)].$$

These conclusions may be stated in a different form as follows:

The derivative of a convolution is the convolution of either of the functions with the derivative of the other.

Thus

$$(f * g)' = f' * g = f * g',$$

and again

$$f * g = f' * \int^x g\,dx$$
$$= f'' * \iint^x g\,dx\,dx$$
$$\cdots\cdot$$

Exercise Investigate the question of what lower limits are appropriate for the integrals. Investigate the formula $f' * g = f * g'$ by integration by parts.

In terms of the derivative of the impulse symbol we may write

$$h = \delta * h.$$

Therefore

$$h' = \delta' * h = \delta' * f * g = (\delta' * f) * g = f' * g$$
$$= f * (\delta' * g) = f * g',$$
$$f * g = \delta * f * g = (\delta' * H) * f * g = (\delta' * f) * (H * g).$$

The formulas quoted here are applicable to the evaluation of particular convolution integrals analytically or numerically but are principally of theoretical value (for example, in the deduction of the uncertainty relation and in deriving the formulas for the response of a filter in terms of its impulse and step responses).

The convenient algebra permitted by the δ notation in conjunction with the associative and other properties of convolution enables rapid generation of relations which may be needed for some problem under study. Many interesting possibilities arise. For example, starting from

$$\frac{d}{dx}(f * g) \supset i2\pi sFG$$

we may factor the right-hand side to get

$$\frac{d}{dx}(f * g) = \frac{d^{\frac{1}{2}}f}{dx^{\frac{1}{2}}} * \frac{d^{\frac{1}{2}}g}{dx^{\frac{1}{2}}} \supset (i2\pi s)^{\frac{1}{2}}F(i2\pi s)^{\frac{1}{2}}G.$$

The transform of a generalized function

Let $p(x)$ be a generalized function defined as in Chapter 5 by the sequence $p_\tau(x)$. Let the Fourier transforms of members of the defining sequence be $P_\tau(s)$, where

$$P_\tau(s) = \int_{-\infty}^{\infty} p_\tau(x) e^{-i2\pi sx}\, dx.$$

Perhaps this new sequence $P_\tau(s)$ defines a generalized function. We know that the members of the sequence are particularly well-behaved; we test the sequence for regularity by means of an arbitrary particularly well-behaved function $F(x)$ whose Fourier transform is $\bar{F}(s)$. From the energy theorem

$$\int_{-\infty}^{\infty} P_\tau(s)\bar{F}(s)\, ds = \int_{-\infty}^{\infty} p_\tau(x)F(-x)\, dx$$

it follows that

$$\lim_{\tau \to 0} \int_{-\infty}^{\infty} P_\tau(s)\bar{F}(s)\, ds = \lim_{\tau \to 0} \int_{-\infty}^{\infty} p_\tau(x)F(-x)\, dx,$$

and we know the latter limit exists. Hence $P_\tau(s)$ defines a generalized function, to which we give the symbol $P(s)$ and the meaning "Fourier transform of the generalized function $p(x)$."

The following statement can be made about $p(x)$ and $P(s)$,

$$\int_{-\infty}^{\infty} P(s)\bar{F}(s)\, ds = \int_{-\infty}^{\infty} p(x)F(-x)\, ds,$$

where $F(x)$ is any particularly well-behaved function and $\bar{F}(s)$ is its Fourier transform.

Let $\phi(x)$ be a function, such as a polynomial, that has derivatives of all orders at all points but whose behavior as $|x| \to \infty$ is not so stringently controlled as that of particularly well-behaved functions. We allow $\phi(x)$ to go infinite as $|x|^N$ where N is finite. Functions such as $\exp x$ and $\log x$ would be excluded. Then products of $\phi(x)$ and any particularly well-behaved function will eventually be overwhelmed by the latter as $|x| \to \infty$, since the particularly well-behaved factor dies out faster than $|x|^{-N}$. Furthermore, since the product has all derivatives at all points, the product itself is particularly well-behaved.

Consider a sequence

$$\phi(x)p_\tau(x).$$

It is particularly well-behaved, and

$$\lim_{\tau \to 0} \int_{-\infty}^{\infty} [\phi(x)p_\tau(x)]F(x)\, dx = \lim_{\tau \to 0} \int_{-\infty}^{\infty} p_\tau(x)[\phi(x)F(x)]\, dx,$$

which exists since $p_\tau(x)$ is a regular sequence and $[\phi(x)F(x)]$ is particularly well-behaved. The generalized function so defined we write as

$$\phi(x)p(x).$$

In practice we use this notation with functions $\phi(x)$ that have a *sufficient* number of derivatives and include exponentially increasing functions when, as is the case with the pulse sequence $\tau^{-1}\Pi(x/\tau)$, the behavior at infinity is inessential.

Nothing is introduced which could be called the product of two generalized functions; the product of two defining sequences is not necessarily a regular sequence and consequently does not in general define a generalized function.

Proofs of theorems

The numerous theorems of Fourier theory, which have proved so fruitful in the preceding sections, have shown themselves perfectly adaptable to the insertion of the impulse symbol $\delta(x)$, the shah symbol $\text{III}(x)$, the duplicating symbol $\text{II}(x)$, and other familiar nonfunctions. In the course of standard proofs of the theorems it is found necessary to eliminate these cases, and in the outcome we have conditions for the applicability of the theorems which we have found in practice need not be observed. This situation was dealt with by introducing the idea of a transform in the limit, and special ad hoc interpretations as limits were placed on expressions containing impulse symbols.

Having established the algebra for generalized functions, we can also give systematic proofs of the various theorems, free from the awkward conditions that arise when attention is confined to ordinary functions possessing regular transforms. The difficulties associated with functions that do not have derivatives disappear, for generalized functions possess derivatives of all orders. And more intolerable circumstances, such as the lack of a regular spectrum for direct current, also vanish.

Addition theorem The proof of this theorem can be supplied by the reader.

Similarity and shift theorems We prove these theorems simultaneously. Let $p(x)$ be a generalized function with Fourier transform $P(s)$. Then

$$p(ax + b) \supset \frac{1}{|a|} e^{i2\pi bs/a} P\left(\frac{s}{a}\right).$$

Proof: Since

$$\lim_{\tau \to 0} \int_{-\infty}^{\infty} p_\tau(ax + b)F(x)\, dx = \frac{1}{|a|} \lim_{\tau \to 0} \int_{-\infty}^{\infty} p_\tau(x) F\left(\frac{x - b}{a}\right) dx$$

exists, we have a meaning for $p(ax + b)$. Now

$$p_\tau(ax + b) \supset \frac{1}{|a|} e^{i2\pi bs/a} P_\tau\left(\frac{s}{a}\right)$$

by substitution of variables. Hence the two theorems follow.

Derivative theorem The Fourier transform of $p'_\tau(x)$ is $i2\pi s P_\tau(s)$. Hence the Fourier transform of $p'(x)$ is $i2\pi s P(s)$.

Power theorem Since no meaning has been assigned to the product of two generalized functions, the best theorem that can be proved is

$$\int_{-\infty}^{\infty} P(s)\bar{F}(s)\,ds = \int_{-\infty}^{\infty} p(x)F(-x)\,dx,$$

where $F(x)$ is a particularly well-behaved function and $p(x)$ is a generalized function. The theorem follows from the fact that

$$\lim_{\tau\to 0}\int_{-\infty}^{\infty} P_\tau(s)\bar{F}(s)\,ds = \lim_{\tau\to 0}\int_{-\infty}^{\infty}\int_{-\infty}^{\infty} \bar{F}(s)p(x)e^{-i2\pi sx}\,dx\,ds$$
$$= \lim_{\tau\to 0}\int_{-\infty}^{\infty} p_\tau(x)F(-x)\,dx.$$

Summary of theorems

The theorems discussed in the preceding pages are collected for reference in Table 6.1.

Table 6.1 Theorems for the Fourier transform

Theorem	$f(x)$	$F(s)$
Similarity	$f(ax)$	$\frac{1}{\lvert a\rvert}F\left(\frac{s}{a}\right)$
Addition	$f(x) + g(x)$	$F(s) + G(s)$
Shift	$f(x - a)$	$e^{-i2\pi as}F(s)$
Modulation	$f(x)\cos\omega x$	$\frac{1}{2}F\left(s - \frac{\omega}{2\pi}\right) + \frac{1}{2}F\left(s + \frac{\omega}{2\pi}\right)$
Convolution	$f(x) * g(x)$	$F(s)G(s)$
Autocorrelation	$f(x) * f^*(-x)$	$\lvert F(s)\rvert^2$
Derivative	$f'(x)$	$i2\pi sF(s)$
Derivative of convolution	$\frac{d}{dx}[f(x) * g(x)] = f'(x) * g(x) = f(x) * g'(x)$	
Rayleigh	$\int_{-\infty}^{\infty} \lvert f(x)\rvert^2\,dx = \int_{-\infty}^{\infty} \lvert F(s)\rvert^2\,ds$	
Power	$\int_{-\infty}^{\infty} f(x)g^*(x)\,dx = \int_{-\infty}^{\infty} F(s)G^*(s)\,ds$	
(f and g real)	$\int_{-\infty}^{\infty} f(x)g(-x)\,dx = \int_{-\infty}^{\infty} F(s)G(s)\,ds$	

Problems

1 Using the transform pairs given for reference, deduce the further pairs listed below by application of the appropriate theorem. Assume that A and σ are positive.

$$\frac{\sin x}{x} \supset \pi\Pi(\pi s) \qquad \left(\frac{\sin x}{x}\right)^2 \supset \pi\Lambda(\pi s)$$

$$\frac{\sin Ax}{Ax} \supset \frac{\pi}{A}\Pi\left(\frac{\pi s}{A}\right) \qquad \left(\frac{\sin Ax}{Ax}\right)^2 \supset \frac{\pi}{A}\Lambda\left(\frac{\pi s}{A}\right)$$

$$e^{-x^2} \supset \pi^{\frac{1}{2}}e^{-\pi^2 s^2} \qquad \delta(ax) \supset \frac{1}{|a|}$$

$$e^{-Ax^2} \supset \left(\frac{\pi}{A}\right)^{\frac{1}{2}} e^{-\pi^2 s^2/A} \qquad \delta(ax+b) \supset \frac{1}{|a|}e^{i2\pi bs/a}$$

$$e^{-x^2/2\sigma^2} \supset (2\pi)^{\frac{1}{2}}\sigma e^{-2\pi^2\sigma^2 s^2} \qquad e^{ix} \supset \delta\left(s - \frac{1}{2\pi}\right)$$

2 Show that the following transform pairs follow from the addition theorem, and make graphs.

$$1 + \cos \pi x \supset \delta(s) + \mathrm{II}(s)$$
$$1 + \sin \pi x \supset \delta(s) + i\,{}_{\mathrm{I}}\mathrm{I}(s)$$
$$\operatorname{sinc} x + \tfrac{1}{2}\operatorname{sinc}^2 \tfrac{1}{2}x \supset \Pi(s) + \Lambda(2s)$$
$$A^{-\frac{1}{2}}e^{-\pi x^2/A} + A^{\frac{1}{2}}e^{-\pi Ax^2} \supset e^{-\pi s^2/A} + e^{-\pi As^2}$$
$$4\cos^2 \pi x + 4\cos^2 \tfrac{1}{2}\pi x - 3 \supset \delta(s+\tfrac{1}{2}) + \delta(s-\tfrac{1}{2}) + \delta(s) + \tfrac{1}{2}\delta(x-1) + \tfrac{1}{2}\delta(s+1).$$

3 Deduce the following transform pairs, using the shift theorem.

$$\frac{\cos \pi x}{\pi(x-\frac{1}{2})} \supset -e^{-i\pi s}\Pi(s)$$
$$\frac{\sin \pi x}{\pi(x-1)} \supset -e^{-i2\pi s}\Pi(s)$$
$$\Lambda(x-1) \supset e^{-i2\pi s}\operatorname{sinc}^2 s$$
$$\Pi(x-\tfrac{1}{2}) \supset e^{-i\pi s}\operatorname{sinc} s$$
$$\Pi(x)\operatorname{sgn} x \supset -i\sin \tfrac{1}{2}\pi s \operatorname{sinc} \tfrac{1}{2}s$$
$$\Pi\left(\frac{x-\frac{1}{2}a}{a}\right) \supset |a|e^{-i\pi as}\operatorname{sinc} as.$$

4 Use the convolution theorem to find and graph the transforms of the following functions: sinc x sinc $2x$, (sinc x cos $10x)^2$.

5 Let $f(x)$ be a periodic function with period a, that is, $f(x+a) = f(x)$ for all x Since the Fourier transform of $f(x+a)$ is, by the shift theorem, equal to exp $(i2\pi as)F(s)$, which must be equal to $F(s)$, what can be deduced about the transform of a periodic function?

6 Graph the transform of $f(x) \sin \omega x$ for large and small values of ω, and explain graphically how, for small values of ω, the transform of $f(x) \sin \omega x$ is proportional to the derivative of the transform of $f(x)$.

7 Graph the transform of $\exp(-x)H(x) \cos \omega x$. Is it an even function of s?

8 Show that a pulse signal described by $\Pi(x/X) \cos 2\pi f x$ has a spectrum

$$\tfrac{1}{2}X\{\text{sinc}\,[X(s+f)] + \text{sinc}\,[X(s-f)]\}$$

9 Show that a modulated pulse described by $\Pi(x/X)(1 + M \cos 2\pi F x) \cos 2\pi f x$ has a spectrum

$$\tfrac{1}{2}X\{\text{sinc}\,[X(s+f)] + \text{sinc}\,[X(s-f)]\} + \tfrac{1}{4}MX\{\text{sinc}\,[X(s+f+F)] + \text{sinc}\,[X(s+f-F)] + \text{sinc}\,[X(s-f+F)] + \text{sinc}\,[X(s-f-F)]\}.$$

Graph the spectrum to a suitably exaggerated scale for a case where there are 100 modulation cycles and 100,000 radio-frequency cycles in one pulse and the modulation coefficient M is 0.6. Show by dimensioning how the factors 100, 100,000, and 0.6 enter into the shape of the spectrum.

10 A function $f(x)$ is defined by

$$f(x) = \begin{cases} 0 & |x| > 2 \\ 2 - |x| & 1 < |x| < 2 \\ 1 & |x| < 1; \end{cases}$$

show that

$$f(x) = 2\Lambda\left(\frac{x}{2}\right) - \Lambda(x) = \Lambda(x) * [\delta(x+1) + \delta(x) + \delta(x-1)]$$

and hence that

$$F(s) = 4\,\text{sinc}^2\,2s - \text{sinc}^2\,s = \text{sinc}^2\,s(1 + 2\cos 2\pi s).$$

11 Prove that $f * g * h \supset FGH$ and hence that $f^{*n} \supset F^n$.

12 The notation f^{*n} meaning $f(x)$ convolved with itself $n - 1$ times, where $n = 2, 3, 4, \ldots$, suggests the idea of fractional-order self-convolution. Show that such a generalization of convolution is readily made and that, for example, one reasonable expression for $f(x)$ convolved with itself half a time would be

$$f^{*1\frac{1}{2}} \equiv \int e^{i2\pi sx}\left[\int e^{-i2\pi su} f(u)\,du\right]^{1\frac{1}{2}} ds.$$

13 Prove that

$$(f * g)(h * j) \supset (FG) * (HJ)$$

and that

$$(f + g) * (h + j) \supset FH + FJ + GH + GJ.$$

14 Use the convolution theorem to obtain an expression for

$$e^{-ax^2} * e^{-bx^2}.$$

15 Prove that

$$\int_{-\infty}^{\infty} f^*(u) g^*(x - u)\,du \supset F^*(-s) G^*(-s).$$

16 Prove that

$$\int_{-\infty}^{\infty}\int_{-\infty}^{\infty} f^*(u)g^*(u-x)e^{-i2\pi xs}\,du\,dx = F^*(-s)G^*(s).$$

17 Show by Rayleigh's theorem that

$$\int_{-\infty}^{\infty} \operatorname{sinc}^2 x\,dx = 1$$

$$\int_{-\infty}^{\infty} \operatorname{sinc}^4 x\,dx = \int_{-\infty}^{\infty} [\Lambda(x)]^2\,dx = \tfrac{2}{3}$$

$$\int_{-\infty}^{\infty} [J_0(x)]^2\,dx = \infty$$

$$\int_{-\infty}^{\infty} \frac{dx}{(1+x^2)^2} = \frac{\pi}{2}$$

18 Complete the following schemata for reference, including thumbnail sketches of the functions.

function	⊃ transform
autocorrelation $f \star f$	⊃ power spectrum $\lvert F(s)\rvert^2$

$\Pi(x)$	

$\Lambda(x)$	

$\delta(x)$	

$\operatorname{sinc} x$	

$\Pi(x)\cos 2\pi fx$	

19 Show the fallacy in the following reasoning. "The Fourier transform of $\int_{-\infty}^{x} f(x)\,dx$ must be $F(s)/i2\pi s$ because the derivative of $\int_{-\infty}^{x} f(x)\,dx$ is $f(x)$, and hence by the derivative theorem the transform of $f(x)$ would be $F(s)$, which is true."

20 Establish an integral theorem for the Fourier transform of the indefinite integral of a function.

21 Use the derivative theorem to find the Fourier transform of $xe^{-\pi x^2}$.

22 Show that $2\pi x\Pi(x) \supset i\operatorname{sinc}' s$.

23 The following brief derivation appears to show that the area under a derivative is zero. Thus

$$\int_{-\infty}^{\infty} f'(x)\,dx = \overline{f'(x)}\Big|_0 = i2\pi s F(s)\Big|_0 = 0.$$

Confirm that this is so, or find the error in reasoning.

24 Show that

$$f(ax-b) \supset \frac{1}{|a|}\,e^{-i2\pi bs/a}F\left(\frac{s}{a}\right).$$

25 Show from the energy theorem that

$$\int_{-\infty}^{\infty} e^{-\pi x^2}\cos 2\pi a x\,dx = e^{-\pi a^2}.$$

26 Show from the energy theorem that

$$\int_{-\infty}^{\infty} \operatorname{sinc}^2 x \cos \pi x\,dx = \tfrac{1}{2}.$$

27 Show that the function whose Fourier transform is $|\operatorname{sinc} s|$ has a triangular autocorrelation function.

28 As a rule, the autocorrelation function tends to be more spread out than the function it comes from. But show that

$$\frac{1}{\pi x} * \frac{-1}{\pi x} = \delta(x).$$

Show that $(\pi x)^{-1}$ must have a flat energy spectrum, and from that deduce and investigate other functions whose autocorrelation is impulsive.

29 The Maclaurin series for $F(s)$ is

$$F(0) + sF'(0) + \frac{s^2}{2!}F''(0) + \cdots.$$

Consider the case of $F(s) = \exp(-\pi s^2)$, where the series is known to converge and to converge to $F(s)$. Thus, in this particular case,

$$F(s) = \sum_{n=0}^{\infty} \frac{s^n}{n!}F^{(n)}(0).$$

If $F(s)$ is the transform of $f(x)$, then transforming this equation we obtain

$$f(x) = \delta(x)\int_{-\infty}^{\infty} f(x)\,dx - \delta'(x)\int_{-\infty}^{\infty} xf(x)\,dx + \delta''(x)\int_{-\infty}^{\infty}\frac{x^2}{2!}f(x)\,dx + \cdots.$$

How do you explain this result?

Chapter 7 Doing transforms

A number of transforms were introduced earlier to illustrate the basic theorems of the Fourier transformation. Of course, one need not necessarily be aware of any particular Fourier transform pairs to appreciate the meaning of the theorems. Many of the chains of argument in which the Fourier transformation is important are independent of any knowledge of particular examples. Even so, carrying out a general argument with a special case in mind often serves as insurance against surprises.

The examples of Fourier transform pairs chosen for illustration were all introduced without derivation and asserted to be verifiable by evaluation of the Fourier integral. Obviously this does not help when it is necessary to generate new pairs. We therefore consider various ways of carrying out the Fourier transformation. Numerical methods based on the discrete Fourier transform and allowing for use of the fast Fourier transform algorithm are discussed in Chapter 18.

Starting from the given function $f(x)$ whose Fourier transform is to be deduced, one may first contemplate the integral

$$\int_{-\infty}^{\infty} f(x)e^{-i2\pi xs}\,dx.$$

If this integral can be evaluated for all s, the problem is solved.

A number of different approaches from the standpoint of integration are discussed separately in subsequent sections.

In addition to this direct approach we have a powerful resource in the basic theorems which have now been established. Many of the theorems take the form "If f and F are a transform pair then g and G are also." Thus, if one knows a transform pair to begin with, others may be generated. It is indeed possible to build up an extensive dictionary of transforms by means of the theorems, starting from the beginnings already laid down. It may be surmised that some classes of function will never be stumbled on in this way; on the other hand, a variety of physically feasible functions prove to be accessible. Generation from theorems is taken up later with examples. Finally, there is the possibility of extracting a desired transform from tables.

Integration in closed form

It is a propitious circumstance if $f(x)$ is zero over some range of x and if in addition its behavior is simple where it is nonzero. Thus if

$$f(x) = \Pi(x),$$

then

$$\begin{aligned}\int_{-\infty}^{\infty} f(x)e^{-i2\pi xs}\,dx &= \int_{-\frac{1}{2}}^{\frac{1}{2}} e^{-i2\pi xs}\,dx \\ &= \int_{-\frac{1}{2}}^{\frac{1}{2}} \cos 2\pi xs\,dx \\ &= \frac{\sin \pi s}{\pi s} \\ &= \operatorname{sinc} s.\end{aligned}$$

If $f(x)$ is nonzero on several segments of the abscissa and constant within each segment, the integration in closed form can likewise be done. Thus if we let

$$f(x) = a\Pi\left(\frac{x-b}{c}\right),$$

then

$$\begin{aligned}F(s) &= \int_{b-\frac{1}{2}c}^{b+\frac{1}{2}c} ae^{-i2\pi xs}\,dx \\ &= a\int_{-\frac{1}{2}c}^{\frac{1}{2}c} e^{-i2\pi(u+b)s}\,du \\ &= ae^{-i2\pi bs}\int_{-\frac{1}{2}c}^{\frac{1}{2}c} \cos 2\pi us\,du \\ &= ace^{-i2\pi bs} \operatorname{sinc} cs.\end{aligned}$$

It follows that if

$$f(x) = \sum_n a_n\Pi\left(\frac{x-b_n}{c_n}\right),$$

then

$$F(s) = \sum_n a_n c_n e^{-i2\pi b_n s} \operatorname{sinc} c_n s.$$

Functions built up segmentally of rectangle functions include staircase functions and functions suitable for discussing Morse code, teleprinter signals, and on/off servomechanisms. They not only occur frequently in engineering but can also simulate, as closely as may be desired, any kind of variation.

An even simpler case arises if $f(x)$ is zero almost everywhere; for example, suppose that $f(x)$ is a set of impulses of various strengths at various values of x:

$$f(x) = \sum_n a_n \, \delta(x - b_n).$$

Then only the values of $a_n \exp(-i2\pi xs)$ at the points $x = b_n$ can matter. By the sifting theorem for the impulse symbol,

$$\begin{aligned} F(s) &= \int_{-\infty}^{\infty} \sum a_n \, \delta(x - b_n) e^{-i2\pi xs} \, dx \\ &= \int_{b_1-}^{b_1+} a_1 \, \delta(x - b_1) e^{-i2\pi xs} \, dx + \int_{b_2-}^{b_2+} a_2 \, \delta(x - b_2) e^{-i2\pi xs} \, dx + \cdots \\ &= a_1 e^{-i2\pi b_1 s} + a_2 e^{-i2\pi b_2 s} + \cdots. \end{aligned}$$

If $f(x)$ has some special functional form, it may prove possible to perform the integration, but no general rules can be given. Some simple examples follow.

1. Let $f(x) = \operatorname{sinc} x$. Then

$$\begin{aligned} F(s) &= \int_{-\infty}^{\infty} \operatorname{sinc} x \, e^{-i2\pi xs} \, dx \\ &= \int_{-\infty}^{\infty} \frac{\sin \pi x \cos 2\pi xs}{\pi x} \, dx \\ &= \int_{-\infty}^{\infty} \left[\frac{\sin (\pi x + 2\pi xs)}{2\pi x} + \frac{\sin (\pi x - 2\pi xs)}{2\pi x} \right] dx \\ &= \int_{-\infty}^{\infty} \left\{ \frac{1 + 2s}{2} \operatorname{sinc} [(1 + 2s)x] + \frac{1 - 2s}{2} \operatorname{sinc} [(1 - 2s)x] \right\} dx \\ &= \frac{1 + 2s}{2|1 + 2s|} + \frac{1 - 2s}{2|1 - 2s|} \\ &= \Pi(s). \end{aligned}$$

We have used the result that

$$\int_{-\infty}^{\infty} \operatorname{sinc} ax \, dx = \frac{1}{|a|}.$$

2. Let $f(x) = e^{-|x|}$. Then

$$\begin{aligned} F(s) &= \int_{-\infty}^{\infty} e^{-|x|} e^{-i2\pi xs} \, dx \\ &= 2 \int_0^{\infty} e^{-x} \cos 2\pi xs \, dx \\ &= 2 \operatorname{Re} \int_0^{\infty} e^{-x} e^{i2\pi xs} \, dx \\ &= 2 \operatorname{Re} \int_0^{\infty} e^{(i2\pi s - 1)x} \, dx \\ &= 2 \operatorname{Re} \frac{-1}{i2\pi s - 1} \\ &= \frac{2}{4\pi^2 s^2 + 1}. \end{aligned}$$

3. Let $f(x) = e^{-\pi x^2}$. Then

$$\begin{aligned} F(s) &= \int_{-\infty}^{\infty} e^{-\pi x^2} e^{-i2\pi xs}\,dx \\ &= \int_{-\infty}^{\infty} e^{-\pi(x^2+i2xs)}\,dx \\ &= e^{-\pi s^2} \int_{-\infty}^{\infty} e^{-\pi(x+is)^2}\,dx \\ &= e^{-\pi s^2} \int_{-\infty}^{\infty} e^{-\pi(x+is)^2} d(x+is) \\ &= e^{-\pi s^2}. \end{aligned}$$

In this example we have used the known result that the infinite integral of $\exp(-\pi x^2)$ is unity. The next case illustrates a contour integral.

4. Let $f(x) = x^{-1}$. The infinite discontinuity at the origin causes the standard Fourier integral to diverge; hence we consider instead

$$\lim_{\epsilon \to 0} \left(\int_{-\infty}^{-\epsilon} \frac{e^{-i2\pi sx}}{x}\,dx + \int_{\epsilon}^{\infty} \right).$$

Now consider

$$\int_C \frac{e^{-i2\pi sz}}{z}\,dz,$$

where the contour C is a semicircle of radius R in the complex plane of z, whose diameter lies along the real axis and has a small indentation of radius ϵ at the origin. The contour integral is zero since no poles are enclosed. Thus

$$\int_{-R}^{-\epsilon} \frac{e^{-i2\pi sx}}{x}\,dx + \int_{\pi}^{0} ie^{-i2\pi s\epsilon e^{i\theta}}\,d\theta + \int_{\epsilon}^{R} \frac{e^{-i2\pi sx}}{x}\,dx + \int_{0}^{\pi} ie^{-i2\pi sRe^{i\theta}}\,d\theta = 0.$$

As $R \to \infty$, the fourth integral vanishes and the second equals $\pm i\pi$ according to the sign of s, so that we have

$$\lim_{\epsilon \to 0} \left(\int_{-\infty}^{-\epsilon} \frac{e^{-i2\pi sx}}{x}\,dx + \int_{\epsilon}^{\infty} \right) + i\pi \operatorname{sgn} s = 0.$$

Hence the desired transform pair in the limit is

$$\frac{1}{x} \supset -i\pi \operatorname{sgn} s.$$

5. Let $f(x) = \operatorname{sgn} x$. This is the previous example in reverse, and it is seen that in this case the Fourier integral fails to exist in the standard sense because the function does not possess an absolutely convergent integral. Therefore consider a sequence of transformable functions which approach $\operatorname{sgn} x$ as a limit, for example, the sequence $\exp(-\tau|x|) \operatorname{sgn} x$ as

$\tau \to 0$. The transform will be

$$\int_{-\infty}^{\infty} e^{-\tau|x|} \operatorname{sgn} x \, e^{-i2\pi xs} \, dx = \int_{-\infty}^{0} -e^{(\tau - i2\pi s)x} \, dx + \int_{0}^{\infty} e^{-(\tau + i2\pi s)} \, dx$$

$$= -\frac{1}{\tau - i2\pi s} + \frac{1}{\tau + i2\pi s}.$$

As $\tau \to 0$ this expression has a limit $1/i\pi s$. Hence we have the Fourier transform pair in the limit,

$$\operatorname{sgn} x \supset \frac{1}{i\pi s}.$$

Numerical Fourier transformation

If the values of a function have been obtained by physical measurement and it is necessary to find its Fourier transform, various possibilities exist.

First of all, certain limitations of physical data will influence the result, and we shall begin with this aspect of numerical transformation.

The data will be given at discrete values of the independent variable x. The interval Δx may be so fine that there is no concern about interpolating intermediate values, but in any case we can take the view that the data can hardly contain significant information about Fourier components with periods less than $2\ \Delta x$. Therefore it is not necessary to take the computations to frequencies higher than about $(2\ \Delta x)^{-1}$.

In addition, observational data are given for a finite range of x, let us say for $-X < x < X$. By a corresponding argument, the Fourier transform need not be calculated for values of s more closely spaced than $(2X)^{-1}$. If there were any significant fine detail in $F(s)$ that required a finer tabulation interval than $(2X)^{-1}$ for its description, then measurements of $f(x)$ would have to be extended beyond $x = X$ to reveal it.

These simple but important facts for the computer programmer may be summarized by saying that the function $f(x)$ tabulated at interval Δx over a range $2X$ possesses $2X/\Delta x$ degrees of freedom, and this should be comparable with the number of data computed for the transform. In Chapters 10 and 18 the underlying thought is explored in detail.

A further property of physical data is to contain errors. There is therefore a limit to the precision that is warranted in calculating values of the Fourier transform. This limit is expressed concisely by the power spectrum of the error component (or, what is equivalent, the autocorrelation function of the error component). Sometimes only the magnitude of the errors is available, and not their spectrum; and sometimes the magnitude is not known either. Nevertheless, the errors set a limit to the number of physically significant decimal places in a computed value of the transform.

Let $f(x)$ be represented by values at $x = n$, where the integer n ranges from $-N$ to N. Before the Fourier transform of $f(x)$ can be computed, it is necessary to have information about $f(x)$ over the whole infinite range of x. Therefore we call on our physical knowledge and make some assumption about the behavior outside the range of measurement. In this case suppose that $f(x)$ is zero where $|x| > N$. Then the sum

$$\sum_{-N}^{N} f(n) e^{i2\pi sn}$$

will be an approximation to $F(s)$. In practice one computes the real and imaginary parts separately, forming the sums

$$\sum_{-N}^{N} f(n) \cos 2\pi sn \qquad \text{and} \qquad \sum_{-N}^{N} f(n) \sin 2\pi sn.$$

This summation, over $2N + 1$ terms, must be done for each chosen value of s. For this reason it is convenient in using a desk calculator to possess tables of cosine and sine prepared for suitable values of s; instead of being limited to one quadrant they should run on and on, showing negative values as they occur.

Generation of transforms by theorems

A wide variety of functions, especially those occurring in theoretical work, can be transformed if some property can be found that permits a simplifying application of a theorem. For example, consider a polygonal function, as in Fig. 7.1. If we perceive that it can be expressed as the convolution of the triangle function and set of impulses, then we can handle

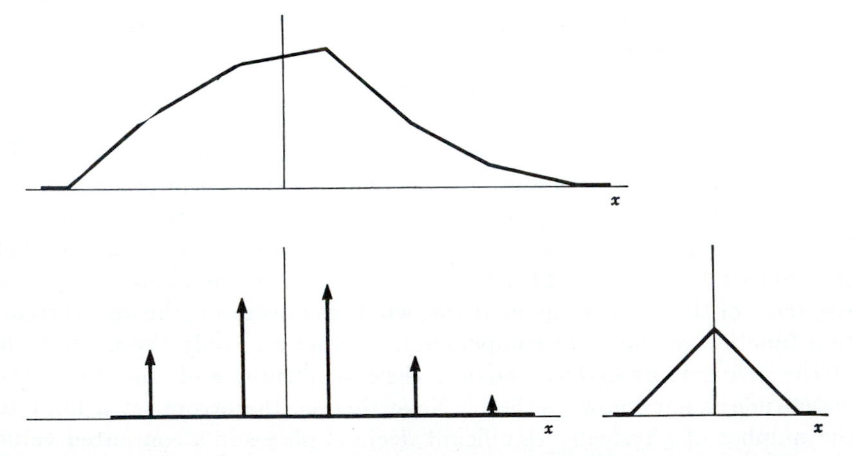

Fig. 7.1 A polygonal function (above) that can be regarded as the convolution of a set of impulses and a triangle function (below).

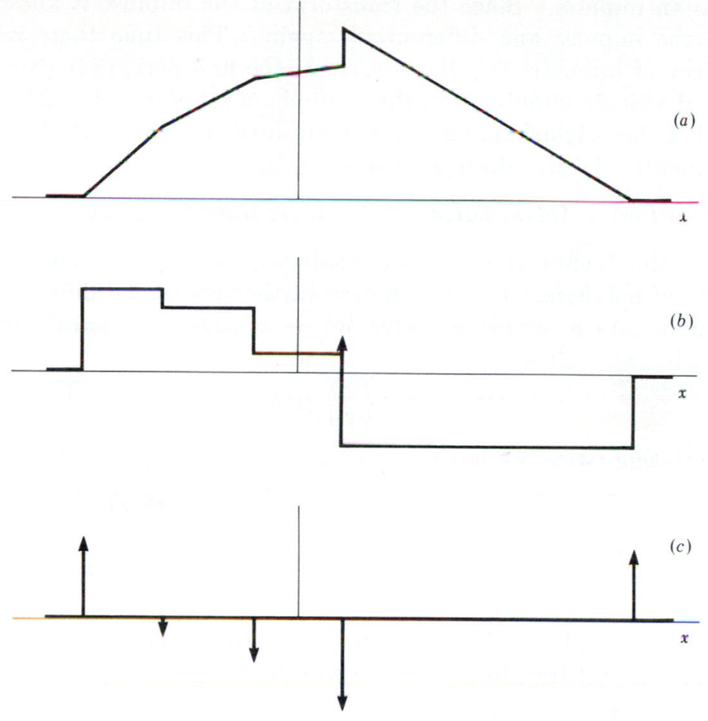

Fig. 7.2 Technique of reduction to impulses by continued differentiation.

the impulses as described above and multiply their transform by the transform of the triangle function. Thus

$$\Lambda(x) * \sum_n A_n \, \delta(x - a_n) \supset \operatorname{sinc}^2 s \sum_n A_n e^{-i2\pi a_n s}.$$

As an example take the trapezoidal pulse $\Lambda(x) * \Pi(x)$. Evidently

$$\Lambda(x) * \Pi(x) \supset \operatorname{sinc}^2 s \cos \pi s.$$

In practice the convolution theorem is frequently applicable for the generation of derived transforms, and many examples of the use of the convolution theorem and other theorems will be given in the problems for this chapter.

Application of the derivative theorem to segmented functions

There is a special application of the derivative theorem that has wide use in connection with switching waveforms. Consider a segmentally linear function such as that of Fig. 7.2*a*. The first derivative, shown in *b*, contains an impulse. Since the transform of the impulse is known, we remove the impulse and differentiate again. This time there remains only a set of impulses $\Sigma C_n\,\delta(x - c_n)$. If the first derivative contained, instead of a single impulse as in the example, a set of impulses $\Sigma B_n\,\delta(x - b_n)$, and if the original function $f(x)$ contained impulses $\Sigma A_n\,\delta(x - a_n)$, then evidently the transform $F(s)$ is given by

$$(i2\pi s)^2F(s) = (i2\pi s)^2\Sigma A_n e^{-i2\pi a_n s} + i2\pi s\Sigma B_n e^{-i2\pi b_n s} + \Sigma C_n e^{-i2\pi c_n s}.$$

Clearly this technique extends immediately to functions composed of segments of polynomials, in which case further continued differentiation is required. As a simple example let us consider the parabolic pulse $(1 - x^2)\Pi(x/2)$. Here

$$(1 - x^2)\Pi\left(\frac{x}{2}\right) \supset F(s).$$

Differentiating twice, we have

$$-2x\Pi\left(\frac{x}{2}\right) \supset i2\pi sF(s)$$

$$2\delta(x + 1) - 2\Pi\left(\frac{x}{2}\right) + 2\delta(x - 1) \supset (i2\pi s)^2F(s).$$

Now the left-hand side has a known transform $4 \cos 2\pi s - 4 \operatorname{sinc} 2s$. Hence the desired transform $F(s)$ is given by

$$F(s) = \frac{4 \cos 2\pi s - 4 \operatorname{sinc} 2s}{(i2\pi s)^2} + K_1\delta(s) + K_2\delta'(s),$$

where K_1 and K_2 are integration constants arising from the fact that a constant K_1 may be added to the original parabolic pulse without changing its first derivative, and similarly for the second derivative. Integration of the given pulse shows that no additive constant or linear ramp is present, so in this case $K_1 = K_2 = 0$ and

$$F(s) = -\frac{\cos 2\pi s}{\pi^2 s^2} + \frac{\sin 2\pi s}{2\pi^3 s^3}.$$

Chapter 8 The two domains

We may think of functions and their transforms as occupying two domains, sometimes referred to[1] as the upper and the lower, as if functions circulated at ground level and their transforms in the underworld. There is a certain convenience in picturing a function as accompanied by a counterpart in another domain, a kind of shadow which is associated uniquely with the function through the Fourier transformation, and which changes as the function changes. In the illustrations given here the uniform practice is to keep the functions on the left and the transforms on the right.

The theorems given earlier can be regarded as a list of pairs of corresponding simultaneous operations, one in the function domain and the other in the transform domain. For example, compression of the abscissas in the function domain means expansion of the abscissas plus contraction of the ordinates in the transform domain; translation in the function domain involves a certain kind of twisting in the transform domain; and convolution in the function domain involves multiplication in the transform domain.

By applying these theorems to everyday problems we are able at will to cross from one domain to the other and to carry out required operations in whichever domain is more advantageous. We may find on reaching the conclusion to a line of argument that we are in the wrong domain; our conclusion is a statement about the transform of the function we are interested in.

We therefore now consider pairs of corresponding properties; the area under a function and the central ordinate of its transform are such a pair.

[1] For example, see G. Doetsch, "Theory and Application of the Laplace Transformation," Dover Publications, New York, 1943.

Suppose that we are interested in the area under a function, but we have only its transform; the desired quantity proves to be equal to the central ordinate of the transform. It is not necessary for us to invert the Fourier transformation in full and to integrate the function so determined.

It very often happens that particular features of a function are all we need to know. The corresponding pairs of properties assembled here then assume an important role in enabling us to sidestep details of transformations and go to a direct answer.

Definite integral

The definite integral of a function from $-\infty$ to ∞ is equal to the value of its transform at the origin; that is,

$$\int_{-\infty}^{\infty} f(x)\, dx = F(0).$$

Derivation:

$$\int_{-\infty}^{\infty} f(x)\, dx = \int_{-\infty}^{\infty} f(x) e^{-i2\pi xs}\, dx \Big|_{s=0} = F(0).$$

We often refer to this integral as the area under a function; thus in Fig. 8.1 it is the area shown shaded. The term "central ordinate" is used for the value at the origin of abscissas and this value is indicated in the figure by a heavy line.

It follows from the symmetry of the Fourier transform that the area under the transform is equal to $f(0)$. From this simple correspondence of properties many very interesting conclusions can be drawn. For instance, any operation on $f(x)$ which leaves its area unchanged—say a translation along the x axis—leaves the central ordinate of the transform unchanged. In the case of translation, this conclusion is verifiable from the shift theorem, for

$$\overline{f(x-a)} \Big|_0 = e^{-i2\pi as} F(s) \Big|_0 = F(0).$$

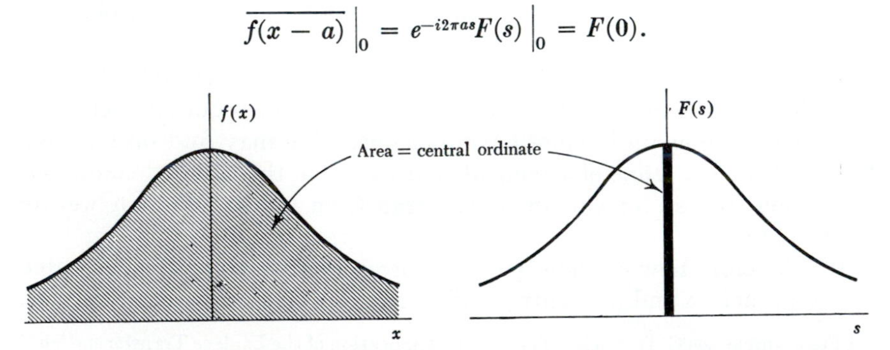

Fig. 8.1 The area under a function is equal to the central ordinate of its transform.

Table 8.1 Some functions and their areas

$f(x)$	$F(0)$	$f(x)$	$F(0)$
$\Pi(x)$	1	$\cos x$	0
$\operatorname{sinc} x$	1	$^{\mathrm{I}}{}_{\mathrm{I}}(x)$	0
$\Lambda(x)$	1	$\sin x$	0
$\operatorname{sinc}^2 x$	1	$\mathrm{III}(x)$	∞
$e^{-\pi x^2}$	1	$\delta(x)$	1
$J_0(x)$	1	e^{ix}	0
$\dfrac{J_1(x)}{x}$	1	$H(x)$	...
$\mathrm{II}(x)$	1	$\operatorname{sgn} x$	...

A more abstruse example is furnished by $f(x - a)H(x - a) + f(x + a)$ $H(-x - a)$, where $f(x)$ is supposed separated at $x = 0$ into two equal parts, which are then pushed apart by an amount $2a$. Since the area has not been changed, the central ordinate of the transform is the same. On the other hand, since the central ordinate of the function is now zero, we can say that the area of the transform must now be zero. These two special facts about the transform are a good example of the powerful reasoning based on corresponding properties.

Exercise If $f(x)$ were a rectangle function, then the foregoing operation of splitting and separating the parts would result in a rectangular pulse pair. Obtain the full transform by use of the modulation theorem, and verify that the infinite integral of the spectrum of a pulse pair is indeed zero.

Exercise What distinctive character can you assign to the spectrum of a signal in Morse code?

In Table 8.1 many special cases are collected, the right-hand column showing the area under the function to the left.

Exercise The functions $\cos x$ and $\sin x$ have infinite integrals which are oscillatory, yet the central ordinates of their transforms have a perfectly definite value of zero. However, the transforms are transforms in the limit. The definite-integral/central-ordinate relation, when applied to transforms in the limit, must thus imply some rule for integrating (co)sinusoids. Investigate this matter. Hint: Consider transformable functions such as $\cos x\, \Pi(x/M)$ and $\cos x\, \Pi[(x - \frac{1}{4}\pi)/M]$, with the aid of the convolution and shift theorems, in the limit as $M \to \infty$.

Exercise Consider what would be appropriate entries in the table for $H(x)$ and $\operatorname{sgn}(x)$.

The first moment

By analogy with mass distribution along a line, the first moment of $f(x)$ about the origin is defined as

$$\int_{-\infty}^{\infty} xf(x)\,dx.$$

The following theorem connects the first moment of a function with the central slope of its transform.

If $f(x)$ has the Fourier transform $F(s)$ then the first moment of $f(x)$ is equal to $-(2\pi i)^{-1}$ times the slope of $F(s)$ at $s = 0$; that is,

$$\int_{-\infty}^{\infty} xf(x)\,dx = \frac{F'(0)}{-2\pi i}.$$

Derivation: By the derivative theorem,

$$\int_{-\infty}^{\infty} (-2\pi ix)f(x)e^{-i2\pi xs}\,dx = F'(s).$$

Therefore
$$-2\pi i\int_{-\infty}^{\infty} xf(x)\,dx = F'(0).$$

In Fig. 8.2 the impulse pair, $-{}^{\mathrm{I}}\!\mathrm{I}(x)$ is shown with its transform $-i\sin\pi s$. Each impulse has a positive moment $\frac{1}{4}$, hence the total moment is $\frac{1}{2}$; thus

$$-\int_{-\infty}^{\infty} x\,{}^{\mathrm{I}}\!\mathrm{I}(x)\,dx = -\int_{-\infty}^{\infty} x\tfrac{1}{2}\delta(x+\tfrac{1}{2})\,dx + \int_{-\infty}^{\infty} x\tfrac{1}{2}\delta(x-\tfrac{1}{2})\,dx = \tfrac{1}{2}.$$

Hence
$$F'(0) = (-2\pi i)\tfrac{1}{2} = -i\pi.$$

In Fig. 8.3 a truncated exponential and its transform are shown. The derivative of the real part of the transform is zero at the origin, so the "slope at the origin" is given by the derivative $-i\tan\theta$ of the imaginary

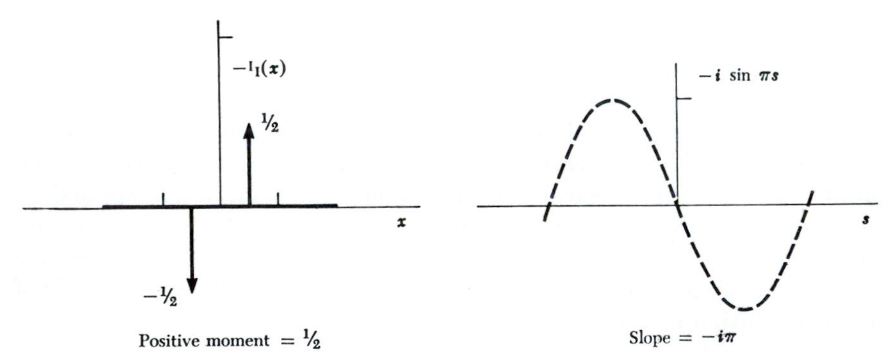

Fig. 8.2 The moment of a function is proportional to the central slope of the transform.

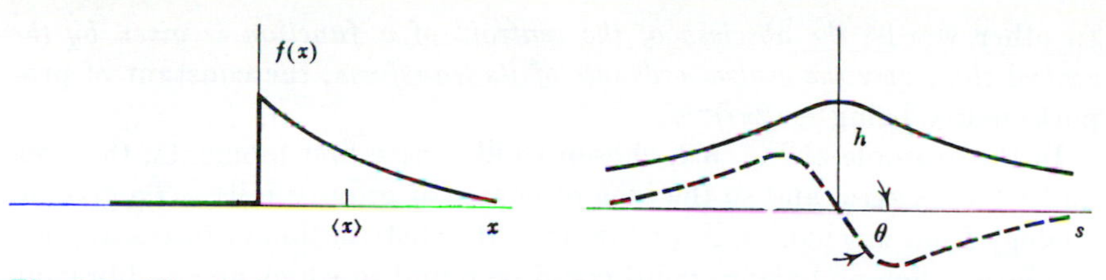

Fig. 8.3 The abscissa of the center of gravity is equal to $(\tan\theta)/2\pi h$.

part alone. The first moment of the exponential distribution is thus given by $(\tan\theta)/2\pi$, a positive real quantity.

As a special case consider a function whose first moment is zero. Its transform has zero slope at its origin. Conversely, *if a function has zero slope at the origin, its transform has zero first moment.* Furthermore, if a function has zero slope at the origin and is not zero there, then its transform has its center of gravity at the origin.

Exercise Discuss the first moment and center of gravity of the transform of $\Pi(x - a)$ as a goes from zero to infinity.

Exercise The function $(x - z)\Pi(x - z)$ is zero and has zero slope at the origin, and it has zero area. What can be said about its transform?

Exercise Under what condition on $f(x)$ will the derivative of $F(s)$ be purely imaginary at $s = 0$?

Centroid

By the centroid of $f(x)$ we mean the point with abscissa $\langle x\rangle$ such that the area of the function times $\langle x\rangle$ is equal to the first moment. Thus

$$\langle x\rangle = \frac{\int_{-\infty}^{\infty} xf(x)\,dx}{\int_{-\infty}^{\infty} f(x)\,dx}.$$

Roughly speaking, $\langle x\rangle$ tells us where a function $f(x)$ is mainly concentrated or, when a signal pulse is described as a function of time, $\langle t\rangle$ bears on the epoch of the signal. In statics, $\langle x\rangle$ is the abscissa of the center of gravity of a rod whose mass density is $f(x)$. Later we shall find that the centroid of $[f(x)]^2$ is a more suitable measure of location or epoch in cases where $f(x)$ can go negative. In the worst case the area of $f(x)$ is zero and $\langle x\rangle$ becomes infinite.

Since $\langle x\rangle$ is the ratio of the first moment to the area of $f(x)$, its connection with $F(s)$ follows immediately from the preceding sections:

$$\langle x\rangle = -\frac{F'(0)}{2\pi iF(0)}.$$

In other words, *the abscissa of the centroid of a function is given by the central slope over the central ordinate of its transform*, the constant of proportionality being $-(2\pi i)^{-1}$.

In the example of Fig. 8.2, chosen to illustrate first moments, the area under $f(x)$ is zero, and so the idea of center of gravity fails. To give an analogy from mechanics, if ${}^{\mathrm{I}}\mathrm{I}(x)$ described a distribution of forces applied to a lever, then no balance point could be found to which an equilibrating force could be applied.

In Fig. 8.3, however, the center of gravity of $f(x)$ falls at a finite positive value of x. Since $F(s)$ is hermitian, its derivative at the origin is purely imaginary, say, $i \tan \theta$; the central ordinate h is given by the value of the real part of $F(s)$ alone; thus

$$\langle x \rangle = \frac{\tan \theta}{2\pi h}.$$

Moment of inertia (second moment)

The moment of inertia or second moment of a function $f(x)$ about the origin is defined as

$$\int_{-\infty}^{\infty} x^2 f(x)\, dx.$$

One finds that the higher the second moment of a function the more curved its transform at the origin (Fig. 8.4). Because of the factor x^2, it is clear that the second moment of a function is sensitive to the behavior

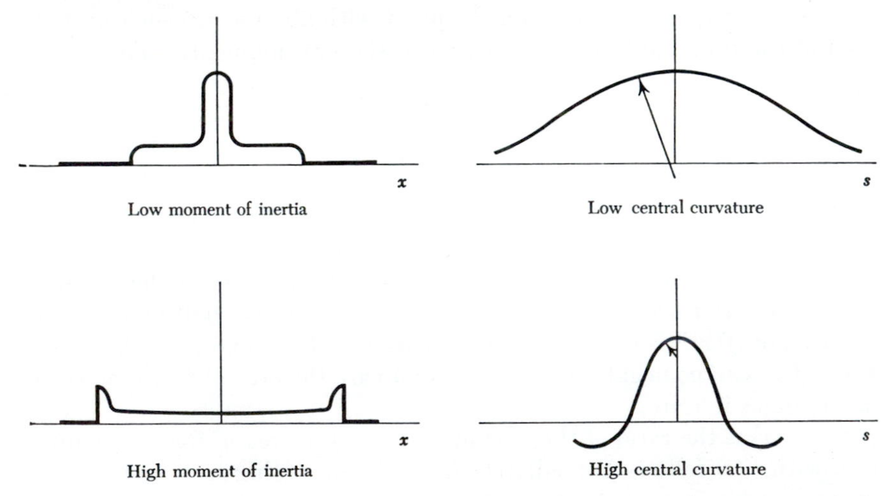

Fig. 8.4 The moment of inertia is proportional to the central curvature of the transform.

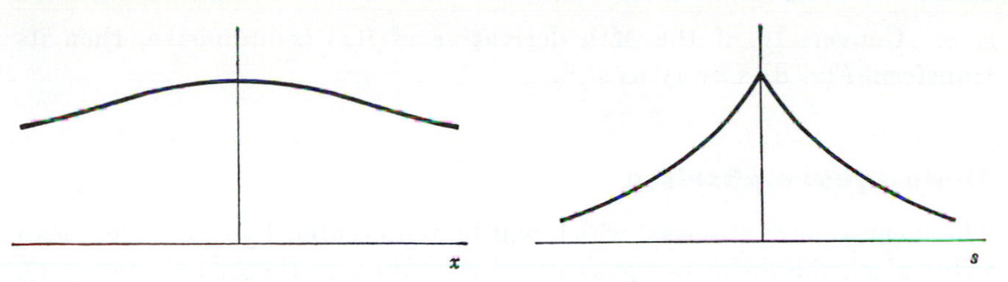

Fig. 8.5 A function with an infinite second moment has a transform with a cusp at the origin.

of $f(x)$ at large values of x; thus what happens at *large* x is reflected in the behavior of the transform at the *origin* of s. This remark also applies to the theorem on the first moment.

The moment of inertia of $f(x)$ is equal to $(4\pi^2)^{-1}$ times the downward curvature of its transform at the origin; that is,

$$\int_{-\infty}^{\infty} x^2 f(x)\, dx = -\frac{1}{4\pi^2} F''(0).$$

Derivation: By the derivative theorem,

$$\int_{-\infty}^{\infty} (-i2\pi x)^2 f(x) e^{-i2\pi xs}\, dx = F''(s).$$

Therefore
$$\int_{-\infty}^{\infty} x^2 f(x)\, dx = -\frac{F''(0)}{4\pi^2}.$$

A special situation of frequent occurrence is shown in Fig. 8.5, where a function whose second moment is infinite has a transform whose central curvature is infinite, that is, there is a cusp at the origin.

Moments

The nth moment of $f(x)$ is equal to $(-2\pi i)^{-n}$ times the nth differential coefficient of $F(s)$ at the origin; that is,

$$\int_{-\infty}^{\infty} x^n f(x)\, dx = \frac{F^{(n)}(0)}{(-2\pi i)^n},$$

provided that the moment exists.

If $f(x)$ behaves as $|x|^{-m}$ for large $|x|$, where m is positive, then only the first $[m]$ moments will exist, and only the first $[m]$ differential coefficients of $F(s)$ at $s = 0$ will exist; there will be a discontinuity of the $[m + 1]$th order at $s = 0$. (The function $[m]$ is the greatest integer less than m.)

For example, if m is a positive integer M, and $f(x)$ dies away as x^{-M}, then the Mth moment is infinite and the Mth derivative of $F(s)$ is impulsive. Conversely, if the Mth derivative of $f(x)$ is impulsive, then its transform $F(s)$ dies away as s^{-M}.

Mean-square abscissa

The mean-square abscissa, which will be represented by $\langle x^2 \rangle$ is the mean value of x^2, the mean being weighted according to the distribution $f(x)$. Thus

$$\langle x^2 \rangle \int_{-\infty}^{\infty} f(x)\, dx = \int_{-\infty}^{\infty} x^2 f(x)\, dx,$$

and hence by previous results

$$\langle x^2 \rangle = -\frac{F''(0)}{4\pi^2 F(0)}.$$

In dynamics, the square of the radius of gyration of a rod whose mass density is $f(x)$ is the same as $\langle x^2 \rangle$. In statistics, $\langle x^2 \rangle$ for a frequency distribution function $f(x)$ is an important concept [the same as the variance of the function $f(x)$] and it is useful to know this relationship it bears to the Fourier transform of $f(x)$ (the "characteristic function"). The mean-square abscissa possesses a very interesting additive property under convolution.

The mean-square abscissa for $f * g$ is equal to the sum of the mean-square abscissas for f and g, provided that f (or g) has its centroid at its origin.

We expect the convolution $f * g$ to be wider, in some sense, than either of its components; this theorem is one quantitative expression of the smoothing-out or diffusing effect of convolution.

Derivation:

$$\begin{aligned}\int_{-\infty}^{\infty} x^2(f * g)\, dx &= \overline{x^2(f * g)}\Big|_0 = -(4\pi^2)^{-1}\overline{(f * g)}''\Big|_0 \\ &= -(4\pi^2)^{-1}(FG)''\Big|_0 \\ &= -(4\pi^2)^{-1}[F''G + 2F'G' + FG'']_0\end{aligned}$$

$$\int_{-\infty}^{\infty} (f * g)\, dx = FG\Big|_0$$

$$\langle x^2 \rangle_{f*g} = \frac{\int_{-\infty}^{\infty} x^2(f * g)\, dx}{\int_{-\infty}^{\infty} (f * g)\, dx} = -(4\pi^2)^{-1}\left[\frac{F''}{F} + 2\frac{F'G'}{FG} + \frac{G''}{G}\right]_0$$

$$= \langle x^2 \rangle_f + \langle x^2 \rangle_g + 2\frac{F'(0)}{2\pi i F(0)}\frac{G'(0)}{2\pi i G(0)}.$$

The final term on the right-hand side, which is twice the product of the abscissas of the centroids of f and g, will vanish if either f or g has its centroid at its origin.

By taking $g(x) = \delta(x - a)$, for which $\langle x^2 \rangle_g = a^2$, we have the theorem which is familiar in relation to the moment of inertia of a mass about an axis not through its center of gravity. For if $f(x)$ has its origin at its centroid,

$$\langle x^2 \rangle_{f(x-a)} = \langle x^2 \rangle_f + a^2.$$

Other ways of weighting the mean of x^2 may be considered. For example, if $f(x)$ has negative-going parts, they will reduce the variance, but there may be interest in measuring the spread of $f(x)$ irrespective of its sign. Two quantities which are met in these circumstances are

$$\langle x^2 \rangle_{|f|} \qquad \text{and} \qquad \langle x^2 \rangle_{|f|^2}.$$

Radius of gyration

This term stands for the root-mean-square value of x—that is, the square root of the mean-square abscissa—and its convenience is largely associated with having the same dimensions as x. Properties such as the relation to the central curvature of the transform, the additive property, and the radius of gyration about a displaced axis, are usually simpler when expressed in terms of the mean-square abscissa rather than in terms of its square root.

Variance

The variance σ^2 is the mean-square deviation referred to the centroid; that is,

$$\sigma^2 = \langle (x - \langle x \rangle)^2 \rangle = \frac{\int_{-\infty}^{\infty} (x - \langle x \rangle)^2 f(x)\, dx}{\int_{-\infty}^{\infty} f(x)\, dx}$$

$$= -\frac{F''(0)}{4\pi^2 F(0)} + \frac{[F'(0)]^2}{4\pi^2 [F(0)]^2},$$

and the statements about $\langle x^2 \rangle$ apply also to variance, since a choice of the origin of abscissas which makes $\langle x \rangle$ zero will mean that $\langle x^2 \rangle$ is the same as the variance.

The variance of $f * g$ is equal to the sum of the variances of f and g.

The proof follows from the earlier result for mean-square abscissas.

Smoothness and compactness

The smoother a function is, as measured by the number of continuous derivatives it possesses, the more compact is its transform; that is, the faster it dies away with increasing s.

If a function and its first $n - 1$ derivatives are continuous, its transform dies away at least as rapidly as $|s|^{-(n+1)}$ for large s; that is,

$$\lim_{|s|\to\infty} |s|^n \bar{F}(s) = 0.$$

In the customary symbology of Landau, $\bar{F}(s) = O(|s|^{-n})$. In the course of the proof of this theorem the Fourier transforms of the first $n - 1$ derivatives appear, and it is therefore necessary to mention, in a rigorous statement of the theorem, that the first n derivatives should possess absolutely convergent integrals.

In Fig. 8.6 several of the common cases are illustrated. The rather familiar fact that $\text{sinc}^2\, x$ dies away much more rapidly than sinc x is seen to be associated with the smoother behavior of its transform. We can say that functions possessing discontinuities in slope ("corners"), and whose second derivatives are therefore impulsive, will have transforms that die away as $|s|^{-2}$. In general one can say that if the kth derivative becomes impulsive, then the transform behaves as $|s|^{-k}$ at infinity. This simplifies statements, such as the one given above, which contain $n - 1$, n, and $n + 1$ in the same sentence.

It is interesting to note that the function $\exp(-\pi x^2)$ is continuous and has all continuous derivatives, and is therefore in the present sense as smooth as possible. Likewise its transform $\exp(-\pi s^2)$ is as compact as possible, since it dies away faster than s^{-n} for all n. This property is not unique to $\exp(-\pi x^2)$, being shared with many other pairs of transforms (including, for example, the self-reciprocal pair sech πx, sech πs).

To prove the above statement, let $f(x)$ have an absolutely integrable derivative [$f(x)$ is of course itself absolutely integrable and differentiable]. Since $f'(x)$ is absolutely integrable,

$$g(x) = \int_{-\infty}^{x} f'(\xi)\, d\xi$$

exists, and since $g'(x) = f'(x)$,

$$f(x) = g(x) + c.$$

As $x \to -\infty$, $g(x) \to 0$. Now if c were not zero, $f(x)$ could not be absolutely integrable. Hence $g(x) = f(x)$, and thus as $x \to -\infty$, we also have $f(x) \to 0$.

Now put

$$\int_{-\infty}^{\infty} f'(\xi)\, d\xi = c_1;$$

then $$f(x) = \int_{-\infty}^{x} f'(\xi)\, d\xi = c_1 - \int_{x}^{\infty} f'(\xi)\, d\xi.$$

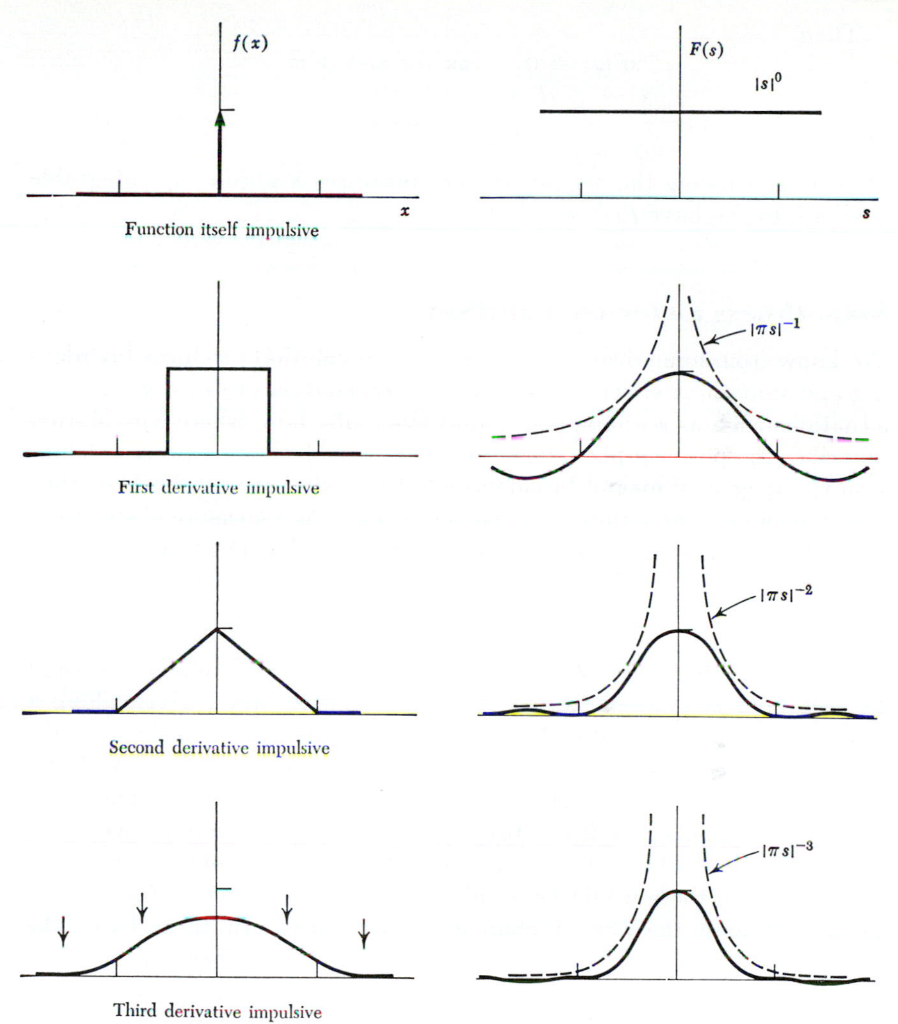

Fig. 8.6 If the kth derivative of a function becomes impulsive, its transform behaves as $|s|^{-k}$ at infinity.

So again $c_1 = 0$ and thus $f(x) \to 0$ for $x \to \infty$. Now

$$F(s) = \int_{-\infty}^{\infty} f(x) e^{-i2\pi sx}\, dx$$
$$= \frac{\int_{-\infty}^{\infty} f'(x) e^{-i2\pi sx}\, dx}{i2\pi s},$$

on integrating by parts and using the fact that $f(x) \to 0$ as $x \to \pm\infty$.

Then

$$sF(s) \to 0 \quad \text{as} \quad s \to \pm\infty$$

$$sF(s) = o(|s|^{-1})$$

or

$$F(s) = o(|s|^{-2}).$$

If then, continuing the argument, $f(x)$ possesses n absolutely integrable derivatives, we have $F(s) = o(|s|^{-n})$.

Smoothness under convolution

We know from numerical experience that convolution produces broadening and smoothing effects. This is also expected on physical grounds in situations such as scanning the sound track of a film, where convolution describes the physical phenomenon. One way of giving a precise definition to the general idea of broadening has already been established; thus it is known that the standard deviation σ_f, which is a measure of spread of a function $f(x)$, is increased under convolution with $g(x)$ according to the result

$$\sigma^2{}_{f*g} = \sigma^2{}_f + \sigma^2{}_g.$$

There are other ways of measuring spread, and smoothness too could be measured in different ways. If an automobile were driven along a road whose height varied as Π^{*2}, or $\Lambda(x)$, the ride would not be smooth; but it would be smoother than driving along a road described by $\Pi(x)$, which has cliffs. A road surface Π^{*3}, composed of two straight and three parabolic segments, free from abrupt changes in height or slope, would be smoother. It would not be perfectly smooth, for at each of the four points of articulation there would be an abrupt change in the centrifugal force, associated with the abrupt change in curvature. In these cases the smoothness increases with the number of derivatives possessed by the function.

Let us say that if the kth derivative of a function is impulsive, then it has smoothness of order k. Now let $f(x)$ have smoothness of order m, and $g(x)$ smoothness of order n. When they are convolved together, is the convolution indeed smoother, as we would wish, under our quantitative definition?

Since

$$F(s) \sim s^{-m}$$

and

$$G(s) \sim s^{-n}$$

it follows that

$$F(s)G(s) \sim s^{-(m+n)}.$$

But $F(s)G(s)$ is the Fourier transform of $f * g$, hence the $(m + n)$th derivative of $f * g$ is impulsive, and it therefore has smoothness of order $m + n$. It is thus smoother than the functions entering into the convolution; in fact the smoothness proves to be additive under convolution just as σ^2 is.

The word "smoothing" is sometimes used as a synonym for convolution. "Inverse smoothing" and "sharpening" are terms used to mean solving the convolution integral equation. It is rather interesting from the standpoint of this section that "sharpening" procedures make use of further smoothing. Consequently the quantitative measure here given to smoothness does not always coincide with qualitative conceptions of smoothness. As a particular example, consider $\Lambda(x) * \tau^{-1} \exp(-\pi x^2/\tau^2)$, where τ is very small. This function is now infinitely smooth but is a close approximation to the rough $\Lambda(x)$. It differs appreciably only at the corners, which are microscopically rounded off to prevent impulsive derivatives. The new road profile might, however, be just as rough to ride over as the old.

Asymptotic behavior

It is well known qualitatively that sharp or abrupt features in a waveform betoken the presence of high-frequency components. Now we can go further and say quantitatively how the spectrum holds up as frequency increases.

We have seen that if a function is continuous but has a discontinuous derivative—that is, it has no jumps but has corners—then its transform dies away as s^{-2} and its power spectrum as s^{-4}. To describe such behavior in electric-circuit language, we would refer to it as attenuation at 12 decibels per octave or 40 decibels per decade.

On a log-log plot, a function varying as s^{-2} would become a straight line of slope -2. Logarithmic units such as the octave or decade, the decibel, the neper, and the magnitude (in astronomy) permit slope, which is dimensionless in itself, to be stated in expressive dimensionless units.

Figure 8.7 shows linear and log-log versions of two familiar functions. (Where the function is negative, the logarithm of the absolute value is shown.) Certain aspects of the functions, especially the nature of their asymptotic behavior and their closeness of approach thereto, are more clearly evidenced by the logarithmic than by the linear plot. For this and other reasons, logarithmic plots are often used to display electrical-filter characteristics and antenna patterns.

The question arises why only slopes of 0, 3, 6, . . . , decibels per octave were encountered when discontinuities of various kinds were discussed in the preceding section. We can study this by considering some transform which dies away at 4.5 decibels per octave, that is, as $s^{-\frac{3}{4}}$, and inquiring what sort of discontinuity the corresponding original function possesses.

Since division by s corresponds to integration, division by $s^{\frac{1}{2}}$ corresponds to half-order integration. We ignore powers of $2\pi i$.

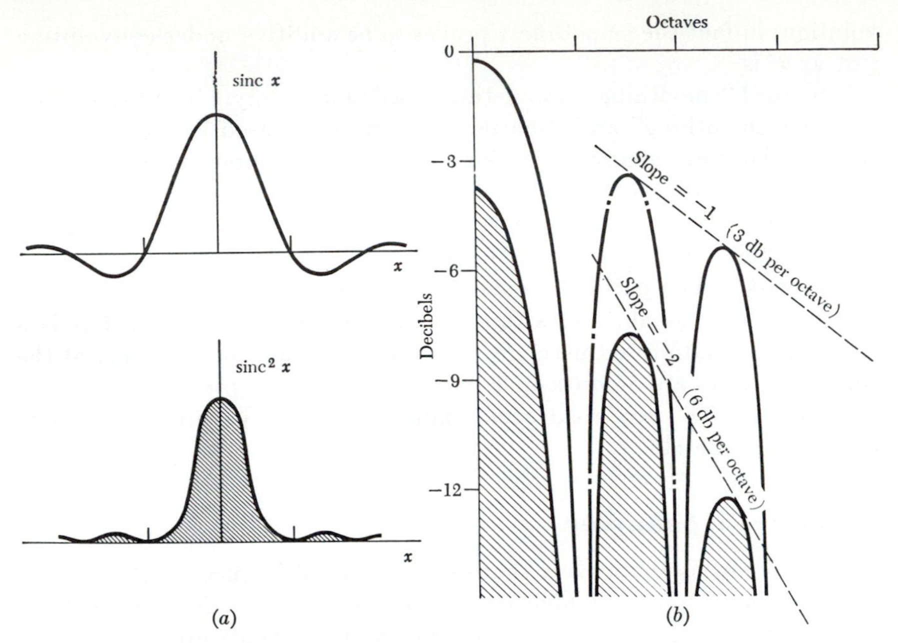

Fig. 8.7 Asymptotic behavior as expressed on a double logarithmic plot in decibels per octave: (a) linear plot for comparison; (b) logarithmic plot (negative lobes shown broken).

The examples tabulated in Table 8.2 show that for the purposes of asymptotic behavior there are indeed discontinuities intermediate in severity between jumps and corners. Thus a corner where a function changes its slope from horizontal to vertical (A) ranks midway between simple corners and simple jumps. A discontinuity (B) midway between a simple jump and an impulsive infinity proves to be less severe.

There are further subtleties to this matter. Thus $\delta(x)$ and x^{-1} prove to have the same kind of infinity at $x = 0$, and the infinite discontinuity of $\log x$, which is of the weakest possible kind, proves to be equivalent to a simple jump.

Equivalent width

A convenient measure of the width of a function, for certain shapes of function, is

$$\frac{\int_{-\infty}^{\infty} f(x)\, dx}{f(0)},$$

that is, the area of the function divided by its central ordinate. Expressed differently, the equivalent width of a function is the width of the rectangle

Table 8.2 Fractional-order discontinuities

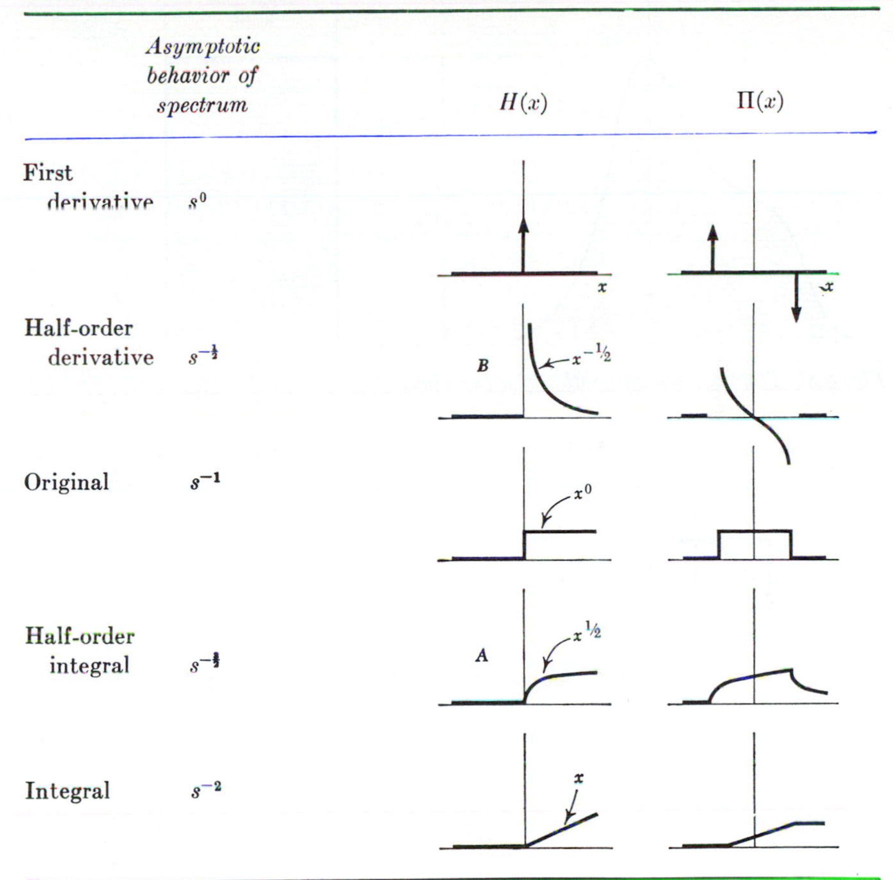

	Asymptotic behavior of spectrum	$H(x)$	$\Pi(x)$
First derivative	s^{0}		
Half-order derivative	$s^{-\frac{1}{2}}$		
Original	s^{-1}		
Half-order integral	$s^{-\frac{3}{2}}$		
Integral	s^{-2}		

whose height is equal to the central ordinate and whose area is the same as that of the function. This is illustrated in Fig. 8.8.

Examples can be drawn from many sources. In spectroscopy the equivalent width of a spectral line is defined as the width of a rectangular profile which has the same central intensity and the same area as the line. In antenna theory the effective beamwidth of an antenna has the character of an equivalent width (Chapter 13). Other examples occur in radiometry (Chapter 16). Some examples of functions, with accompanying rectangles illustrating their equivalent width, are shown in Fig. 8.9; all the examples have the same equivalent width.

If $f(0) = 0$, the equivalent width does not exist. For all cases, however, where $f(0) \neq 0$ and $\int_{-\infty}^{\infty} f(x)\,dx$ exists, there exists an "equivalent width," even though the character of equivalence in the spectroscopic sense is lost.

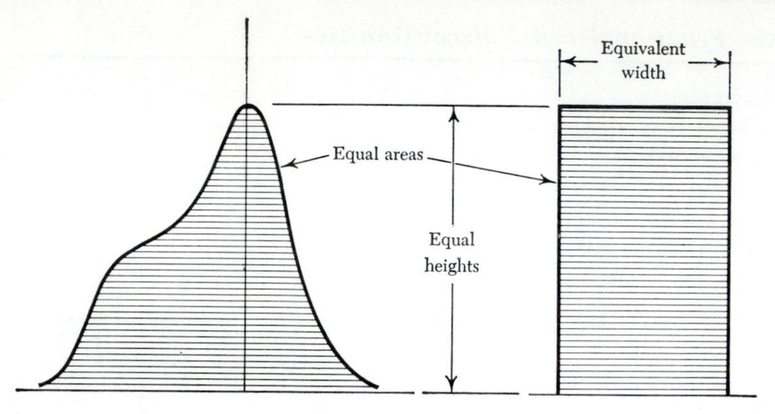

Fig. 8.8 The "equivalent width" of a function.

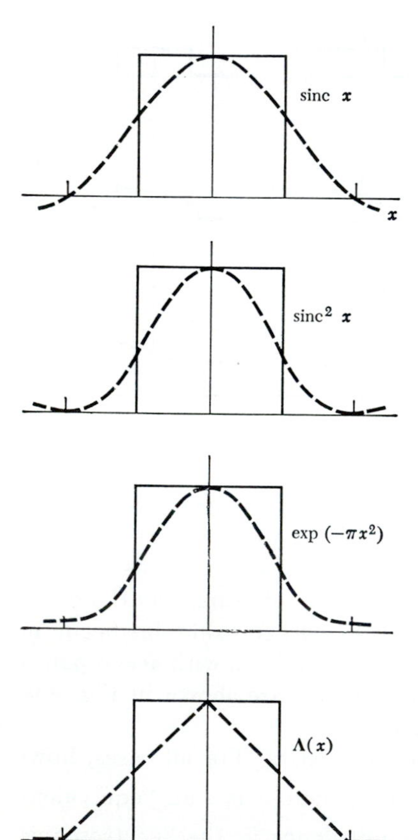

Fig. 8.9 Equivalent widths of some common distributions.

One expects that the wider a function is, the narrower its spectrum will be, but as functions of various shapes are contemplated, can we express this inverse relationship quantitatively? It turns out that a very precise relation exists. Whenever the equivalent width exists, it possesses the following elegant property.

The equivalent width of a function is equal to the reciprocal of the equivalent width of its transform; that is,

$$\frac{\int_{-\infty}^{\infty} f(x)\, dx}{f(0)} = \frac{F(0)}{\int_{-\infty}^{\infty} F(s)\, ds}.$$

This theorem is reminiscent of the similarity theorem, according to which the compression and expansion of a function and its transform behave reciprocally. The similarity theorem, however, is restricted to functions of a given shape. The present theorem says what quantity it is that behaves reciprocally, and it permits unlike functions to be ranked in order of that quantity.

Equivalent width is not always the best measure of width. For example, in expressing the bandwidth of a filter or the beamwidth of an antenna, it is common to adopt the "width to half power." According to this measure the second example illustrated above is narrower, for some purposes, than the first, which agrees with experience.

An interesting paradox arises when a localized turbulent motion spreads outward in space while the turbulence spectrum spreads out to higher wave numbers. In this case equivalent width is the wrong measure of spread.

The effective solid angle of an antenna radiation pattern (which is equal to 4π/directivity) is the "equivalent width" of the pattern, generalized to two dimensions.

Exercise Investigate the behavior of equivalent width under convolution, beginning with some exploratory trials on Gaussian and rectangle functions of various relative widths.

Table 8.3 lists equivalent widths for a number of common functions, together with autocorrelation widths, which are introduced in the next section. In this table the following abbreviations are used:

$$f \star f = \int_{-\infty}^{\infty} f(x+u) f(u)\, du$$

$$W_f = \frac{1}{f(0)} \int_{-\infty}^{\infty} f\, dx$$

$$W_{f\star f} = \frac{1}{f\star f|_0} \int_{-\infty}^{\infty} (f\star f)\, dx$$

Table 8.3 Equivalent widths and autocorrelation widths

$f(x)$	$\int f\,dx$	$f(0)$	W_f	$f \star f$	$\int (f \star f)\,dx$	$f \star f\big\vert_0$	$W_{f\star f}$
$\Pi(x)$	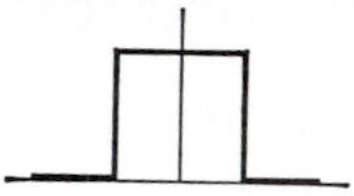 1	1	1	$\Lambda(x)$	1	1	1
$\Lambda(x)$	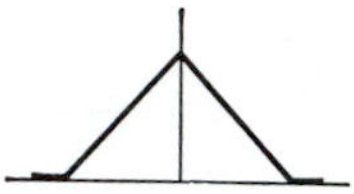 1	1	1	$\Lambda * \Lambda$	1	$\frac{2}{3}$	$\frac{3}{2}$
$\cos \pi x \Pi(x)$	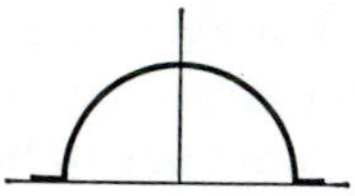 $2/\pi$	1	$2/\pi$	$\left\{\frac{1}{2}(1 - \vert x\vert) \cos \pi x + \frac{1}{2\pi} \sin \pi \vert x\vert \right\} \Pi\left(\frac{x}{2}\right)$	$4/\pi^2$	$\frac{1}{2}$	$8/\pi^2$
$e^{-x}H(x)$	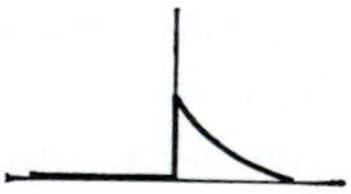 1	1†	1	$\frac{1}{2}e^{-\vert x\vert}$	1	$\frac{1}{2}$	2
$e^{-\pi x^2}$	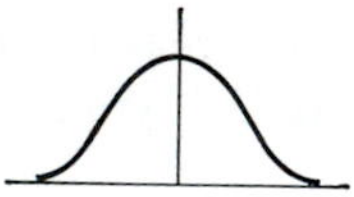 1	1	1	$2^{-\frac{1}{2}}e^{-\pi x^2/2}$	1	$2^{-\frac{1}{2}}$	$2^{\frac{1}{2}}$
$e^{-\vert x\vert}$	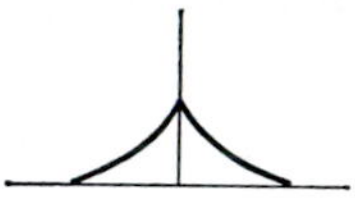 2	1	2	$(\vert x\vert + 1)e^{-\vert x\vert}$	4	1	4
$\frac{1}{1+x^2}$	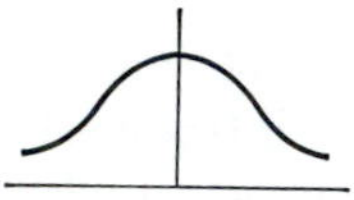 π	1	π	$\frac{2\pi}{4+x^2}$	π^2	$\pi/2$	2π
$\operatorname{sinc} x$	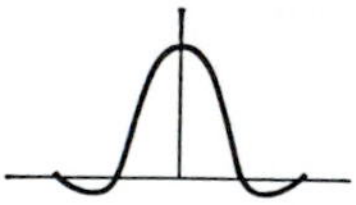 1	1	1	$\operatorname{sinc} x$	1	1	1

† The value given is $f(0+)$.

Table 8.3 Equivalent widths and autocorrelation widths (cont'd)

$f(x)$	$\int f\,dx$	$f(0)$	W_f	$f \star f$	$\int (f \star f)\,dx$	$f \star f \Big\vert_0$	$W_{f \star f}$
$\operatorname{sinc}^2 x$	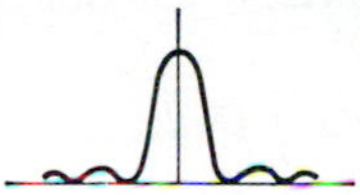 1	1	1	$\frac{1}{\pi x^2}(1 - \operatorname{sinc} 2x)$	1	$\frac{2}{3}$	$\frac{3}{2}$
$\delta(x)$	 1	∞	0	$\delta(x)$	1	∞	0
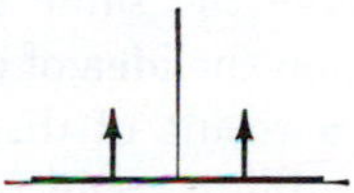 $\text{II}(x)$	1	0	∞	$\frac{1}{4}\delta(x+1) + \frac{1}{2}\delta(x) + \frac{1}{4}\delta(x-1)$	1	∞	0
$2K_0(2\pi x)$				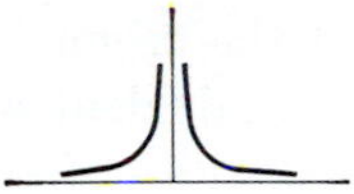 Fourier transform of $(1 + x^2)^{-1}$	1	π	π^{-1}
$\Lambda(x)H(x)$	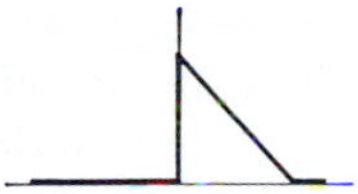 $\frac{1}{2}$	1†	$\frac{1}{2}$	$\frac{1}{6}\Pi\left(\frac{x}{2}\right)(\vert x\vert^3 - 3\vert x\vert + 2)$	$\frac{1}{4}$	$\frac{1}{3}$	$\frac{3}{4}$
$\cos \pi x \Pi(x) H(x)$	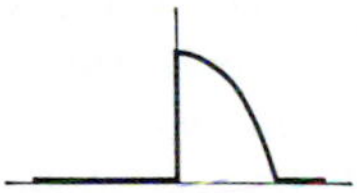 π^{-1}	1†	π^{-1}		π^{-2}	$\frac{1}{4}$	$4/\pi^2$
$\frac{H(x)}{1 + x^2}$	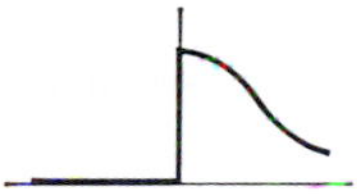 $\pi/2$	1†	$\pi/2$	. . .	$\pi^2/4$	$\pi/4$	π
$x\Pi(x - \frac{1}{2})$	 $\frac{1}{2}$	0	∞	$\frac{1}{6}\Pi\left(\frac{x}{2}\right)(\vert x\vert^3 - 3\vert x\vert + 2)$	$\frac{1}{4}$	$\frac{1}{3}$	$\frac{3}{4}$
$\frac{1}{1 + i2\pi a x}$	$1/2a$	1	$2a$	. . .	. . .	. . .	. . .
$\frac{1}{(1 + x^2)^2}$	$\pi/2$	1	$\pi/2$	. . .	$\pi^2/4$	$5\pi/16$	$4\pi/5$

† The value given is $f(0+)$.

Autocorrelation width

Another interesting "width" is the equivalent width of the autocorrelation function, which we shall call the autocorrelation width:

$$W_{f\star f^*} = \frac{\int_{-\infty}^{\infty} (f \star f^*)\,dx}{f \star f^*\Big|_0} = \frac{\int f\,dx \int f^*\,dx}{\int ff^*\,dx}.$$

Autocorrelation width so defined does not necessarily refer to concentration about the origin. For example, the two functions shown in Fig. 8.10 would have the same autocorrelation width because they have the same autocorrelation function, but they would not have the same equivalent width. If a rectangular pulse were considered, then the idea of equivalent width could completely break down simply as a result of displacement along the axis of abscissas, because the central ordinate could fall to zero. This possibility is eliminated when one uses the equivalent width of the autocorrelation function, which is a maximum at the origin, never zero.

It is clear from the reciprocal property of the equivalent widths of a function and its transform that the autocorrelation width of a function is the reciprocal of *the equivalent width of its power spectrum.* Thus the idea of autocorrelation width breaks down if the equivalent width of the power spectrum loses meaning. This happens when the power spectrum has a zero central ordinate, which in turn means that the function (and its autocorrelation function) have zero area.

Hence, in relation to functions such as those used to represent wave packets which have little or no direct-current component, the autocorrelation width is not an appropriate measure of "extent" or duration.

The reciprocal of the autocorrelation width of the transform of a function is the equivalent width of the squared modulus of the function:

$$\frac{\int ff^*\,dx}{f(0)f^*(0)}.$$

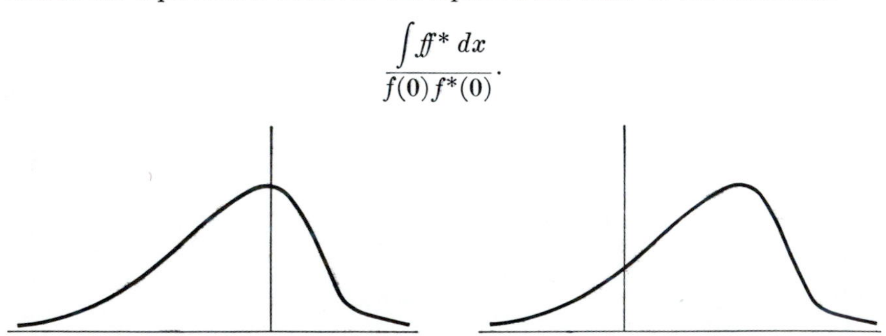

Fig. 8.10 Functions with different equivalent widths but the same autocorrelation width.

The advantage of autocorrelation width as a measure of dispersion comes from the removal of the sensitivity to the value of the central ordinate exhibited by the equivalent width. Two functions that have the same power spectrum, but whose spectral components are slid about into any relative phase, have the same autocorrelation function. Therefore the quantity $W_{f \star f^*}$ copes with (1) displacement of the function away from the origin and (2) internal phase interference of the spectral components, which happens to produce a small central ordinate. However, of two functions with the same power spectrum, one can be narrow and another wide. For example, the Fresnel diffraction radiation fields on planes parallel to an illuminated aperture have the same autocorrelation, but on any intuitive view the width of the illumination increases with distance from the source aperture.

Difficulty (1) can be removed by considering equivalent widths about the centroid. Difficulty (2) indicates that (*a*) internal phase relations of the components cannot be ignored and that (*b*) a width measure hinging on some one ordinate will always be sensitive as a consequence.

This critical approach to width measures should not obscure the fact that equivalent width and autocorrelation width both occur as the appropriate quantity in particular physical circumstances. Thus it would be pointless to criticize the sensitive dependence of antenna directivity on one ordinate of the radiation pattern.

A situation where the autocorrelation width occurs is rectification of white noise passed through a filter. It is well known qualitatively that the average interval between effectively independent output values is inversely proportional to the "width" of the pass band. In fact it is the autocorrelation width of the power pass characteristic for which this statement is quantitatively accurate.

An important property of the autocorrelation width which fits it for many applications is its invariance under shuffling (see Problems).

Mean-square widths

Since the equivalent width and concentration just described do not fill all needs, other measures of width are also encountered. For example, the mean-square value of x defined by

$$\langle x^2 \rangle = \frac{\int_{-\infty}^{\infty} x^2 f(x)\, dx}{\int_{-\infty}^{\infty} f(x)\, dx}$$

is widely appropriate as a width measure, as is also the variance in functions not centered on the origin of x:

$$\langle (x - \langle x \rangle)^2 \rangle = \langle x^2 \rangle - \langle x \rangle^2$$
$$= \frac{F''(0)}{-4\pi^2 F(0)} + \frac{1}{4\pi^2}\left[\frac{F'(0)}{F(0)}\right]^2.$$

These measures break down when the area of the function is zero and are therefore inappropriate for measuring the width of an oscillatory signal or wave packet. To cope with this case we think in terms of the energy density and the centroid and variance of the energy distribution. We shall refer to the mean-square departure from the centroid of $|f(x)|^2$ as the variance of the squared modulus, and it is given by

$$(\Delta x)^2 = \frac{\int_{-\infty}^{\infty} x^2 |f(x)|^2 \, dx}{\int_{-\infty}^{\infty} |f(x)|^2 \, dx} - \left[\frac{\int_{-\infty}^{\infty} x |f(x)|^2 \, dx}{\int_{-\infty}^{\infty} |f(x)|^2 \, dx}\right]^2.$$

This may seem an elaborate expression for the simple physical idea of duration or bandwidth of a signal packet, but it behaves more reasonably for many purposes than the simpler measures considered before.

Of course, the square of a given function may not have a finite variance; for example, sinc x and $(x^2 + 1)^{-\frac{1}{2}}$, two concentrated peaked functions having reasonable "equivalent widths" and "autocorrelation widths," show up as infinitely broad on a variance basis. Caution is therefore necessary before assuming that widths based on variance will have any relation to intuitive ideas of width in particular cases.

In the rigorous expression of the uncertainty relation referred to below, the squared uncertainties are variances of squared moduli. Suppose an observed quantity had the form sinc x, as would happen if an impulsive wave were sampled with receiving equipment of limited high-frequency response, or if an infinite plane wave were sampled through an aperture of limited spatial extent. Then that spread which is subject to the quantitative uncertainty relation would evidently be by no means as well in agreement with one's intuitive notion of the spread as it is in the case of a Gaussian packet exp $(-ax^2)$ sin x.

Table 8.4 lists the centroid and the mean-square abscissa for a variety of functions.

Some inequalities

The discontinuities, cusps, or other sharp behavior characteristics of a function reveal the presence of high frequencies in its spectrum; in fact a quantitative measure of the sharpness of the behavior is linked with the decay of the spectrum in another section. Now if a function is quite

Table 8.4 Centroid and mean-square abscissa

$f(x)$	$\langle x \rangle$	$\langle x^2 \rangle$	$F''(0)$	$f''(0)$	$(\Delta x)^2$
$\Pi(x)$	0	$\frac{1}{12}$	$-\pi^2/3$	0	$\frac{1}{12}$
$\Lambda(x)$	0	$\frac{1}{6}$	$-2\pi^2/3$	∞	$\frac{1}{4}$
$\operatorname{sinc} x$	0	osc	0	$-\pi^2/3$	∞
$\operatorname{sinc}^2 x$	0	∞	∞	$-2\pi^2/3$	...
$H(x)$	∞	∞	∞	∞	∞
$e^{-x}H(x)$	1	2	$-8\pi^2$	∞	$\frac{1}{4}$
$e^{-\|x\|}$	0	4	$-16\pi^2$	∞	1
$\dfrac{1}{1+x^2}$	0	∞	∞	-2	...
$\cos \pi x \Pi(x)$	0	$\dfrac{\pi^2 - 8}{4\pi^2}$	$-\dfrac{2}{\pi}(\pi^2 - 8)$	$-\pi^2$	...
$(1 - x^2)\Pi\left(\dfrac{x}{2}\right)$	0	$\frac{1}{5}$	$-\dfrac{16\pi^2}{15}$	-2	...
$e^{-\pi x^2 a^2}$	0	$\dfrac{1}{2\pi a^2}$	$-\dfrac{2\pi}{a^3}$	$-2\pi a^2$	$\dfrac{1}{4\pi a^2}$
$e^{-\pi x^2 a^2} \cos \omega x$	0	...	$\dfrac{-2\pi}{a^3}$	$-(\omega^2 + 2\pi a^2)$	...
$\text{III}(x)$	0	$\frac{1}{4}$	$-\pi^2$	0	∞
$\operatorname{sech} \pi x$	0	$\frac{1}{4}$	$-\pi^2$	$-\pi^2$	...

continuous, with all derivatives continuous, there is still a limit to the slope, or curvature, which it can have.

If such a function undergoes a large change in a short interval of abscissa as in Fig. 8.11 (that is, it has a very large slope in this interval), it may for some practical purposes be effectively discontinuous. Its spectrum will indeed be found to behave appropriately. The fact that ultimately the asymptotic behavior appropriate to a continuous function sets in is then not of practical concern, since only periodic components of period short compared with the above-mentioned short interval exhibit the expected behavior, and they are beyond the limit of interest, for we have said that for the purpose the change is effectively discontinuous.

Upper limits to ordinate and slope Just how steep can the slope of a function be? Consider first the problem of how great a function can become. We can say that

$$|f(x)| \leqslant \int_{-\infty}^{\infty} |F(s)|\, ds$$

as a direct consequence of the relation

$$f(x) = \int_{-\infty}^{\infty} F(s) e^{i2\pi xs}\, ds.$$

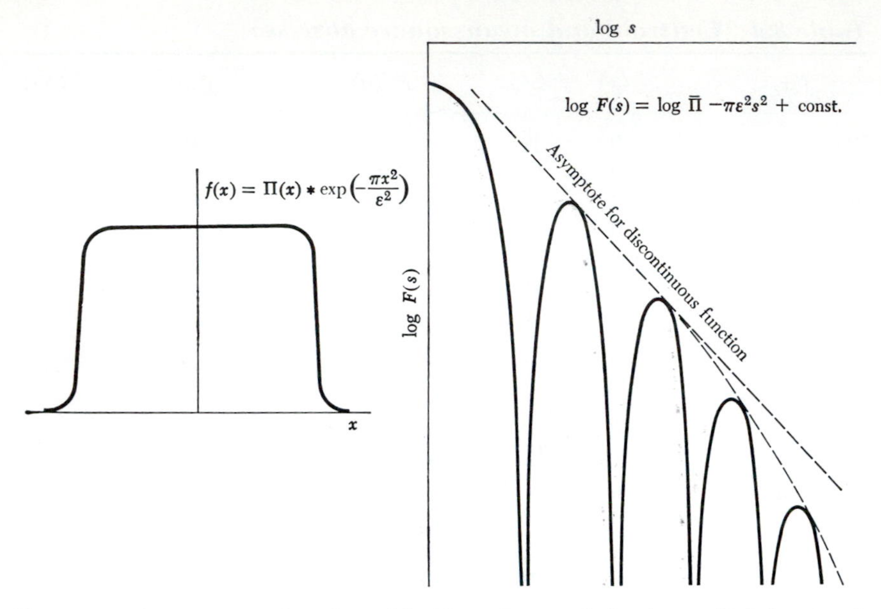

Fig. 8.11 A continuous function with all continuous derivatives, which for certain purposes is effectively discontinuous, and its transform.

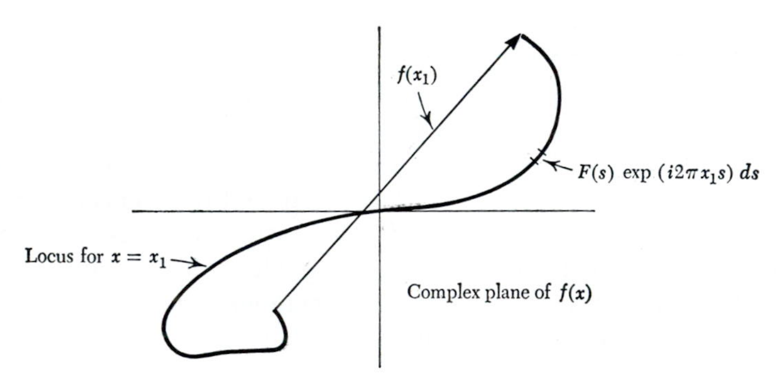

Fig. 8.12 Locus on the complex plane of $f(x)$.

On the complex plane of $f(x)$ a Fourier integral for a particular value of x is illustrated in Fig. 8.12. For other values of x the locus would be different but the length of the arc, which is equal to

$$\int_{-\infty}^{\infty} |F(s)|\, ds,$$

must remain the same. This arc length is the maximum value which $|f(x)|$ could assume; and it would do so if there were a value of x for which the arc extended into a straight line. In other words, the maximum value of a function cannot exceed the resultant of all its components at that point, if it exists, where they all come into phase together.

The slope of the component $F(s)\exp(i2\pi xs)ds$ is $2\pi isF(s)\exp(i2\pi xs)ds$, and by considering the possibility of all components having their maximum slopes together, we have

$$|f'(x)| \leqslant 2\pi \int_{-\infty}^{\infty} |sF(s)|\, ds.$$

The two inequalities we have derived both stem from the geometrical phenomenon that an arc joining two points is equal to or greater than the chord. This in turn can be further distilled and depends ultimately on the property that two sides of a triangle are together greater than or equal to the third. In vector notation

$$|\mathbf{A} + \mathbf{B}| \leqslant |\mathbf{A}| + |\mathbf{B}|.$$

It appears that inequalities are often reducible to one or the other of a few geometrical inequalities. This particular one is referred to as Bessel's inequality.

Schwarz's inequality If two real functions $f(x)$ and $g(x)$ are defined in the interval $a < x < b$, then

$$\left[\int_a^b f(x)g(x)\, dx\right]^2 \leqslant \int_a^b [f(x)]^2\, dx \int_a^b [g(x)]^2\, dx.$$

A related vector inequality states that the scalar product of two vectors cannot exceed the product of their absolute values; thus

$$\mathbf{A} \cdot \mathbf{B} \leqslant \mathrm{AB}.$$

If F and G are complex functions of x, we have the Schwarz inequality in the form most suited to the present work:

$$\left|\int (F^*G + FG^*)\, dx\right|^2 \leqslant 4 \int FF^*\, dx \int GG^*\, dx.$$

To prove Schwarz's inequality, let ϵ be a real constant. Then

$$0 < \int (F + \epsilon G)(F + \epsilon G)^*\, dx,$$

because the integrand is a positive quantity (unless by accident F happens to be proportional to G, in which case the value of ϵ that makes $F + \epsilon G$ identically equal to zero must be avoided). By expanding the above inequality we can write

$$0 < \int FF^*\, dx + \epsilon \int (F^*G + FG^*)\, dx + \epsilon^2 \int GG^*\, dx.$$

The right-hand side is a quadratic expression in ϵ^2; call it $c + b\epsilon + a\epsilon^2$. The condition that it should not reduce to zero for any real value of ϵ is $b^2 - 4ac \leqslant 0$.

When the Schwarz inequality is applied, definite limits may be assigned to the integrals. For example, in connection with the uncertainty relation infinite limits are adopted. The inequality holds, however, for any limits.

The uncertainty relation

It is well known that the bandwidth-duration product of a signal cannot be less than a certain minimum value. This is essentially a mathematical phenomenon bound in with the interdependence of time and frequency, which prevents arbitrary specification of signals on the time-frequency plane. One may specify arbitrary functions of time or arbitrary spectra, but not both together. Thus a finite area of the time-frequency plane can contain only a finite number of independent data. The bandwidth-duration product of a signal cannot be less than the value for an elementary signal containing only one datum. Such a signal may be brief and wideband or quasi-monochromatic and persistent.

Since the equivalent widths of a function and its transform are reciprocals, it follows that

$$\text{Equivalent duration} \times \text{equivalent bandwidth} = 1.$$

The product of the autocorrelation widths is deducible from a previous section as

$$\frac{|f(0)|^2 \left| \int f\,dx \right|^2}{\left[\int |f|^2\,dx \right]^2},$$

and if we deal in mean-square widths,

$$\langle x^2 \rangle \langle s^2 \rangle = - \frac{f''(0) \int x^2 f\,dx}{4\pi^2 f(0) \int f\,dx} = \frac{f''(0)F''(0)}{16\pi^4 f(0)F(0)}.$$

Neither of these last two width products is a constant, and the second can range from zero to infinity.

If we consider $(\Delta x)^2$, the variance of $|f(x)|^2$, we find the usual expression of the uncertainty relation, namely,

$$\Delta x\,\Delta s \geqslant \frac{1}{4\pi}.$$

Proof of uncertainty relation In this section all integrals are to be taken between infinite limits. We need the theorem

$$\int f'f'^*\,dx = 4\pi^2 \int s^2 FF^*\,ds,$$

which may be verified by combining the derivative theorem with Rayleigh's theorem.

We also need the Schwarz inequality in the form

$$4 \int ff^* \, dx \int gg^* \, dx \geqslant \left| \int (f^*g + fg^*) \, dx \right|^2,$$

and the formula for integration by parts between infinite limits in the form

$$\left| \int xf' \, dx \right| = \left| \int f \, dx \right|.$$

Now, taking $f(x)$, and thus also $F(s)$, to be centered on its centroid,

$$\begin{aligned}
(\Delta x)^2(\Delta s)^2 &= \frac{\int x^2 ff^* \, dx \int s^2 FF^* \, ds}{\int ff^* \, dx \int FF^* \, ds} \\
&= \frac{\int xf.xf^* \, dx \int f'f'^* \, dx}{4\pi^2 \left(\int ff^* \, dx \right)^2} \\
&\geqslant \frac{\left| \int (xf^*.f' + xf.f'^*) \, dx \right|^2}{16\pi^2 \left(\int ff^* \, dx \right)^2} \\
&= \frac{\left| \int x \frac{d}{dx} (ff^*) \, dx \right|^2}{16\pi^2 \left(\int ff^* \, dx \right)^2} \\
&= \frac{\left| \int ff^* \, dx \right|^2}{16\pi^2 \left(\int ff^* \, dx \right)^2} \\
&= \frac{1}{16\pi^2}.
\end{aligned}$$

Therefore $\Delta x \, \Delta s \geqslant \frac{1}{4\pi}$.

Example of uncertainty relation Take $f(x)$ to be a voltage $V(t)$ whose Fourier transform is $S(f)$. We select $V(t)$ so that its spectrum is real for the purpose of illustration but in general $S(f)$ is complex. Voltages or the fields of electromagnetic waves, on the other hand, can only be real. The signal energy distributions and the energy spectrum are shown in Fig. 8.13, together with Δt and Δf the root-mean-square departures from the centroids of the two energy distributions. The uncertainty relation says that the product $\Delta t \, \Delta f$ must exceed or equal $(4\pi)^{-1}$. In terms of angular frequency ω,

$$\Delta t \, \Delta \omega \geqslant \tfrac{1}{2}.$$

In the case illustrated we can make a much stronger statement. Suppose that a radar transmitter emits a 1-microsecond pulse at a fre-

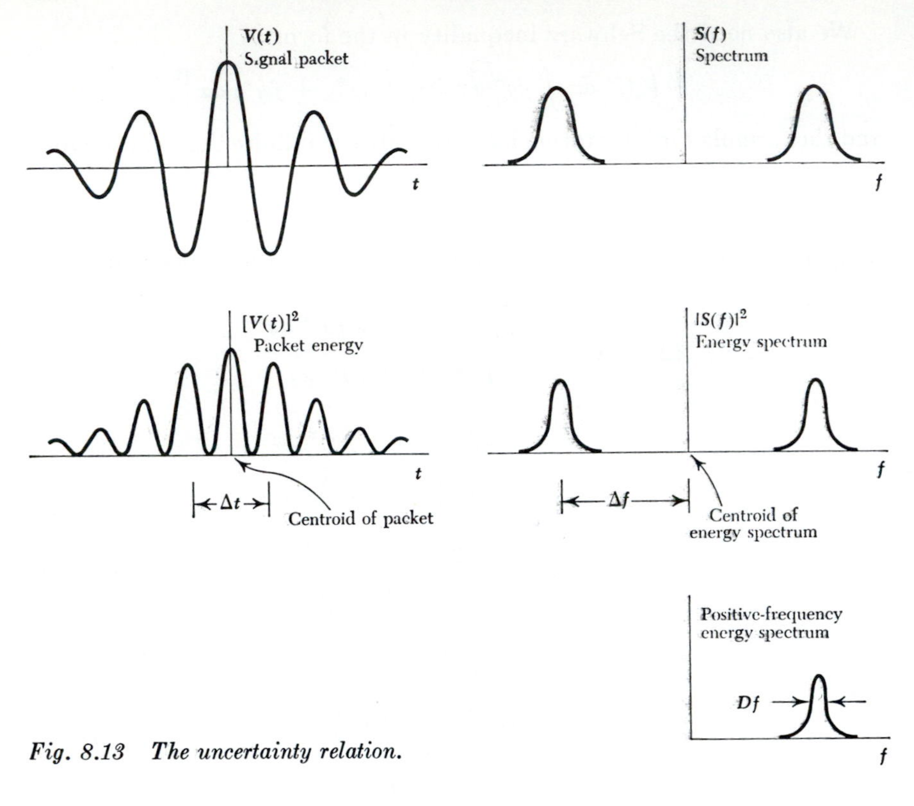

Fig. 8.13 The uncertainty relation.

quency of 10,000 megacycles per second. Then the uncertainty principle says that

$$10^{-6} \geqslant \frac{1}{4\pi \times 10^{10}}.$$

Now if Df is the width of the positive-frequency power spectrum defined by

$$2|S(f)|^2 \qquad f > 0,$$

we know from experience with radio receivers that

$$\Delta t \sim \frac{1}{Df},$$

since in the case of a 1-microsecond pulse a receiver bandwidth of 1 megacycle per second is an optimum choice for detection. The uncertainty relation thus sets a lower limit to the duration-bandwidth product, which in the case illustrated fails by a factor of 10^5 to be practical.

While the positive-frequency energy spectrum is not quite as amenable to mathematical handling as $S(f)$, the conventional spectrum in terms of which Δf is defined, it is closely connected with what would be observed if the energy spectrum were explored experimentally with tunable resonators equipped for energy detection.

The next step is to consider the product

$$\Delta t \, Df,$$

which can be shown to agree with experience in cases such as that quoted but which is not subject to the lower limit of $(4\pi)^{-1}$ applicable to $\Delta t \, \Delta f$. In fact, as far as is known at present, it may not have a positive lower limit at all.

Calling $\Delta x \, \Delta s$ the uncertainty product, we may investigate its numerical value for a particular case.

Let $f(x) = \exp(-\pi a^2 x^2)$. Then $ff^* = \exp(-2\pi a^2 x^2)$, for which the variance of x is $1/4\pi a^2$, and so $\Delta x = (4\pi a^2)^{-\frac{1}{2}}$. Similarly, for the transform, $FF^* = [a^{-1} \exp(-\pi s^2/a^2)]^2$ with variance $a^2/4\pi$ and $\Delta s = (4\pi/a^2)^{-\frac{1}{2}}$. Hence, for this case,

$$\Delta x \, \Delta s = \frac{1}{4\pi},$$

which is the minimum value permitted by the uncertainty relation in its quantitative form.

The finite difference

We define the finite difference of $f(x)$, taken over the interval a, to be

$$\Delta_a f(x) = f(x + \tfrac{1}{2}a) - f(x - \tfrac{1}{2}a).$$

The finite difference is often encountered in connection with functions that are tabulated at discrete intervals and so is defined only at discrete values of x. However, we may also take the finite difference of a function that is defined for all x, and in that case the finite difference is also a function of the continuous variable x. When the differencing interval a is small, then $\Delta_a f(x)$ is small but approaches proportionality to the derivative; when a is large, $\Delta_a f(x)$ may separate out into a displaced $f(x)$ followed by an inverted $f(x)$. These possibilities are illustrated in Figs. 8.14 and 8.15.

Since finite differencing in the x domain corresponds to a simple multiplying operation in the transform domain, it pays to be alert to the possibility of expressing functions as finite differences.

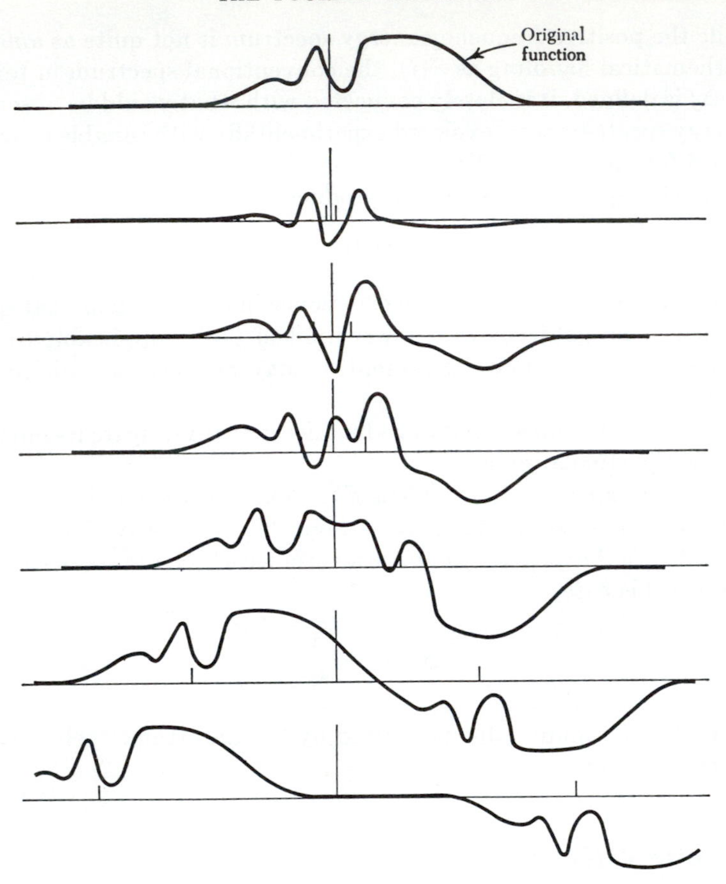

Fig. 8.14 Finite differences taken over different intervals.

Exercise Deduce the original functions of which the examples in Fig. 8.16 are the finite differences (if possible). Can you say whether your solutions are unique?

Exercise Discuss the question of whether $\Delta_a f(x)$ is an odd function.

The process of taking the finite difference will be recognized as the equivalent of convolution; that is,

$$\Delta_1 f(x) = 2\mathrm{I_I}(x) * f(x).$$

The Fourier transform of the odd impulse pair has already been discussed; hence we know that

$$\frac{1}{|a|}\,\mathrm{I_I}\left(\frac{x}{a}\right) \supset i \sin \pi as.$$

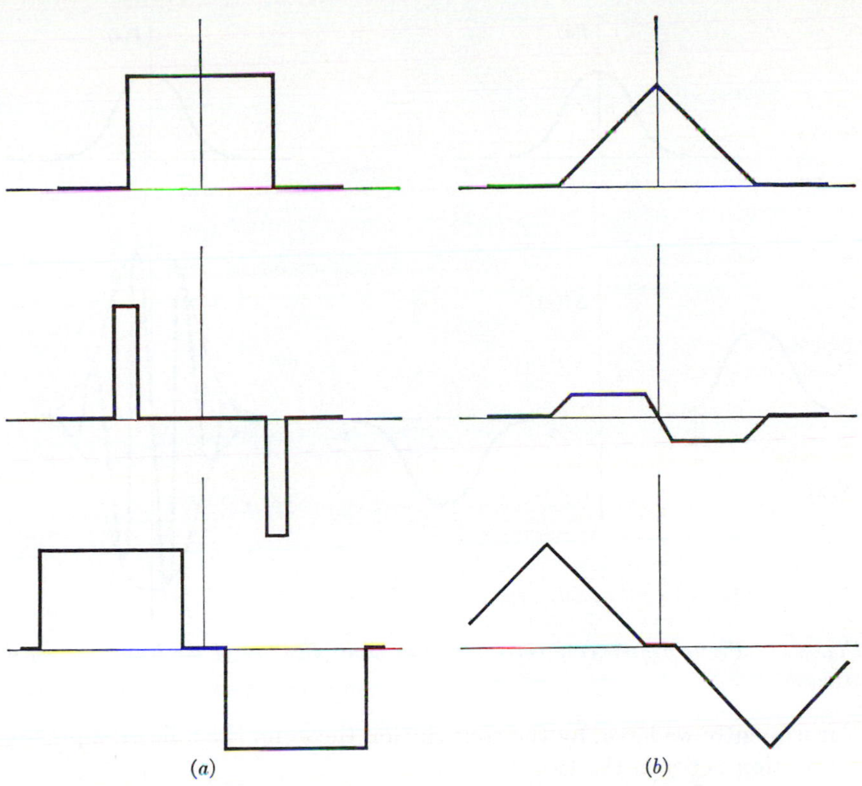

Fig. 8.15 Finite differences taken over short and long intervals of (a) rectangle function, (b) triangle function.

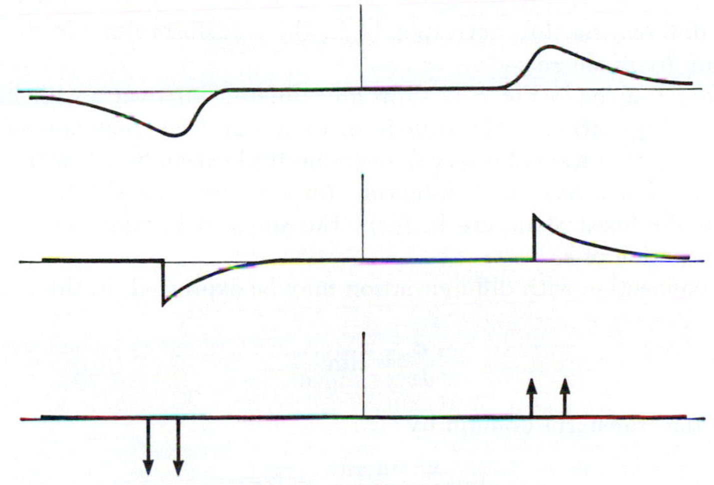

Fig. 8.16 Functions to be expressed as finite differences.

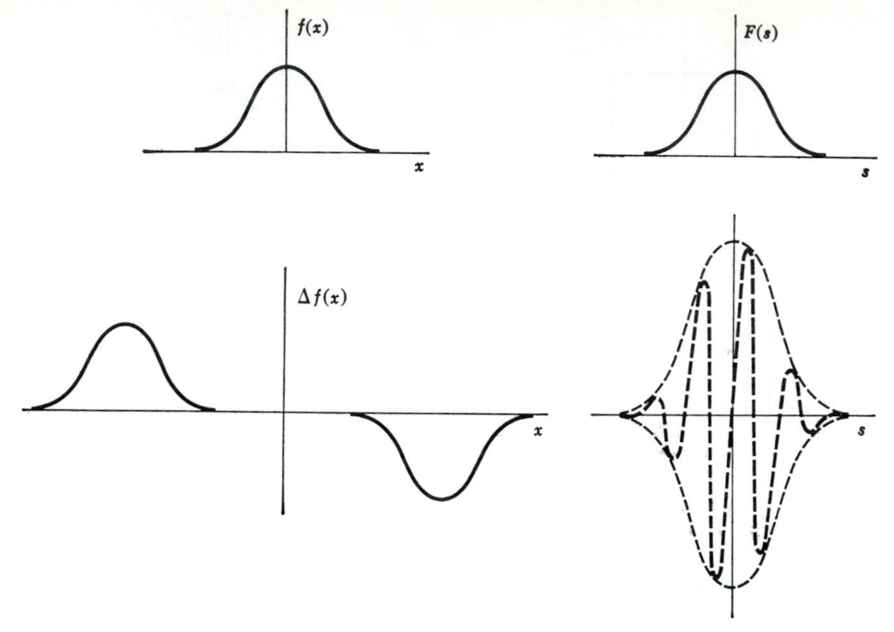

Fig. 8.17 Finite differencing corresponds to multiplication of the transform by a sinusoid.

Consequently we have, by the convolution theorem, the following precise connection between the two domains.

If $\Delta_a f(x)$ is the finite difference of $f(x)$, taken over the interval a, then the Fourier transform of $\Delta_a f(x)$ is $2i \sin \pi as\, F(s)$.

Finite differencing thus corresponds, in the transform domain, to multiplication by $2i \sin \pi as$.

Examples given in Fig. 8.17 show how differencing over a wide interval incurs multiplication of the transform by a high-frequency sinusoid, but differencing over a small interval incurs multiplication by a low-frequency sinusoid. For a *very* small interval—for example, one shorter than the scale of the finest structure in $f(x)$—the sinusoid becomes effectively a linear function of s.

The connection with differentiation may be expressed, in the x domain by

$$\frac{d}{dx} \equiv \lim_{a \to 0} \frac{\Delta_a}{a},$$

and in the transform domain by

$$\lim \frac{2i \sin \pi as}{a} = i2\pi s.$$

The second difference $\Delta^2_{aa}f(x)$ is defined by

$$\Delta^2_{aa}f(x) = \Delta_a[\Delta_a f(x)] = f(x + a) - 2f(x) + f(x - a).$$

Since $\Delta^2_{aa}f(x)$ is obtained by taking the finite difference of the finite difference, it follows that it corresponds in the transform domain to two successive multiplications by $2i \sin \pi as$. Hence the Fourier transform of $\Delta^2 f(x)$ is $-4 \sin^2 \pi as\, F(s)$.

Geometrical interpretations of the finite differences are given in Fig. 8.18. The first difference is proportional to the average slope and the second difference to the average curvature over finite intervals.

Exercise Show that $d^2/dx^2 \equiv a^{-2} \lim_{a\to 0} \Delta^2$.

Running means

In the reduction of meteorological data, such as rainfall data, the unimportant day-to-day fluctuations are customarily smoothed out to reveal the meaningful seasonal trend by taking running means. The running mean of $f(x)$ over the interval a is defined by

$$\frac{1}{a}\int_{x-\frac{1}{2}a}^{x+\frac{1}{2}a} f(x')\,dx'.$$

It is quite general practice, when recording data of any sort automatically, to smooth out unwanted rapid fluctuations; for example, a signal to be fed to a recording milliammeter is often passed through a smoothing filter. The running mean is one kind of smoothed value; it is generated by electronic counters that integrate an input voltage and print out the result at the end of fixed time intervals.

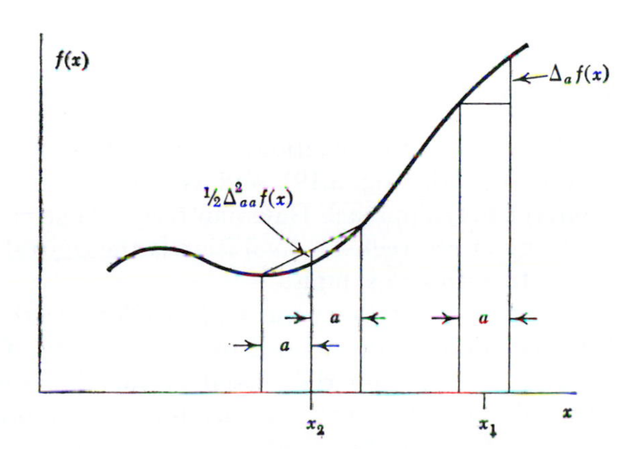

Fig. 8.18 Geometrical interpretation of the first difference of $f(x)$ at $x = x_1$ and the second difference at $x = x_2$.

It is evident that the running mean may be expressed in the form of a convolution with a rectangle function of base a and unit area:

$$a^{-1}\Pi\left(\frac{x}{a}\right) * f(x).$$

Since the Fourier transform of $a^{-1}\Pi(x/a)$ is sinc as, it follows from the convolution theorem that *the transform of the running mean of $f(x)$, taken over the interval a, is* sinc $as\, F(s)$. Thus the process of taking running means corresponds, in the transform domain, to multiplication by sinc as.

Exercise Interpret the properties

$$\lim_{a\to 0} \text{sinc}\, as = 1 \qquad \text{and} \qquad \lim_{a\to 0} \text{sinc}\, as = a^{-1}\delta(s)$$

in terms of running means.

Running means of higher order result from successive applications of the same process; thus the second-order running mean

$$a^{-1}\Pi\left(\frac{x}{a}\right) * a^{-1}\Pi\left(\frac{x}{a}\right) * f(x)$$

has a Fourier transform

$$\text{sinc}^2\, as\, F(s).$$

Recognizing sinc^2 as as the transform of $a^{-1}\Lambda(x/a)$, we note that the second-order running mean may be regarded as a *weighted* running mean generated by convolution with $a^{-1}\Lambda(x/a)$.

A number of successive running means of an original rectangle function, shown in Fig. 8.19 together with the successive transforms, illustrate the relation

$$a^{-1}\Pi\left(\frac{x}{a}\right) * \ . \ . \ . \ n \text{ times } . \ . \ . \ * a^{-1}\Pi\left(\frac{x}{a}\right) \supset \text{sinc}^n\, as.$$

Central-limit theorem

If a large number of functions are convolved together, the resultant may be very smooth (Fig. 8.19), and as the number increases indefinitely, the resultant may approach Gaussian form. The rigorous statement of this tendency of protracted convolution is the central-limit theorem. Let us illustrate with an example.

In Fig. 8.19 two sequences of profiles are shown, each approaching Gaussian form in the central part as $n \to \infty$. Over a small range of s about $s = 0$ the top right-hand profile sinc s is approximated by the parabolic curve $1 - ms^2$, and therefore the nth profile is approximated by

$(1 - ms^2)^n$. Now

$$(1 - ms^2)^n = \left[1 - nms^2 + \frac{n(n-1)}{1.2} m^2 s^4 - \cdots\right]$$
$$= \left[1 - nms^2 + \frac{n^2 m^2 s^4}{1.2} - \cdots\right] + \text{remainder}$$
$$= e^{-nms^2} + \text{remainder}.$$

Keeping s fixed and letting n tend to infinity we see that we have a Gaussian term which continually gets narrower, its width varying as $n^{-\frac{1}{2}}$, plus a remainder of diminishing relative importance. The central curvature of the right-hand profile increases proportionally to n.

The Gaussian term has a Gaussian transform

$$\left(\frac{\pi}{nm}\right)^{\frac{1}{2}} e^{-\pi^2 x^2/nm},$$

which continually gets wider, the variance increasing proportionally to n; this, of course, is a consequence of the property that variances add under convolution.

The interesting phenomenon is the tendency toward Gaussian form under successive self-convolution and also under successive self-multiplication.

Exercise Graph $\cos^{10} s$ and $[J_0(s)]^{10}$ for small s.

How the remainder term behaves in general is discussed by Khinchin,[2] and obviously not all original profiles lead to the Gaussian result under continued self-convolution; for example, $\Pi(x) \sin 2\pi x$ does not. However, the functions which do manage to smooth themselves out into regular Gaussian form are numerous and include unsymmetrical functions such as $e^{-x}H(x)$ and rather unlikely looking discontinuous functions, such as $\Pi(x)$. Even $\text{II}(x)$, which has two infinite discontinuities, comes close. The sometimes surprising rapidity with which the approach to Gaussian form sets in is interesting to investigate numerically by trial. [For example, form the successive serial products $(1\ 1)^{*n}$, $(1\ 1\ 1\ 1)^{*n}$, $(3\ 2\ 1)^{*n}$, and test by plotting the logarithms of the terms against the square of the distance from the center.]

The central-limit theorem in its general form states that under applicable conditions the convolution of n functions (not necessarily all the same as in the discussion above) is equal to a Gaussian function whose variance is the sum of the variances of the different functions, plus a remainder which diminishes with increasing n in a certain way.

Without studying at length the conditions of applicability, we can see

[2] A. Ya. Khinchin, "Statistical Mechanics," Dover Publications, Inc., New York, 1949.

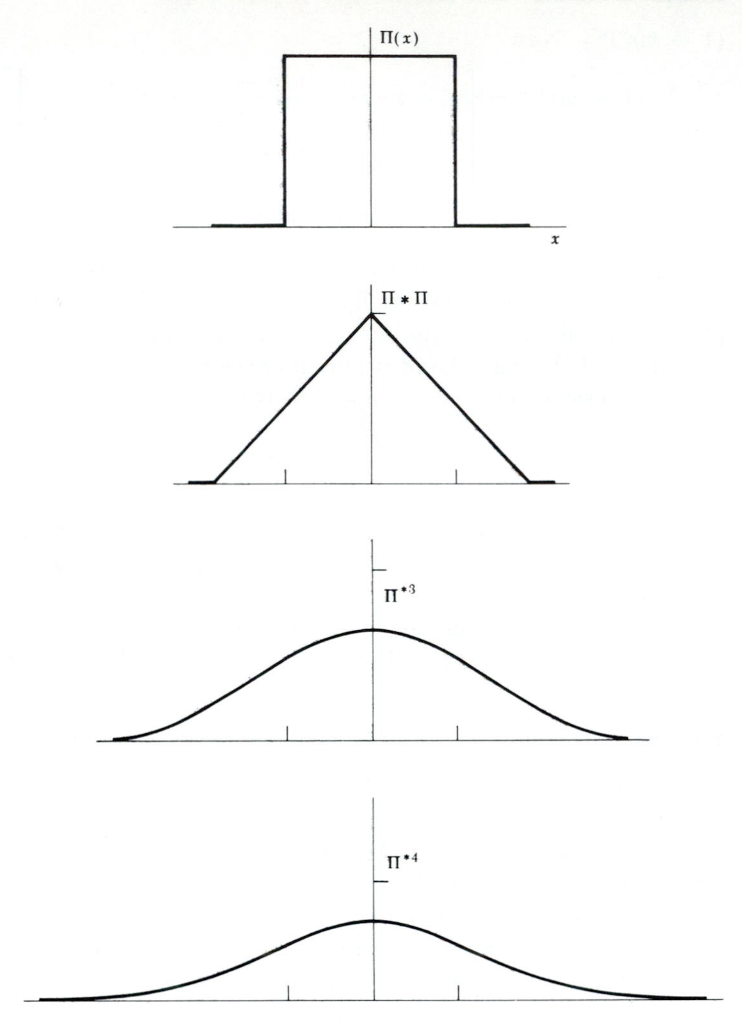

that the essential thing is that the Fourier transform of each of the functions entering into the convolution should exhibit the humped kind of behavior at the origin which is possessed by our example; that is, it should be of the form $a - bs^2$. The quantity a must not be zero; that is, no one of the n functions may have zero area, or else the product of the transforms will be zero at $s = 0$, and the total convolution will be reduced to zero area. Similarly, it must not be infinite. It does not matter if b is occasionally zero, but it should in no case be infinite. Clearly, nothing resembling $\Lambda(s)$ at the origin should be included, for one such factor would leave a permanent corner at the origin, thus preventing approach to the Gaussian form.

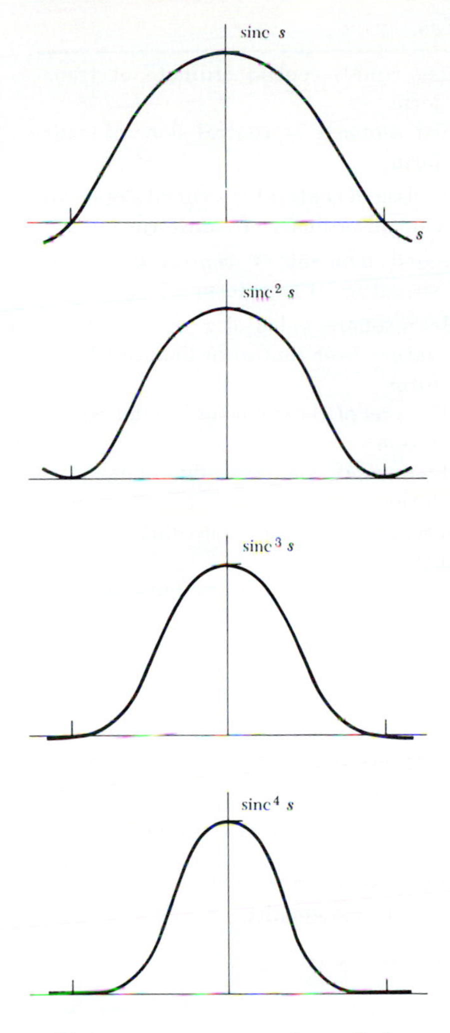

Fig. 8.19 Successive running means and their transforms.

It is not necessary that all the transforms have zero slope at the origin, but this could always be ensured by referring each function to its centroid, provided that the slope was not infinite.

The conditions may be expressed in the x domain as follows. First consider the case of identical real functions $f(x)$,

$$f(x) * f(x) * \ . \ . \ . \ * f(x),$$

and eliminate the possibility of $f(x)$ having negative area. Then the first condition ($a \neq 0$ or ∞) becomes

$$0 < \int_{-\infty}^{\infty} f(x) \, dx < \infty.$$

Table 8.5 Table of correspondences

$\int_{-\infty}^{\infty} f(x)\,dx = F(0)$	Area equals central ordinate of transform
$\int_{-\infty}^{\infty} xf(x)\,dx = \dfrac{F'(0)}{-2\pi i}$	First moment $\propto$ central slope of transform
$\langle x \rangle = -\dfrac{F'(0)}{2\pi i F(0)}$	Abscissa of centroid $\propto$ central slope over central ordinate of transform
$\int_{-\infty}^{\infty} x^2 f(x)\,dx = -\dfrac{F''(0)}{4\pi^2}$	Second moment $\propto$ central downward curvature of transform
$\langle x^2 \rangle = -\dfrac{F''(0)}{4\pi^2 F(0)}$	Mean-square value of $x \propto$ central curvature over central ordinate of transform
$\langle x \rangle_{f*g} = \langle x \rangle_f + \langle x \rangle_g$	Abscissas of centroids add under convolution
$\langle x^2 \rangle_{f*g} = \langle x^2 \rangle_f + \langle x^2 \rangle_g$	Mean-square abscissas add under convolution
$\sigma^2_{f*g} = \sigma^2_f + \sigma^2_g$	Variances add under convolution

If the kth derivative of $f(x)$ becomes impulsive, $F(s) \sim |s|^{-k}$.

If the mth derivative of $f(x)$ and the nth derivative of $g(x)$ become impulsive, the $(m+n)$th derivative of $f * g$ becomes impulsive.

$W_f = \dfrac{\int_{-\infty}^{\infty} f(x)\,dx}{f(0)}$	The equivalent width of $f(x)$ is its area over its central ordinate
$W_f = \dfrac{1}{W_F}$	The equivalent width of a function is the reciprocal of the equivalent width of its transform
$\lvert f(x)\rvert \leqslant \int_{-\infty}^{\infty} \lvert F(s)\rvert\,ds$	Maximum possible magnitude
$\lvert f'(x)\rvert \leqslant 2\pi \int_{-\infty}^{\infty} \lvert sF(s)\rvert\,ds$	Maximum possible slope
$[\int f(x)g(x)\,dx]^2 \leqslant \int [f(x)]^2\,dx \int [g(x)]^2\,dx$	Schwarz's inequality
$\sigma_{\lvert f\rvert^2}\sigma_{\lvert F\rvert^2} \geqslant \dfrac{1}{4\pi}$	Uncertainty relation
$\Delta f(x) = 2\mathrm{I_I}(x) * f(x)$	Finite differencing is convolution with odd impulse pair
$\Delta_a f(x) \supset 2i \sin \pi as\, F(s)$	Finite differencing over interval a corresponds, in the transform domain, to multiplication by $2i \sin \pi as$. (Note: Differentiation corresponds to multiplication by $i2\pi s$.)
$a^{-1}\Pi\left(\dfrac{x}{a}\right) * f(x) \supset \operatorname{sinc} as\, F(s)$	Taking running means over interval a corresponds to multiplication by sinc as
$\lim_{n\to\infty} f_1 * f_2 * \cdots * f_n = e^{-ax^2}$	The convolution of many functions approaches Gaussian form, provided that their transforms are humped at the origin (central-limit theorem)

The condition on $F'(0)$ becomes

$$\int_{-\infty}^{\infty} xf(x)\,dx = 0.$$

The requirement that $F(s)$ have a finite nonpositive second derivative (b finite) becomes

$$0 \leqslant \int_{-\infty}^{\infty} x^2 f(x)\,dx < \infty.$$

The possibility should also be considered that there is a discontinuity in the second derivative; but if $f(x)$ is real, the real part of $F(s)$, being even, must have the same curvature to each side of the origin. Therefore any discontinuity must be confined to the imaginary part of $F''(s)$. But the imaginary part of $F(s)$ is zero at $s = 0$; therefore any effect of a discontinuity in $F''(s)$ at $s = 0$ will die out in the limit. It is therefore sufficient to require $f(x)$ to have finite area, finite mean, and finite variance.

In the event of nonidentical functions being convolved, a finite absolute third moment is required plus a more elaborate condition due to Lyapunov to ensure that the third moments are not too strong.

Exercise Consider the behavior of $(\text{sinc}\, x)^{*n}$, $(\text{sinc}^2\, x)^{*n}$, $[(1 + x^2)^{-1}]^{*n}$, $[x\Pi(x)]^{*n}$, $[\Pi(x)\sin x]^{*n}$, $[e^{-\alpha x}\sin \beta x H(x)]^{*n}$.

Summary of correspondences in the two domains

The results of the preceding discussion are tabulated in Table 8.5 for reference.

Problems

1 Deduce a simple expression for $\exp(-x^2) * \exp(-x^2)$.

2 Show that the self-convolution of $(1 + x^2)^{-1}$ is identical with itself, except for scale factors. Show that the self-convolution is twice as wide as the original function, that is, that the width is additive under convolution in this case, and reconcile this with the known general fact that variance is additive.

3 Investigate the functional form of $(1 + x^2/a^2)^{-1} * (1 + x^2/b^2)^{-1}$ and its width in terms of the widths of the convolved functions.

4 Show by direct integration that

$$\Lambda(x)H(x) \supset \frac{1}{i2\pi s} - \frac{e^{-i\pi s}\,\text{sinc}\, s}{i2\pi s};$$

verify the result by applying first the addition and shift theorems to find the

Fourier transform of $(d/dx)[\Lambda(x)H(x)]$, and then the derivative theorem in reverse.

5 By separation of the preceding transform into real and imaginary parts, show that

$$\Lambda(x)H(x) \supset \tfrac{1}{2}\operatorname{sinc}^2 s + \frac{1 - \cos \pi s \operatorname{sinc} s}{i2\pi s};$$

as checks on the algebra, verify that the central ordinate and slope are connected in the appropriate way with the area and first moment of $\Lambda(x)H(x)$, respectively, and note whether the transform is indeed hermitian. Break $\Lambda(x)H(x)$ into its even and odd parts, and obtain the transform of each separately.

6 Because of the irregular variation in the number of sunspots, the sequence of daily sunspot numbers is smoothed by taking five-day running totals; that is, for each day we add the sunspot number for the preceding and following two days. Here is a sequence of five-day running totals beginning Jan. 1, 1900:

45, 35, 25, 15, 5, 0, 0, 0, 0, 15, 50, 80, 100, 125, 125, 100, 80, 70, 45, 30, 30, 30, 35, 60, 80, 90, 95, 100, 90, 85, 75.

From this smoothed sequence, what can be deduced about the actual daily values?

7 Show that

$$W_{f\star g} = \frac{W_f W_g}{W_{fg}},$$

where W_f is the equivalent width of $f(x)$.

8 Show that squares of equivalent widths are additive under convolution of Gaussian functions.

9 Show that the equivalent width of $4 \operatorname{sinc}^2 2s - \operatorname{sinc}^2 s$ is $\frac{1}{3}$, and calculate the equivalent width of $\operatorname{sinc} s + \operatorname{sinc}^2 2s$.

10 Investigate the properties of the Jones bandwidth[3] of $f(x)$, which is defined as

$$\frac{\int_0^\infty FF^*\, ds}{F_{\max}}.$$

11 Show that the autocorrelation width of $f(x)H(x)$ is twice that of $f(x)$.

12 Verify the following autocorrelation widths.

Function:	$x\Pi(x - \frac{1}{2})$	$\Pi(2x + \frac{3}{4}) + \Pi(2x - \frac{3}{4})$	$e^{-x^2/2\sigma^2}$
Width:	$\frac{3}{4}$	1	$2\pi^{\frac{1}{2}}\sigma$

13 The abscissa x of a function $f(x)$ is divided into a finite number of finite segments (plus two semi-infinite end segments). The finite segments are rearranged without overlapping, thus defining a new function which we may describe as

[3] R. C. Jones, Phenomenological Discussion of the Response and Detecting Ability of Radiation Detectors, *Proc. IRE*, vol. 47, p. 1496, 1959.

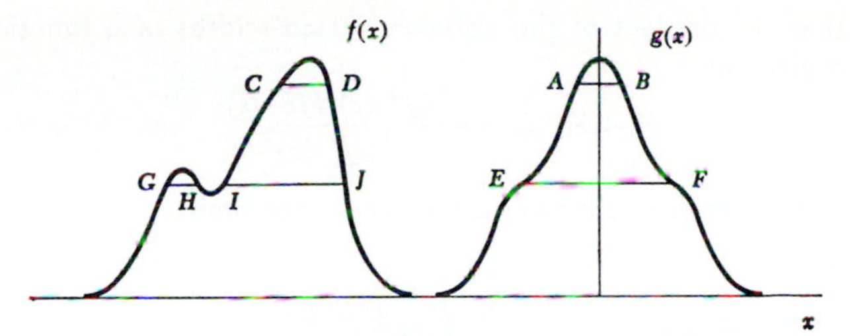

Fig. 8.20 Steiner symmetrization.

derivable from $f(x)$ by shuffling. For example, the string of 11 pulses

$$\sum_{n=-5}^{n=+5} \Pi(11x - n)$$

and the function $|x|\Pi(x/2)$ are derivable respectively from $\Pi(x)$ and $\Lambda(x)$ by shuffling, and vice versa. Show that the equivalent width is unaffected by shuffling if the segment containing $x = 0$ is not dislodged in the shuffle, and that the autocorrelation width is not affected in any case. Consider the effect of shuffling on the total energy of a waveform, and mention several parameters of its power spectrum which are invariant under shuffling.

14 An even function $g(x)$ is derived from a function $f(x)$ by the process of symmetrization illustrated in Fig. 8.20, where $AB = CD$, $EF = GH + IJ$, and so on. This process, known as Steiner symmetrization, was used by Jacob Steiner to prove that the circle is the figure of minimum perimeter for a given area. Show that the autocorrelation functions of $f(x)$ and $g(x)$ have the same equivalent width.

15 State the relationship between the functions $\exp(-x)H(x)$ and $\exp(-|x|)$ in the light of symmetrization, and say how their respective power spectra are related.

16 Show that the convolution of two odd functions is even.

17 Establish the following relations between functions and their autocorrelation functions.

Function	*Autocorrelation*
$\Pi(x)$	$\Lambda(x)$
$e^{-\pi x^2}$	$2^{-\frac{1}{2}}e^{-\frac{1}{2}\pi x^2}$
$\delta(x)$	$\delta(x)$
$e^{-x}H(x)$	$\frac{1}{2}e^{-\lvert x\rvert}$
$e^{-\lvert x\rvert}$	$e^{-\lvert x\rvert}(1 + \lvert x\rvert)$

18 Show by a simple argument in the Fourier transform domain that the autocorrelation function of $\exp(-\pi x^2)\cos\omega x$ is $2^{-\frac{1}{2}}\exp(-\frac{1}{2}\pi x^2)\cos\omega x$ when ω is large.

19 Some difficulty arises over the autocorrelation of $\cos x$. Show that there is a sense in which the autocorrelation of a cosine function is also a cosine function.

20 Show that the product of the autocorrelation widths of a function and its transform is given by

$$W_{f\star f^*} W_{F\star F^*} = \frac{(\int f\, dx)^2 (\int F\, ds)^2}{(\int f^2\, dx)^2}$$

and that the product does not have a nonzero lower limit.

21 Show that

$$|f * g| \leqslant \int_{-\infty}^{\infty} |FG|\, ds.$$

Chapter 9 Electrical waveforms, spectra, and filters

In this chapter the response of a linear system, such as an electrical filter, is considered from the point of view of resolution of the input into exponential, or harmonic, components, and an equivalent way of looking at filter action in the time domain is exhibited. The fundamental connections between linearity and convolution are also derived.

Electrical waveforms and spectra

An electrical waveform $V(t)$ is a single-valued real function of time t subject only to the general requirement of representing physically possible time dependence. Yet it is firmly established usage to speak as though steps, impulses, and absolutely monochromatic signals were electrical waveforms, even though it is clearly not possible to generate discontinuous, infinite, or eternal currents or voltages. When we make statements about the behavior of circuits in the presence of steps or impulses, we imply that a more rigorous statement exists in which that behavior would appear as the limiting form of behavior under a sequence of stimuli approaching a discontinuous step or infinite pulse. The same applies to statements about behavior in the presence of alternating or direct current.

The spectrum $S(f)$ of the electrical waveform $V(t)$ is defined as its Fourier transform (or transform in the limit as necessary):

$$S(f) = \int_{-\infty}^{\infty} V(t)e^{-i2\pi ft}\,dt$$

whence

$$V(t) = \int_{-\infty}^{\infty} S(f)e^{i2\pi ft}\,df.$$

Since $V(t)$ is by definition restricted to real functions, $S(f)$ is also subject to a restriction on its generality. The real part of a spectrum func-

Table 9.1 Corresponding restrictions

Restriction on waveform			Restriction on spectrum
Real	$\text{Im } V(t) = 0$		$\text{Re } S(f)$ even and $\text{Im } S(f)$ odd
Switches on	$V(t) = 0$	$t < 0$	$\text{Im } S(f)$ is Hilbert transform of $\text{Re } S(f)$
Nonnegative	$V(t) \geqslant 0$		Complicated
Finite energy	$\int_{-\infty}^{\infty} V^2 \, dt = W$		$\int_{-\infty}^{\infty} SS^* \, df = W$
Finite duration	$V(t) = 0$	$\|t\| > T$	$S(f)$ fully determined by $\text{III}(Tf)S(f)$
Band-limited	$V(t)$ fully determined by $\text{III}(Mt)V(t)$		$S(f) = 0, \ \|f - f_0\| \geqslant M$

tion $S(f)$ must always be even and its imaginary part odd (see Chapter 2); that is, $S(f)$ is hermitian.

Waveforms may also be subject on occasion to special restrictions of which some of the most common are given in Table 9.1.

The transform formulas for waveforms and spectra have been written in system 1 of Chapter 2. They are sometimes written in system 2 and then appear as

$$S(\omega) = \int_{-\infty}^{\infty} V(t)e^{-i\omega t} \, dt$$

$$V(t) = \frac{1}{2\pi} \int_{-\infty}^{\infty} S(\omega)e^{i\omega t} \, d\omega.$$

The substitution of ω for $2\pi f$ preserves the physical character of the formulas, factor by factor, except for the coefficient $1/2\pi$, which, of course, may be written in front of either integral according as x and s in the system 2 formulas are identified with t and ω, or vice versa. The form given here is widely used, some users remembering the location of the unsymmetrical coefficient by the fact that if the factor $d\omega/2\pi$ were replaced by df the formulas would be symmetrical. The fact that the system 1 formulas are unchanged on generalization to more than one dimension carries no weight in circuit theory, where t has no higher-order generalization.

One also encounters the symmetrical system 3, in the form

$$S(\omega) = \frac{1}{(2\pi)^{\frac{1}{2}}} \int_{-\infty}^{\infty} V(t)e^{-i\omega t} \, dt$$

$$V(t) = \frac{1}{(2\pi)^{\frac{1}{2}}} \int_{-\infty}^{\infty} S(\omega)e^{i\omega t} \, d\omega.$$

An interesting pair of formulas with total symmetry was proposed by Hartley:

$$\psi(\omega) = \frac{1}{(2\pi)^{\frac{1}{2}}} \int_{-\infty}^{\infty} V(t)(\cos \omega t + \sin \omega t)\, dt$$

$$V(t) = \frac{1}{(2\pi)^{\frac{1}{2}}} \int_{-\infty}^{\infty} \psi(\omega)(\cos \omega t + \sin \omega t)\, d\omega.$$

It is evident that $\cos \omega t + \sin \omega t$ is a Fourier kernel, as defined in Chapter 12, but $\psi(\omega)$ is not the Fourier transform of $V(t)$ and does not obey all the theorems; for example, it does not obey Rayleigh's theorem.

Filters

We shall use the term "filter" here to denote a system having an input and an output. Other equivalent terms are transducer, fourpole, four-terminal network, and two-port element. Although electrical terminology will be used, the present considerations will usually apply to mechanical and acoustical transducers and to equivalent devices in other fields where vibrations or oscillations are transmitted. Of course, even within the electrical field, filters may assume a wide variety of physical embodiments, from clusters of wire coils and capacitors to intricate geometrical structures in waveguide.

When a waveform $A \cos 2\pi ft$ is fed into a linear time-invariant electrical filter, the output is also harmonic, as will be proved later, but has in general a different amplitude and phase; let it be $B \cos (2\pi ft + \phi)$. Then the filter is completely specified by a certain frequency-dependent complex quantity $T(f)$, whose amplitude is given by B/A and whose phase is ϕ; thus

$$T(f) = \frac{B}{A} e^{i\phi}.$$

We refer to $T(f)$ as the transfer factor of the filter.

When an input $V_1(t)$ is applied to the filter and it is desired to calculate the output $V_2(t)$, one analyzes $V_1(t)$ into its spectrum, multiplies each spectral component by the corresponding transfer factor to obtain the spectrum of $V_2(t)$, and then synthesizes $V_2(t)$ from its spectrum. Thus

$$S_2(f) = T(f)S_1(f),$$

and then

$$V_2(t) = \int_{-\infty}^{\infty} T(f)S_1(f)e^{i2\pi ft}\, df. \tag{1}$$

Since multiplication of transforms corresponds to convolution of original functions, it follows that $V_2(t)$ may be derived directly from $V_1(t)$

by smoothing with a certain function which would be characteristic of the filter, that is,

$$V_2(t) = I(t) * V_1(t). \tag{2}$$

where $I(t)$ is the Fourier transform of $T(f)$.

These two procedures are shown schematically as follows:

$$\begin{array}{rcl}
V_1(t) \text{ — transform} \rightarrow S_1(f) & \qquad & V_1(t) \\
| & & | \\
\text{multiply by} & & \text{convolve with} \\
T(f) & & I(t) \leftarrow \text{transform — } T(f) \\
\downarrow & & \downarrow \\
V_2(t) \leftarrow \text{transform — } S_2(f) & & V_2(t)
\end{array}$$

Combining all the quantities into one diagram with time functions on the left and transforms on the right in accordance with the convention, we can summarize the interrelations concisely as follows:

$$\begin{array}{ccc}
V_1(t) & \supset & S_1(f) \\
\text{convolution} & & \text{multiplication} \\
I(t) & \supset & T(f) \\
\downarrow & & \downarrow \\
V_2(t) & \supset & S_2(f)
\end{array}$$

The characteristic waveform $I(t)$ associated with the filter is seen from (1) to be the output obtained when $S_1(f) = 1$, that is, when $V_1(t) = \delta(t)$. It is known in communications as the "impulse response" of the filter. For many purposes it is as useful as the frequency characteristic $T(f)$ as a means of specifying a filter, and it may be much easier to obtain experimentally. Equation (2) simply breaks up $V_1(t)$ into a sequence of impulses and expresses $V_2(t)$ as a sum of the responses for each component impulse.

A third means of specifying a filter is by its "step response," that is, the output corresponding to the switching on of a constant signal $V_1(t) = H(t)$. Call this response $A(t)$. Then, regarding $V_1(t)$ as broken up into a sequence of steps $V_1'(\tau)H(t - \tau)$, we have

$$V_2(t) = A(t) * V_1'(t). \tag{3}$$

Transforming this relation and comparing with (1), we have the relation of the step-function response to the transmission characteristic of the filter:

$$S_2(f) = \bar{A}(f)i2\pi f S_1(f); \tag{4}$$

therefore

$$T(f) = i2\pi f \bar{A}(f),$$

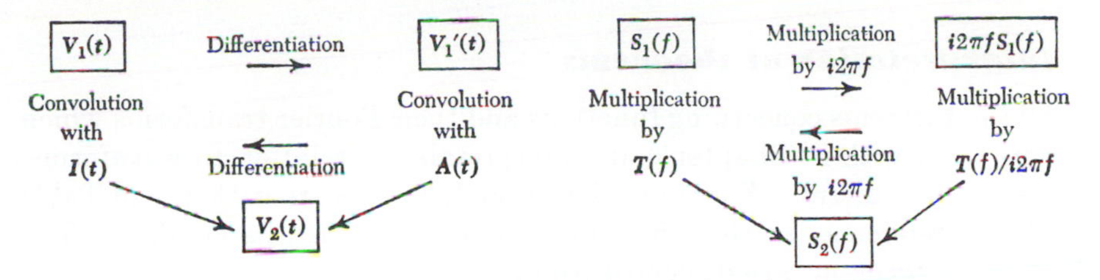

Fig. 9.1 Operations involving the step and impulse responses, and corresponding operations in the transform domain.

whence by a further transformation we have the relation between the step-function response and the impulse response:

$$I(t) = \frac{d}{dt} A(t).$$

Retransforming (4) with a different choice of factors, we may obtain

$$V_2(t) = A'(t) * V_1(t),$$

or formulas involving higher derivatives and integrals, including those of fractional order.

Figure 9.1 shows the relationships involving the step response $A(t)$.

The procedure with filters is summarized in the following table.

Analyze input into	*Specify filter by*	*which has the Fourier transform*
Sine and cosine waves	Transmission characteristic $T(f)$	$I(t)$
Impulses	Impulse response $I(t)$	$T(f)$
Steps	Step response $A(t)$	$T(f)/2\pi i f$

The Fourier integral thus arises in filter theory in any of the following forms:

$$V(t) = \int_{-\infty}^{\infty} S(f) e^{i2\pi ft}\, df$$
$$I(t) = \int_{-\infty}^{\infty} T(f) e^{i2\pi ft}\, df$$
$$A(t) = \frac{1}{2\pi i} \int_{-\infty}^{\infty} \frac{T(f)}{f} e^{i2\pi ft}\, df,$$

and the convolution integral occurs in

$$V_2(t) = \int_{-\infty}^{\infty} I(t-\tau) V_1(\tau)\, d\tau = I(t) * V_1(t) = A'(t) * V_1(t)$$
$$V_2(t) = \int_{-\infty}^{\infty} A(t-\tau) V_1'(\tau)\, d\tau = A(t) * V_1'(t).$$

Interpretation of theorems

All the theorems concerning functions and their Fourier transforms which were dealt with in Chapter 6 are interpretable in relation to waveforms, spectra, and filters. For convenience the theorems are gathered in Table 9.2 for reference, in the symbology appropriate to this chapter. The first four theorems are discussed below.

Similarity theorem The reciprocal relationship of period and frequency is evident in this theorem, for compression of the time scale by a given factor compresses the periods of all harmonic components equally and therefore raises the frequency of every component by the same factor. Since $V(0)$ remains unaffected by time-scale changes, the area under the spectrum must, by the definite-integral theorem, remain constant, hence the compensating factor $|a|^{-1}$, which appropriately weakens the spectrum if it spreads to higher frequencies.

Addition theorem Even when translated into the language of electrical waveforms and spectra, this theorem is simply an expression of the linearity of the Fourier transformation. It has nothing to do with the linearity of the systems in which the waveforms are found; it must, of course, be true even of waveforms in nonlinear circuits. The linear property of the transformation makes it suitable, however, for dealing with linear problems.

Shift theorem When a waveform $V(t)$ is delayed by a given time interval T, its harmonic components are affected in different ways; for example, a component whose frequency is equal to or is an integral multiple of T^{-1} is not affected at all. Components whose period is much greater than T are not affected much, but those with periods short compared with T may be seriously altered in phase. In general, we can say that the component of period $T_1 = f_1^{-1}$ will be unchanged in amplitude but delayed in phase by $2\pi T/T_1$. Hence each component $S(f_1)$ becomes $\exp(-i2\pi f_1 T)S(f_1)$.

Modulation theorem In the ordinary method of imposing an audio-frequency tone of angular frequency ω on a carrier wave of angular frequency Ω, the modulated waveform is

$$(1 + M \cos \omega t) \cos \Omega t,$$

where M is the depth of modulation. This waveform differs from the unmodulated carrier $\cos \Omega t$ by the addition of $M \cos \Omega t \cos \omega t$, a quantity in the form to which the modulation theorem applies.

The modulation theorem states that if a waveform $V(t)$ has a transform $S(f)$, then the waveform $V(t) \cos \omega t$ is derivable by splitting $S(f)$ into two

Table 9.2 Theorems for waveforms and their spectra

Theorem	*Waveform V(t)*	*Spectrum S(f)*
Similarity	$V(at)$	$\frac{1}{\lvert a\rvert} S\left(\frac{f}{a}\right)$
Addition	$V_1(t) + V_2(t)$	$S_1(f) + S_2(f)$
Shift	$V(t - T)$	$e^{-i2\pi fT}S(f)$
Modulation	$V(t) \cos \omega t$	$\frac{1}{2} S\left(f - \frac{\omega}{2\pi}\right) + \frac{1}{2} S\left(f + \frac{\omega}{2\pi}\right)$
Convolution	$I(t) * V_{\text{in}}$	$T(f)S_{\text{in}}(f)$
Autocorrelation	$V_1(t) * V_1(-t)$	$\lvert S(f)\rvert^2$
Differentiation	$V'(t)$	$i2\pi fS(f)$
	$-i2\pi tV(t)$	$S'(f)$
Finite difference	$\Delta V(t) = V(t + \frac{1}{2}T) - V(t - \frac{1}{2}T)$	$2i \sin \pi Tf\, S(f)$
Second difference	$\Delta^2 V(t)$	$-4 \sin^2 \pi Tf\, S(f)$
Running means	$\frac{1}{T} \Pi \left(\frac{t}{T}\right) * V(t)$	$\frac{\sin \pi fT}{\pi fT} S(f)$
Rayleigh	$\int_{-\infty}^{\infty} [V(t)]^2\, dt = \int_{-\infty}^{\infty} SS^*\, df$	
Energy	$\int_{-\infty}^{\infty} V_1(t)V_2(t)\, dt = \int_{-\infty}^{\infty} S_1S_2^*\, df$	
Definite integral	$\int_{-\infty}^{\infty} V(t)\, dt = S(0)$	
Center of gravity	$\langle t\rangle = \frac{\int_{-\infty}^{\infty} tV(t)\, dt}{\int_{-\infty}^{\infty} V(t)\, dt} = -\frac{S'(0)}{2\pi iS(0)}$	
First moment	$\int_{-\infty}^{\infty} tV(t)\, dt = \frac{S'(0)}{-2\pi i}$	
Moment of inertia	$\int_{-\infty}^{\infty} t^2V(t)\, dt = \frac{S''(0)}{-4\pi^2}$	
Moment of nth order	$\int_{-\infty}^{\infty} t^nV(t)\, dt = \frac{S^{(n)}(0)}{(-2\pi i)^n}$	
Equivalent width	$\int_{-\infty}^{\infty} \frac{V(t)}{V(0)}\, dt \int_{-\infty}^{\infty} \frac{S(f)}{S(0)}\, df = 1$	
Inequalities	$\lvert V(t)\rvert \leq \int_{-\infty}^{\infty} \lvert S(f)\rvert\, df$	
	$V'(t) \leq 2\pi \int_{-\infty}^{\infty} \lvert fS(f)\rvert\, df$	

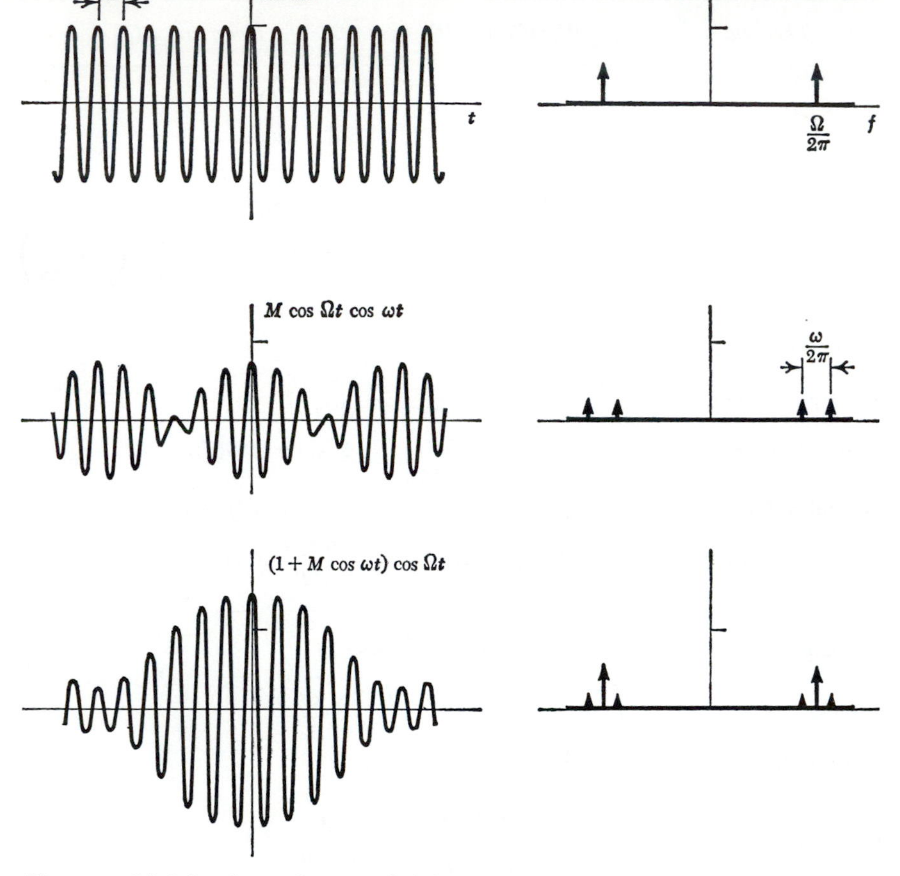

Fig. 9.2 Modulated waveforms and their spectra.

halves, one of which is slipped to the right by an amount $\omega/2\pi$, the other going an equal distance to the left. Hence if

$$V(t) = M \cos \Omega t,$$

for which

$$S(f) = \frac{1}{2} M \delta\left(f + \frac{\Omega}{2\pi}\right) + \frac{1}{2} M \delta\left(f - \frac{\Omega}{2\pi}\right),$$

then the transform of $M \cos \Omega t \cos \omega t$ is

$$\frac{1}{4} M \delta\left(f + \frac{\Omega}{2\pi} + \frac{\omega}{2\pi}\right) + \frac{1}{4} M \delta\left(f + \frac{\Omega}{2\pi} - \frac{\omega}{2\pi}\right) + \frac{1}{4} M \delta\left(f - \frac{\Omega}{2\pi} + \frac{\omega}{2\pi}\right) + \frac{1}{4} M \delta\left(f - \frac{\Omega}{2\pi} - \frac{\omega}{2\pi}\right),$$

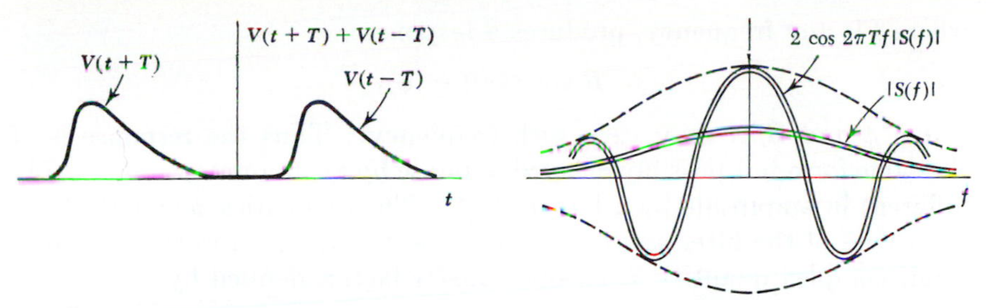

Fig. 9.3 The converse of the modulation theorem.

and this latter expression must be added to the spectrum of the unmodulated carrier cos Ωt (addition theorem) to give the spectrum of the amplitude-modulated wave $(1 + M \cos \omega t) \cos \Omega t$. In this way the familiar sidebands of simple modulation theory are reproduced (see Fig. 9.2).

Converse of modulation theorem Two identical signals are sent out in succession; the spectrum of the composite signal is obtained from the spectrum of one signal alone by multiplication by a cosine function of frequency. Thus

$$V(t + T) + V(t - T) \supset 2 \cos 2\pi Tf \, S(f).$$

In this expression of the converse theorem the time origin has been chosen midway between the origins of the separate signals, but one could also write (see Fig. 9.3)

$$V(t) + V(t - 2T) \supset 2e^{-i2\pi Tf} \cos 2\pi Tf \, S(f).$$

Both forms are derivable directly from the shift theorem.

Linearity and time invariance

Suppose that $V_2(t)$ is the response of a filter to a stimulus $V_1(t)$, and that $W_2(t)$ is the response to $W_1(t)$. Then the filter is said to be linear if the response to $V_1(t) + W_1(t)$ is $V_2(t) + W_2(t)$, irrespective of the choice of $V_1(t)$ and $W_1(t)$. Sometimes a condition is added that $aV_1(t)$ shall have a response $aV_2(t)$ for all a and $V_1(t)$, but one can deduce this relation from the superposition property (proving it first where a is an integer, then a ratio of integers[1]).

Time invariance means that the response to $V_1(t - T)$ is $V_2(t - T)$ for all T and $V_1(t)$.

As a consequence of linearity and time invariance a stimulus $A \cos 2\pi ft$,

[1] For a guide to the mathematical considerations that arise when the possibility of a voltage being irrational is not excluded *ab initio*, see R. W. Newcomb, Distributional Impulse Response Theorems, *Proc. IEEE.*, vol. 51, p. 1157, 1963.

where f is any frequency, produces a response

$$B \cos (2\pi ft - \phi),$$

where ϕ and B/A may vary with frequency. Thus the response is of the same form but possibly delayed in phase by an amount ϕ and possibly different in amplitude by a factor B/A. The quantities ϕ and B/A are properties of the filter and may be compactly expressed in the form of a single complex quantity $T(f)$, the transfer factor, defined by

$$T(f) = \frac{B}{A} e^{i\phi}.$$

The essential feature of the above assertion, which will be now proved, is that a stimulus of harmonic form at a given frequency produces a response that is of the *same* form and the *same* frequency, regardless of the choice of frequency. This property is sometimes loosely referred to as "harmonic response to harmonic input."

Let the input stimulus be of unit strength and let it be the real part of a complex time-dependent function $\hat{V}_1(t)$ defined by

$$\hat{V}_1(t) = e^{i2\pi ft}.$$

Then the response can be shown to be given by

$$\hat{V}_2(t) = T(f)e^{i2\pi ft}.$$

The circumflex accent reminds us that, contrary to custom, we are using complex instantaneous values instead of real ones. This is done for algebraic simplicity.

To prove the assertion, let the response to $\hat{V}_1(t) = \exp(i2\pi ft)$ be $K(f,t) \exp(i2\pi ft)$, a quite general function of f and t, from which the exponential factor is extracted for convenience. Now apply a delayed stimulus $\exp [i2\pi f(t - T)]$; because of time invariance the response will be $K(f, t - T) \exp [i2\pi f(t - T)]$. However, the delayed stimulus can be expressed as the product of the original stimulus with a complex constant $\exp(-i2\pi fT)$; that is,

$$e^{i2\pi f(t-T)} = e^{-i2\pi fT}\hat{V}_1(t);$$

Hence the response can, because of linearity, be expressed as the product of the original response with this same complex factor. Thus

$$K(f, t - T)e^{i2\pi f(t-T)} = e^{-i2\pi fT}K(f,t)e^{i2\pi ft},$$

and therefore

$$K(f, t - T) = K(f,t);$$

that is, $K(f,t)$ is time-invariant and may be represented simply by $T(f)$.

The basic property of harmonic response to harmonic stimulus is thus shown to be a consequence of linearity plus time invariance.

We now show that the existence of a convolution relation

$$V_2(t) = I(t) * V_1(t)$$

between the stimulus and response is an equivalent condition.

Since the filter is linear, the response may be expressed in the form of the most general linear functional of the stimulus $V_1(t)$; that is, let

$$V_2(t) = \int_{-\infty}^{\infty} J(t,t')V_1(t')\,dt'.$$

Given also time invariance, under which $V_1(t - T)$ produces $V_2(t - T)$ for all T, it follows that

$$V_2(t - T) = \int_{-\infty}^{\infty} J(t,t')V_1(t' - T)\,dt'$$

or

$$V_2(t) = \int_{-\infty}^{\infty} J(t + T, t' + T)V_1(t')\,dt'.$$

Hence

$$J(t,t') = J(t + T, t' + T), \text{ for all } T,$$

is the condition for time invariance. It follows that $J(t,t')$ is a function purely of $t - t'$. Thus

$$J(t,t') = I(t - t')$$

and

$$V_2(t) = \int_{-\infty}^{\infty} I(t - t')V_1(t')\,dt'.$$

Hence linearity plus time invariance implies the convolution relation.

Problems

1 When a voltage $t\Pi(t)$ is applied to a circuit, the response $R(t)$ is called the ramp response. Show that the response to a general applied voltage $V_1(t)$ is given by

$$\int_{-\infty}^{\infty} R(t - u)V_1''(u)\,du$$

and that this expression is particularly suited to situations where the input voltage waveform is a polygon.

2 A sound track on film is fed into a high-fidelity reproducing system at twice the correct speed. It is physically obvious, and the similarity theorem confirms that the frequency of a sinusoidal input will be doubled. Ponder what the similarity theorem says about amplitude until this is also obvious physically.

3 A transparent band (at $x = 0$) in an otherwise opaque sound track is scanned by a rectangular slit. Show that the equivalent width of the response is equal either to the width of the band or to the width of the slit.

4 A signal of finite duration is applied to a filter whose impulse response is brief in duration compared with the signal. Show that the output waveform has a longer duration than the input waveform but that the amount of stretching, as measured by equivalent width, is short compared with the duration of the impulse response.

5 The (complex) electrical length of a uniform transmission line is θ. Show that the transfer factor relating the output voltage to the input voltage is given by $T(f) = \operatorname{sech} \theta$ when no load impedance is connected.

6 A filter consists of a tee-section whose series impedances are Z_1 and Z_2 and the shunt impedance is Z_3. Show that the voltage transfer function is given by

$$T(f) = \frac{Z_3}{Z_1 + Z_3}.$$

7 In the tee-section of the previous problem, let $Z_2 = 0$ and $Z_3 = R$. The element Z_1 is an open-circuited length of loss-free transmission line of characteristic impedance R, so that we may write $Z_1 = -iR \cot 2\pi Tf$, where T is a constant. Show that the output voltage response to an input voltage step is a rectangle function of time, and hence that in general the output voltage is the finite difference of the input voltage.

8 A very large number of identical passive two-port networks are cascaded, and a voltage impulse is applied to the input, causing a disturbance to propagate down the chain. The disturbance at a distant point is expressible by repeated self-convolution of the impulse response of a single network. Does the disturbance approach Gaussian form?

9 Show that a *linear* system can be imagined whose response to a modulated signal $(1 + M \cos \omega t) \cos \Omega t$ is proportional to the audio signal $M \cos \omega t$.

Chapter 10 Sampling and series

Suppose that we are presented with a function whose values were chosen arbitrarily, and suppose that no connection exists between the neighboring values chosen for the dependent variable. Thus if the independent variable were time, we would have to expect jumps of arbitrary magnitude and sign from one instant to the next.

In nature such a function would never be observed. Because of limitations of the measuring instrument, or for other reasons, there is always a limit to the rate of change. One might thus surmise that there is at least a little interdependence between waveform values at neighboring instants, and consequently that it might be possible to predict from past values over a certain brief time interval. This suggests that it might be possible to dispense with the values of a function for intervals of the same order and yet preserve essentially all the information by noting a set of values spaced at fine, but not infinitesimal, intervals. From this set of samples it would seem reasonable that the intervening values could be recovered. if only to some degree of approximation.

Sampling theorem

The sampling theorem states that, under a certain condition, it is in fact possible to recover the intervening values with full accuracy; in other words, the sample set can be fully equivalent to the complete set of function values. The condition is that the function should be "band-limited"; that is, its Fourier transform is nonzero over a finite range of the transform variable.

Clearly, the interval between samples is crucial in deciding the utility of the theorem; if the samples had to be very close together, not much would be gained. As an illustration of the fineness of sampling we take a simple band-limited function, namely sinc x, whose spectrum is flat up to a cutoff value $s = \frac{1}{2}$. The sampling interval, deduced as explained below, is 1. Figure 10.1a shows the sample values that define sinc x, and it will be seen that the interval is extremely coarse in comparison, for example, with the interval that would be chosen for numerical integration. The sampling intervals indicated by this theorem often seem, at first, to be surprisingly wide. Also shown in Fig. 10.1b are a set of samples for $\text{sinc}^2 \frac{1}{2}x$, a function whose spectrum cuts off at the same point ($s = \frac{1}{2}$) as that of sinc x. Figures 10.1c and 10.1d provide some samples for experiment.

With any given waveform, there is always a frequency beyond which spectral contributions are negligible for some purposes. However, on the other hand, the transform probably never cuts off absolutely; consequently, in applications of the sampling theorem, the error incurred by taking a given waveform to be band-limited must always be estimated.

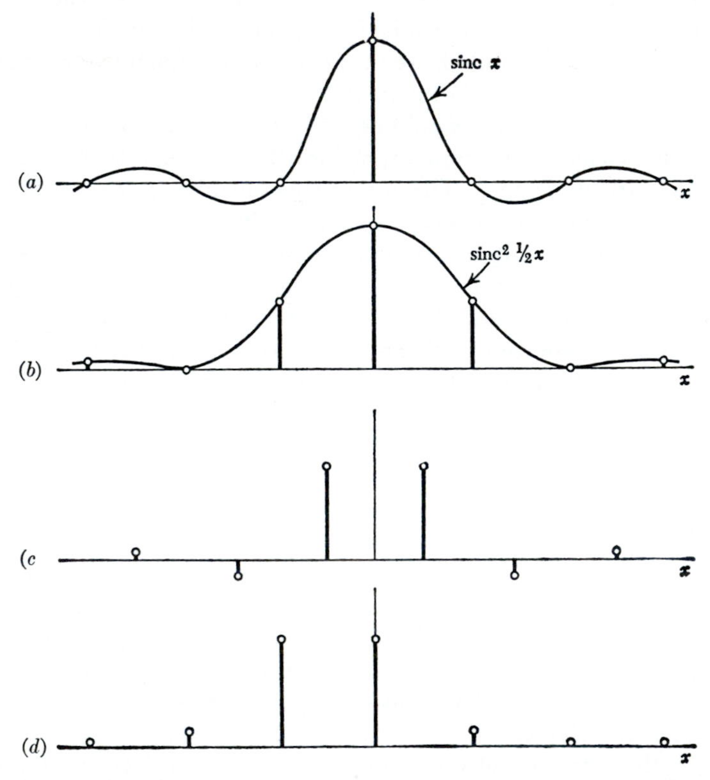

Fig. 10.1 Two functions and their samples, and two sets of samples for practice.

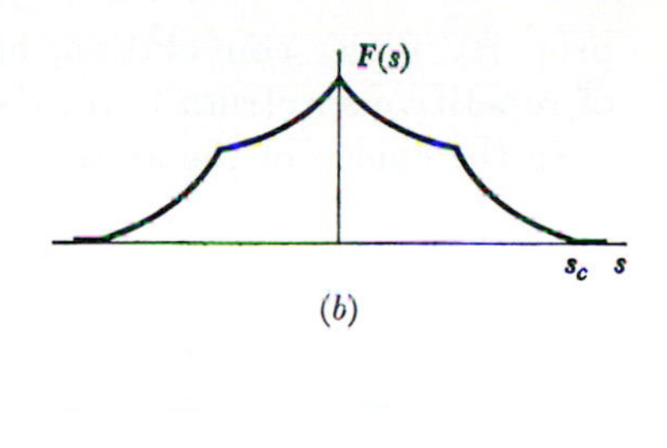

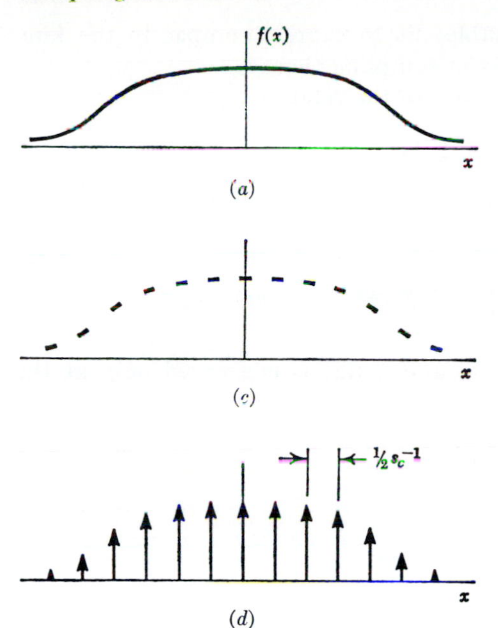

Fig. 10.2 (*a*) *A band-limited function* $f(x)$ *with a cutoff spectrum* (*b*). *The samples* (*c*) *suffice to reconstitute* $f(x)$ *accurately in full detail and are equivalent in content to the train of impulses* (*d*).

Consider a function $f(x)$, whose Fourier transform $F(s)$ is zero where $|s| > s_c$ (see Fig. 10.2). Evidently $f(x)$ is a band-limited function; in this case the band to which the Fourier components are limited is centered on the origin of s. Such a function is representative of a wide class of physical distributions which have been observed with equipment of limited resolving power. We shall refer to such transforms as "cutoff transforms" and describe them as being cut off beyond the "cutoff frequency" s_c.

In general a cutoff transform is of the form $\Pi(s/2s_c)G(s)$, where $G(s)$ is arbitrary, and therefore the general form of functions whose transforms are cut off is

$$\text{sinc } 2s_cx * g(x),$$

where $g(x)$ is arbitrary.

Of course, if the original function is cut off, then it is the transform which is band-limited.

With the exception noted below, band-limited functions have the peculiar property that they are fully specified by values spaced at equal intervals not exceeding $\frac{1}{2}s_c^{-1}$ (see Fig. 10.2*c*).

In the derivation that follows, the introduction of the *shah* symbol proves convenient, because multiplication by $\text{III}(x)$ is equivalent to sampling, in the sense that information is retained at the sampling points and abandoned in between.

As an additional bonus, the *shah* symbol, as a result of its replicating property under convolution, enables us to express compactly the kind of repetitive spectrum that arises in sampling theory.

In the course of the argument we use the relation

$$\mathrm{III}(x) \supset \mathrm{III}(s),$$

which is discussed at greater length at the end of this chapter.

Consider the function

$$\tau^{-1}f(x)\mathrm{III}\left(\frac{x}{\tau}\right) = \sum_n f(n\tau)\ \delta(x - n\tau)$$

shown in Fig. 10.2c. Information about $f(x)$ is conserved only at the

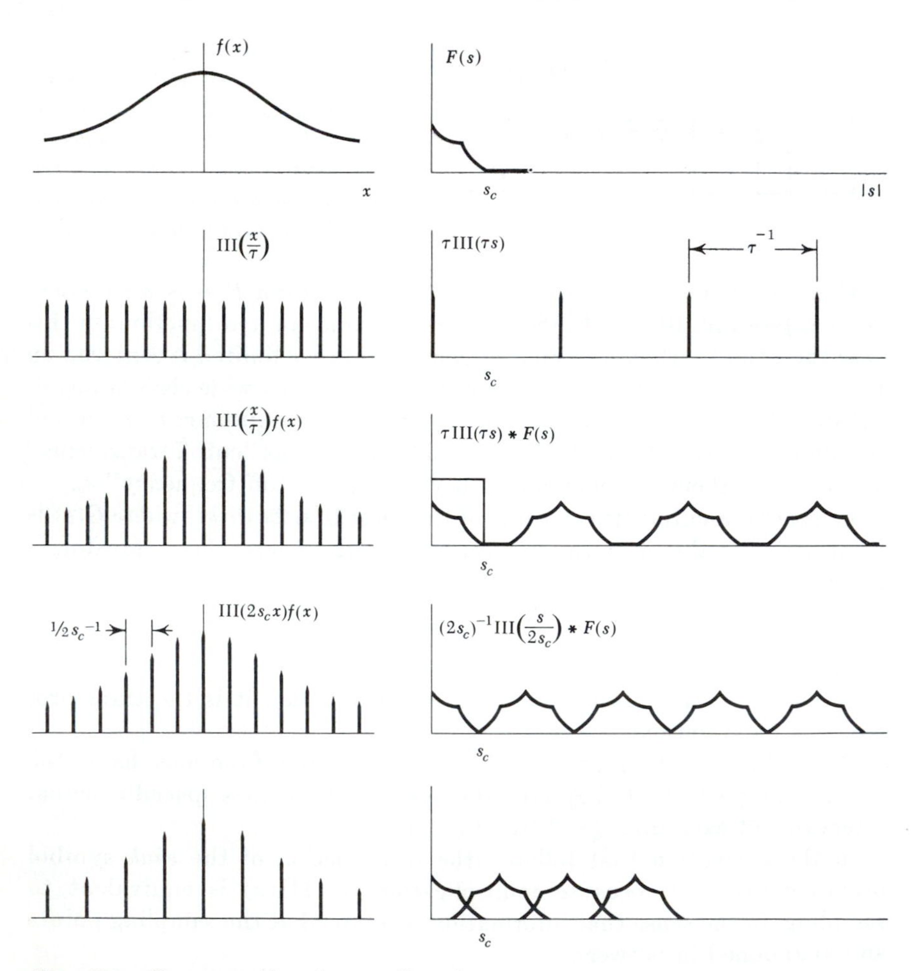

Fig. 10.3 Demonstrating the sampling theorem.

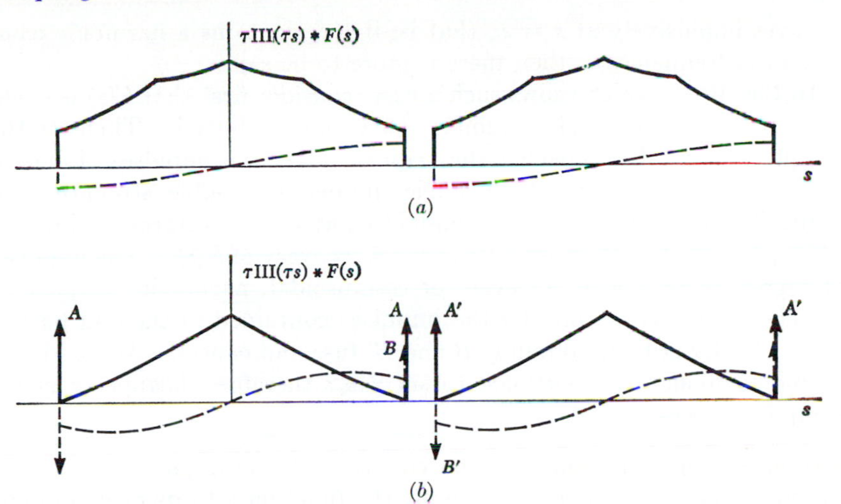

Fig. 10.4 Critical sampling.

sampling points where x is an integral multiple of the sampling interval τ. The intermediate values of $f(x)$ are lost. Therefore, if $f(x)$ can be reconstructed from $f(x)\text{III}(x/\tau)$, the theorem is proved. The transform of $\text{III}(x/\tau)$ is $\tau\text{III}(\tau s)$, Fig. 10.3*b*, which is a row of impulses at spacing τ^{-1}. Therefore

$$\overline{f(x)\text{III}(x/\tau)} = \tau\text{III}(\tau s) * F(s),$$

and we see that multiplication of the original function by $\text{III}(x/\tau)$ has the effect of replicating the spectrum $F(s)$ at intervals τ^{-1} (see Fig. 10.3*c*). We can reconstruct $f(x)$ if we can recover $F(s)$, and this can evidently be done, in the case illustrated in Fig. 10.3*c*, by multiplying $\tau\text{III}(\tau s) * F(s)$ by $\Pi(s/2s_c)$. Except for cases of singular behavior at $s = s_c$ (to be considered below), this is sufficient to demonstrate the sampling theorem.

At the same time a condition for sufficiently close sampling becomes apparent, for recovery will be impossible if the replicated islands overlap as shown in Fig. 10.3*e*, and this will happen if the spacing of the islands τ^{-1} becomes less than the width of an island $2s_c$. Hence the sampling interval τ must not exceed $\frac{1}{2}s_c^{-1}$, the semiperiod of a sinusoid of frequency s_c, and for critical sampling we shall have the islands just touching, as in Fig. 10.3*d*.

A small refinement must now be considered before the sampling theorem can be enunciated with strictness. In the illustration $F(s_c)$ is shown equal to zero. If $F(s_c)$ is not zero, the islands have cliffs which, under conditions of sampling at precisely the critical interval, make butt contact. In Fig. 10.4*a* this is on the point of happening, but careful examination, taking into account also the imaginary part of $F(s)$, reveals that multiplication by $\Pi(s/2s_c)$ permits exact recovery of $F(s)$. However, if $F(s)$

behaves impulsively at $s = s_c$, that is, if $f(x)$ contains a harmonic component of frequency s_c, then there is more to be said.

In Fig. 10.4*b*, which shows such a case, consider first that $F(s)$ is even; that is, ignore the odd imaginary part shown dotted. Then as the sampling interval approaches the critical value, the impulses A and A' tend to fuse at $s = s_c$ into a single impulse of double strength, and multiplication by $\Pi(s/2s_c)$ taken equal to $\frac{1}{2}$ at $s = s_c$, restores the impulse to its proper value, thus permitting exact recovery of $F(s)$. The impulses A represent, of course, an even, or *co*sinusoidal, harmonic component: $2A \cos 2\pi s_c x$. Now consider the impulses contained in the odd part of $F(s)$. Under critical sampling, B and B' fuse and cancel. Any odd harmonic component proportional to $\sin 2\pi s_c x$ therefore disappears in the sampling process.

Exercise The harmonic function $\cos(\omega t - \phi)$ is sampled at its critical interval (the semiperiod π/ω). Split the function into its even and odd parts and note that the sample values are precisely those of the even part alone. Note that the odd part is sampled at its zeros.

The sampling theorem can now be enunciated for reference:

A function whose Fourier transform is zero for $|s| > s_c$ is fully specified by values spaced at equal intervals not exceeding $\frac{1}{2}s_c^{-1}$ save for any harmonic term with zeros at the sampling points.

In this statement of the sampling theorem there is no indication of how the function is to be reconstituted from its samples, but from the argument given in support of the theorem it is clear that it is possible to reconstruct the function from the train of impulses equivalent to the set of samples, using some process of filtration. This procedure, which is envisaged as filtering in the transform domain, evidently amounts in the function domain to interpolation.

Interpolation

The numerical process of calculating intermediate points from samples does not of course depend on calculating Fourier transforms. Since the process of recovery was to multiply a transform by $\Pi(s/2s_c)$, the equivalent operation in the function domain, namely, convolution with $2s_c$ sinc $2s_c x$, will yield $f(x)$ directly from $\text{III}(x/\tau)f(x)$. Convolution with a function consisting of a row of impulses is an attractive operation numerically because the convolution integral reduces exactly to a summation (serial product).

For midpoint interpolation we can permanently record a table (see

Table 10.1 Midpoint interpolation

$\|x\|$	sinc x	$\|x\|$	sinc x	$\|x\|$	sinc x	$\|x\|$	sinc x
$\frac{1}{2}$	0.6366	$9\frac{1}{2}$	−0.0335	$18\frac{1}{2}$	0.0172	$27\frac{1}{2}$	−0.0116
$1\frac{1}{2}$	−0.2122	$10\frac{1}{2}$	0.0303	$19\frac{1}{2}$	−0.0163	$28\frac{1}{2}$	0.0112
$2\frac{1}{2}$	0.1273	$11\frac{1}{2}$	−0.0277	$20\frac{1}{2}$	0.0155	$29\frac{1}{2}$	−0.0108
$3\frac{1}{2}$	−0.0909	$12\frac{1}{2}$	0.0255	$21\frac{1}{2}$	−0.0148	$30\frac{1}{2}$	0.0104
$4\frac{1}{2}$	0.0707	$13\frac{1}{2}$	−0.0236	$22\frac{1}{2}$	0.0141	$31\frac{1}{2}$	−0.0101
$5\frac{1}{2}$	−0.0579	$14\frac{1}{2}$	0.0220	$23\frac{1}{2}$	−0.0135	$32\frac{1}{2}$	0.0098
$6\frac{1}{2}$	0.0490	$15\frac{1}{2}$	−0.0205	$24\frac{1}{2}$	0.0130	$33\frac{1}{2}$	−0.0095
$7\frac{1}{2}$	−0.0424	$16\frac{1}{2}$	0.0193	$25\frac{1}{2}$	−0.0125	$34\frac{1}{2}$	0.0092
$8\frac{1}{2}$	0.0374	$17\frac{1}{2}$	−0.0182	$26\frac{1}{2}$	0.0120	$35\frac{1}{2}$	−0.0090

Table 10.1) of suitably spaced values of sinc x, and it proves practical when further interpolation is required to repeat the midpoint process, using the same array.

Rectangular filtering

Suppose that it is required to remove from a function spectral components whose frequencies exceed a certain limit, that is, to multiply the transform by a rectangle function, which we shall take to be $\Pi(s)$. We are assuming that $s_c = \frac{1}{2}$ and that the critical sampling interval is 1. This is just the operation which has already been carried out for the purpose of interpolation. However, in general the function to be filtered will not consist solely of impulses, and the convolution integral giving one filtered value does not reduce exactly to a summation. However, when it is evaluated numerically it will have to be approximated by a summation,

$$\Sigma_\tau = f(x) * \left[\text{III}\left(\frac{x}{\tau}\right) \text{sinc } x \right],$$

and we may ask how coarse the tabulation interval may be and still sufficiently approximate the desired integral

$$f(x) * \text{sinc } x.$$

Beginning with $\tau = 1$, we find $\Sigma_1 = f(x)$; that is, no filtering at all has resulted. Now trying $\tau = \frac{1}{2}$, we have

$$\Sigma_{\frac{1}{2}} = f(x) * \text{III}(2x) \text{ sinc } x$$

and

$$\Sigma_{\frac{1}{2}} = F(s)[\tfrac{1}{2}\text{III}(\tfrac{1}{2}s) * \Pi(s)],$$

since

$$\Sigma_\tau = F(s)[\tau\text{III}(\tau s) * \Pi(s)].$$

Hence $\bar{\Sigma}_{\frac{1}{2}}$ consists of a central part $F(s)\Pi(s)$ plus remoter parts. For many purposes this simple operation would suffice (for example, when the components to be rejected lie chiefly just beyond the central region).

Adopting an idea from the method of interpolation, where we economize on interpolating arrays by repeated use of the one midpoint interpolation array, we now consider the effect of repeated approximate filtering of the one kind. By using the same filtering array at $\tau = \frac{1}{4}$ we have the filter characteristic $\frac{1}{4}\text{III}(s/4) * \Pi(s/2)$, which when multiplied by $\frac{1}{2}\text{III}(s/2) * \Pi(s)$ gives the bottom line of Fig. 10.5. In other words, repeated application of the process has pushed down more of the outer islands of response. The same result is obtained by taking $\tau = \frac{1}{4}$ initially.

To summarize, approximate rectangular filtering with a cutoff at s_c is carried out by reading off $f(x)$ at half the critical sampling interval (that is, at intervals of $\frac{1}{4}s_c^{-1}$) and taking the convolution (more precisely, the serial product) with $2s_c \operatorname{sinc} 2s_cx$, where $2s_cx$ assumes all half-integral values, including 0. This filtering array (see Table 10.2) contains precisely the values tabulated for interpolation plus interleaved zeros and a central value of unity.

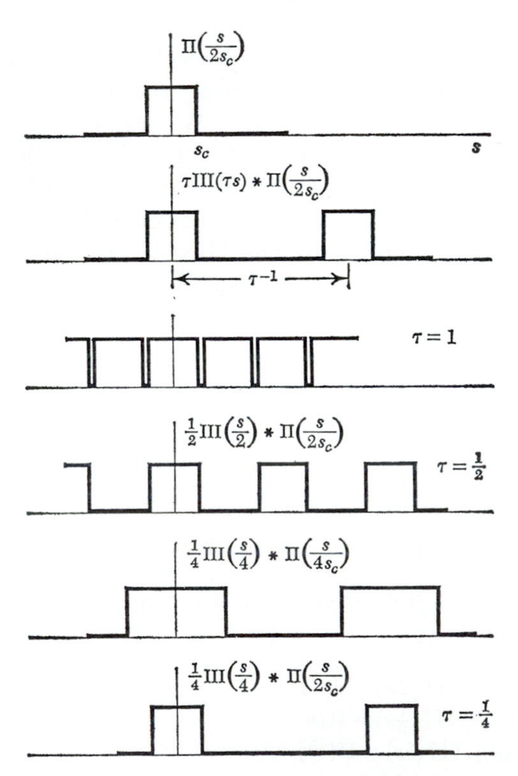

Fig. 10.5 Numerical procedure for achieving the desired filter characteristic $\Pi(s)$.

Table 10.2 Array for approximate rectangular filtering

.
.
.
−0.0909
0
0.1273
0
−0.2122
0
0.6366
⟶1.0000
0.6366
0
−0.2122
0
0.1273
0
−0.0909
.
.
.

Undersampling

Suppose that $f(x)$ (see Fig. 10.6*a*) is read off at intervals corresponding to a desired cutoff for rectangular filtering. Then the band-limited function $g(x)$ (see Fig. 10.6*c*) defined by this set of values (see Fig. 10.6*b*) has a cutoff spectrum of the desired extent and may at first sight appear to be a product of rectangular filtering. But the process is not the same as rectangular filtering, since the result depends on high-frequency components in $f(x)$; for example, one of the sample values may fall at the peak of a narrow spike; furthermore, the *phase* of the coarse sampling points will clearly affect the result. However, the effect may often be a good approximation to rectangular filtering.

By examining the process in terms of Fourier transforms, we see that the band-limited function $g(x)$ derived from too-coarse sampling contains contributions from high-frequency components of $f(x)$, impersonating low frequencies in a way described by reflection of the high-frequency part in the line $s = s_c$. The effect has been referred to as "aliasing." If this high-frequency tail is not too important, then the coarse sampling procedure gives a fair result.

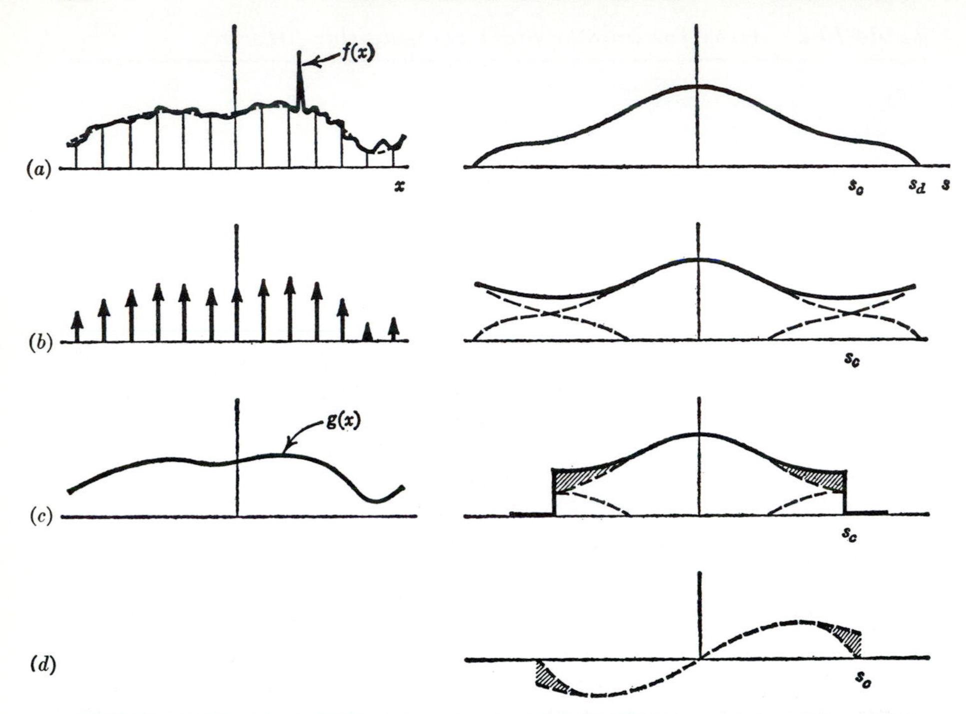

Fig. 10.6 A band-limited function $g(x)$ derived from $f(x)$ by undersampling; high-frequency components of $f(x)$ shift down inside the band limits. Shaded areas indicate high frequencies masquerading as low.

While the effect of undersampling is to reinforce the even part of the spectrum, Fig. 10.6*d* shows that the odd part is diminished. It follows that $g(x)$ will be evener than $f(x)$, which is, of course, to be expected, for the sampling procedure discriminates against the (necessarily odd) components with zeros at the sampling points.

Ordinate and slope sampling

Let $f(x)$ be a band-limited function that is fully specified by ordinates at a spacing of 0.5; but suppose that only IIIf is given; that is, only every second ordinate is given. Then from the overlapping islands (see Fig. 10.7) composing $\overline{\text{III}f}$, it would not be possible to recover $F(s)$ or, consequently, $f(x)$. But if further partial information were given, it might become possible.

If the slope is given, in addition to the ordinate, recovery proves to be possible. Thus, given IIIf and IIIf', one can express $f(x)$ as a combination of linear functionals of IIIf and IIIf':

$$f(x) = a(x) * (\text{III}f) + b(x) * (\text{III}f'),$$

where $a(x)$ and $b(x)$ are solving functions that have to be found. Just as the sinc function, which is the solving function for ordinary sampling, must be zero at all its sample points save the origin, where it must be unity, so must $a(x)$. And $b(x)$ must be zero at all sample points but have unit slope at the origin.

To find $a(x)$ and $b(x)$, note that in the interval $-1 \leqslant s \leqslant 1$

$$\overline{\text{III}f} = F(s+1) + F(s) + F(s-1)$$

and that

$$\overline{\text{III}f'} = i2\pi(s+1)F(s+1) + i2\pi sF(s) + i2\pi(s-1)F(s-1).$$

These two equations can be solved for $F(s)$, for although there appear to be three unknowns, namely, $F(s+1)$, $F(s)$, $F(s-1)$, in fact, for any value of s, one of them is always known to be zero. Thus for positive s we have $F(s+1) = 0$, and on eliminating $F(s-1)$ we have

$$i2\pi F(s) = \overline{\text{III}f'} - i2\pi(s-1)\overline{\text{III}f},$$

and for negative s we have

$$-i2\pi F(s) = \overline{\text{III}f'} - i2\pi(s+1)\overline{\text{III}f}.$$

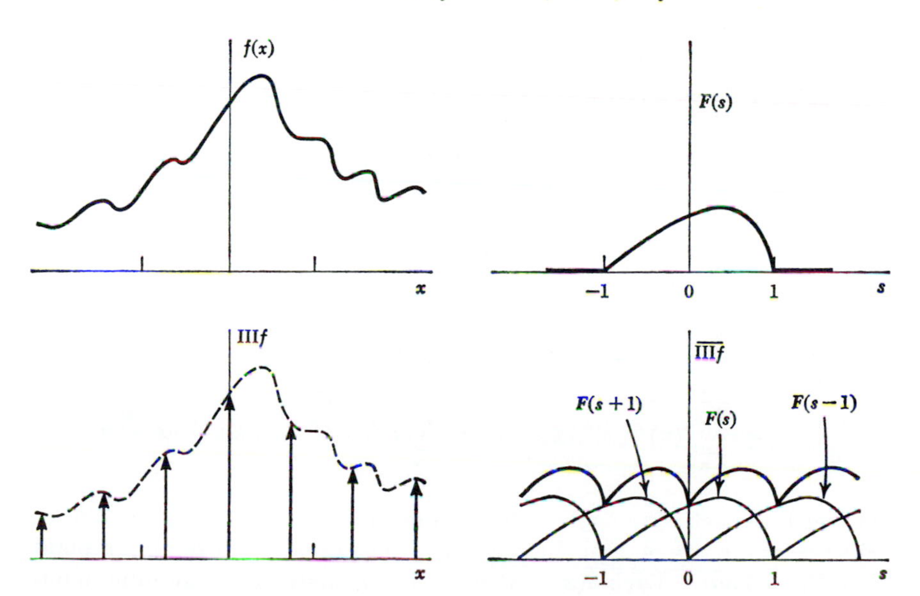

Fig. 10.7 Ordinate sampling at half the rate necessary for full definition.

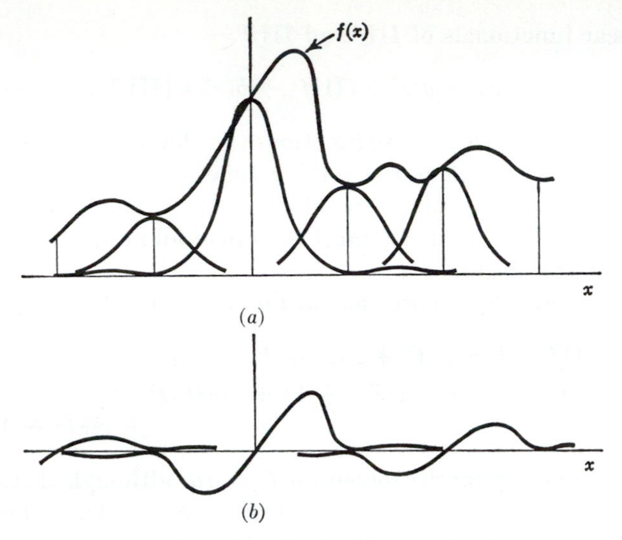

Fig. 10.8 (*a*) *The ordinate-dependent constituents* $f(n)a(x-n)$; (*b*) *The slope-dependent constituents* $f'(n)b(x-n)$.

For all s, both positive and negative, we have concisely

$$F(s) = \frac{i}{2\pi}\Lambda'(s)\overline{\text{III}f'} + \Lambda(s)\overline{\text{III}f}.$$

Hence
$$f(x) = \text{sinc}^2\, x * (\text{III}f) + x\,\text{sinc}^2\, x * (\text{III}f').$$

The solving functions are

$$a(x) = \text{sinc}^2\, x$$
$$b(x) = x\,\text{sinc}^2\, x,$$

and the convolution integrals reduce to a sum of spaced a's and b's of suitable amplitudes,

$$\begin{aligned} f(x) &= \sum_{-\infty}^{\infty} f(n)a(x-n) + \sum_{-\infty}^{\infty} f'(n)b(x-n) \\ &= \sum_{-\infty}^{\infty} f(n)\,\text{sinc}^2\,(x-n) + \sum_{-\infty}^{\infty} f'(n)(x-n)\,\text{sinc}^2\,(x-n). \end{aligned}$$

We see that each $a(x-n)$ is zero at all sampling points (integral values of x) save where $x = n$, and that each has zero slope at all sampling points (see Fig. 10.8*a*). Each $b(x-n)$ has zero ordinate at all sampling points and likewise zero slope, save where $x = n$ (see Fig. 10.8*b*).

Interlaced sampling

As in the previous example, let IIIf represent every second ordinate of the set that is necessary to specify $f(x)$, and let a supplementary set $\mathrm{III}(x - a)f(x)$, interlaced with the first as illustrated in Fig. 10.9, also be available. Will it be possible to reconstitute $f(x)$? It is known that equispaced samples, separated by just more than the critical interval, do not suffice, and in the case of interlaced sampling every second jump exceeds this critical interval. On the other hand, it has been shown that ordinate-and-slope sampling suffices, and this is clearly equivalent to extreme interlacing as a approaches zero.

If there is a solution, it should be in the form of a sum of two linear functionals of IIIf and $\mathrm{III}_a f$, where $\mathrm{III}_a \equiv \mathrm{III}(x - a)$:

$$f(x) = a(x) * (\mathrm{III}f) + b(x) * (\mathrm{III}_a f).$$

The solving function $a(x)$ must be equal to unity at $x = 0$, and zero at all other sampling points, and $b(x)$ must be the mirror image of $a(x)$; that is, $b(x) = a(-x)$.

In the interval $-1 < s < 1$,

$$\overline{\mathrm{III}f} = F(s + 1) + F(s) + F(s - 1)$$

and

$$\overline{\mathrm{III}_a f} = e^{i2\pi a}F(s + 1) + F(s) + e^{-i2\pi a}F(s - 1).$$

For positive s we have $F(s + 1) = 0$, and on eliminating $F(s - 1)$ we have

$$F(s) = -\frac{e^{-i2\pi a}}{1 - e^{-i2\pi a}}\overline{\mathrm{III}f} + \frac{1}{1 - e^{-i2\pi a}}\overline{\mathrm{III}_a f},$$

and for negative s we have

$$F(s) = -\frac{e^{i2\pi a}}{1 - e^{i2\pi a}}\overline{\mathrm{III}f} + \frac{1}{1 - e^{i2\pi a}}\overline{\mathrm{III}_a f}.$$

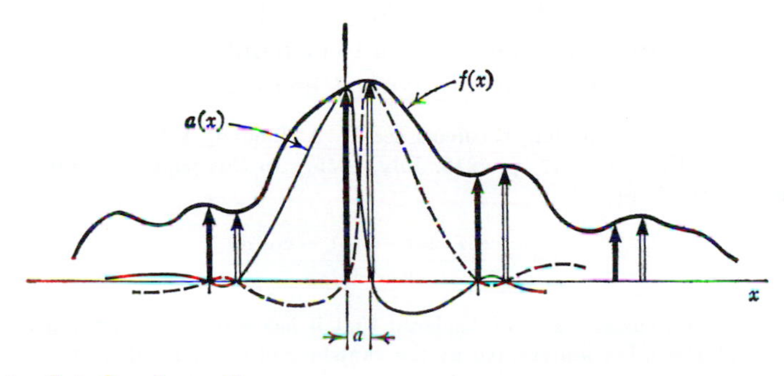

Fig. 10.9 Interlaced samples.

For all s we have

$$F(s) = A(s)\overline{\mathrm{III}f} + A^*(s)\overline{\mathrm{III}_a f},$$

where

$$A(s) = \begin{cases} -\dfrac{e^{-i2\pi a}}{1 - e^{-i2\pi a}} & 0 < s < 1 \\ -\dfrac{e^{i2\pi a}}{1 - e^{i2\pi a}} & -1 < s < 0 \\ 0 & |s| > 1 \end{cases}$$

$$= \tfrac{1}{2}\Pi\left(\frac{s}{2}\right) + \tfrac{1}{2}i \cot a\pi\Lambda'(s).$$

Hence

$$a(x) = \operatorname{sinc} 2x - (\pi \cot a\pi)x \operatorname{sinc}^2 x,$$

as graphed in Fig. 10.9.[1]

It may seem strange that the equidistant samples may be regrouped in pairs, even to the extreme of close spacing. However, it is also possible to bunch the samples in groups of any size separated by such wide intervals as maintain the original average spacing. The bunching may be indefinitely close; see Linden for a proof that the ordinates and first n derivatives, at points spaced by $n + 1$ times the usual spacing, suffice to specify a band-limited function. In the limit, as n approaches infinity, the formula for reconstituting the function becomes the Maclaurin series. This introduces doubt of the practical applicability of higher-order sampling theorems, for it is well known that the Maclaurin series

$$f(0) + xf'(0) + \frac{x^2}{2!}f''(0) + \cdots + \frac{x^n}{n!}f^{(n)}(0)$$

does not usually converge to $f(x)$. (Consider the functions $\Pi(ax)$, which all have the same Maclaurin series.)

In practice, higher-order sampling breaks down at some point because small amounts of noise drastically affect the determination of high-order derivatives or finite differences. In the total absence of noise, trouble would still be expected to set in at some stage because of the impossibility of ensuring perfectly band-limited behavior. In applications of sampling theorems, the claim of a given function to be band-limited must always be scrutinized, and any error resulting must be estimated.

[1] For a discussion of sampling theorems see D. A. Linden, A Discussion of Sampling Theorems, *Proc. IRE.*, vol. 47, p. 1219, July, 1959. In this paper the solving function $a(x)$ is given in the form

$$a(x) \equiv \frac{\cos(2\pi x - a\pi) - \cos a\pi}{2\pi x \sin a\pi}.$$

The numerator is a cosine wave so displaced that it has zeros at $x = n$ and $x = n + a$, but the zero at $x = 0$ is counteracted by the vanishing of the denominator in such a way that $a(0) = 1$.

Sampling in the presence of noise

Suppose that the samples $f(n)$ of a certain function $f(x)$ cannot be obtained without some error being made; that is, the observable quantity is

$$f(n) + \text{error}.$$

Then when an attempt is made to reconstruct $f(x)$ from the observed samples by applying the same procedure used for true samples, the reconstructed values will differ from the true values of $f(x)$.

Consider the case of midpoint interpolation, supposing that samples have been taken at $x = \pm\frac{1}{2}, \pm 1\frac{1}{2}, \pm 2\frac{1}{2}, \ldots .$ Then

$$f(0) = 0.6366[f(\tfrac{1}{2}) + f(-\tfrac{1}{2})] - 0.2122[f(1\tfrac{1}{2}) + f(-1\tfrac{1}{2})] + 0.1273[f(2\tfrac{1}{2}) + f(-2\tfrac{1}{2})] - \cdots .$$

The errors affecting $f(\frac{1}{2})$ and $f(-\frac{1}{2})$ will have the greatest effect on $f(0)$; each error is reduced by 0.6366, and the resulting net error at $x = 0$, due to the errors at the nearest two sampling points, could be anywhere from zero, if the two errors happened to cancel, up to 1.27 times either error. Clearly, the error involved in using the sampling theorem to interpolate will be of the same order of magnitude as the errors affecting the data. More than this could not be expected, and so it may be concluded that the interpolating procedure is tolerant to the presence of noise.

It is not the purpose here to go into statistical matters, but a simple result should be pointed out that arises when the errors are of such a nature that they are independent from one sample to the next, and all come from a population with zero mean value and a certain variance σ^2. Then the variance of the error contributed at $x = 0$ by the error at $x = \frac{1}{2}$ is $(0.6366)^2\sigma^2$, and the variance of the total error at $x = 0$ due to the errors at all the sampling points is

$$[\cdots + (0.1273)^2 + (0.2122)^2 + (0.6366)^2 + (0.6366)^2 + (0.2122)^2 + (0.1273)^2 + \cdots]\sigma^2.$$

Now the terms of the series within the brackets are values of $\operatorname{sinc}^2 x$ at unit intervals of x and hence add up to unity. Therefore, in this simple error situation, the interpolated value is subject to precisely the same error as the data.

Now we apply the same reasoning to interlaced sampling, especially to the extreme situation where the narrow interval is small compared with the wide one, that is, where $a \ll 1$. At the point $x = \frac{1}{2}a + \frac{1}{2}$, which is in the middle of the wide interval, the four nearest sample values enter with coefficients

$$a(\tfrac{1}{2}a + \tfrac{1}{2}),\ b(-\tfrac{1}{2}a + \tfrac{1}{2}),\ a(\tfrac{1}{2}a - \tfrac{1}{2}),\ b(-\tfrac{1}{2}a - \tfrac{1}{2}).$$

The first and last of these are positive and the others negative, and the presence of the factor cot $a\pi$ in the formula for $a(x)$ shows that the numerical values may be large. Thus the interpolated value may result from the cancellation of large terms, and the total error may be large.

In another way of looking at this, a pair of terms

$$f(0)a(x) + f(a)b(x - a)$$

may be reexpressed in the form

$$[a(x) + b(x - a)]\left[\frac{f(0) + f(a)}{2}\right] + \frac{b(x - a) - a(x)}{2}[f(a) - f(0)].$$

Here the first term represents the mean of a sample pair multiplied by a certain coefficient, and the second represents the difference between two close-spaced samples multiplied by a certain other coefficient. In the limit as $a \to 0$, the solving functions for the ordinate-and-slope sampling theorem would result. Now the coefficient of the difference term can be large; for example, when $a < 0.2$, the value adopted in the illustration of interlaced sampling, the coefficient exceeds unity. Thus errors in the difference term may be amplified.

It thus appears that interlaced sampling, where the sample spacing is alternately narrow and wide, is not tolerant to errors, and therefore the magnitude of the errors would have to be estimated carefully in an application. In a full study it would be essential to take account of any correlation between the errors in successive samples since it is clear that the error in $f(a) - f(0)$ would be reduced if both $f(a)$ and $f(0)$ were subject to about the same error.

Fourier series

It is well known that a periodic waveform, such as the acoustical waveform associated with a sustained note of a musical instrument, is composed of a fundamental and harmonics. Exploration of such an acoustical field by means of tunable resonators reveals that the energy is concentrated at frequencies which are integral multiples of the fundamental frequency. There is nothing here that should exclude this case from treatment by the Fourier transform methods so far used. However, insistence on strict periodicity, a physically impossible thing, will clearly lead to an impulsive spectrum and to the refined considerations that are needed in connection with impulses. We now proceed to do this, using the *shah* symbol for convenient handling of the sets of impulses that arise, in connection both with the replication inherent in periodicity and with the sampling associated with harmonic spectra.

The Fourier series will be exhibited as an extreme situation of the Fourier transform, even though the opposite procedure, taking the Fourier series as a point of departure for developing the Fourier transform, is more usual.

For reference let it be stated that the Fourier series associated with the periodic function $g(x)$, with frequency f and period T, is

$$a_0 + \sum_1^\infty (a_n \cos 2\pi nfx + b_n \sin 2\pi nfx),$$

where
$$a_0 = \frac{1}{T}\int_{-\frac{1}{2}T}^{\frac{1}{2}T} g(x)\, dx$$
$$a_n = \frac{2}{T}\int_{-\frac{1}{2}T}^{\frac{1}{2}T} g(x) \cos 2\pi nfx\, dx$$
$$b_n = \frac{2}{T}\int_{-\frac{1}{2}T}^{\frac{1}{2}T} g(x) \sin 2\pi fx\, dx.$$

It is necessary for $g(x)$ to have been chosen so that the integrals exist; otherwise $g(x)$ is arbitrary.

The purpose of a good deal of theory dealing with the Fourier series has been to show that the series associated with a periodic function $g(x)$ does in fact often converge, and furthermore, that when it converges, it often converges to

$$\tfrac{1}{2}[g(x + 0) + g(x - 0)].$$

The rigorous development of this topic was initiated by Dirichlet in 1829, following a controversial period dating back to D. Bernoulli's success in 1753 in expressing the form of a vibrating string as a series

$$y = A_1 \sin x \cos at + A_2 \sin 2x \cos 2at + \cdots .$$

Euler, who had been working on this problem and had just obtained the general solution in terms of traveling waves, said that if Bernoulli was right, an arbitrary function could be expanded as a sine series. This, he said, was impossible. In 1807, when Fourier made this same claim in his paper presented to the Paris Academy, Lagrange rose and said it was impossible. This exciting subject led to many important developments in pure mathematics, including the invention of the Riemann integral. It must be remembered that the expressions for the Fourier constants a_0, a_n, b_n were given long before modern analysis developed.

For the present purpose let us take $T = 1$ and note that the complex constant $a_n - ib_n$ is related to the one-period segment $g(x)\Pi(x)$ by the Fourier transform

$$a_n - ib_n = 2\int_{-\frac{1}{2}}^{\frac{1}{2}} g(x)e^{-i2\pi nx}\, dx$$
$$= 2\int_{-\infty}^{\infty} g(x)\Pi(x)e^{-i2\pi nx}\, dx.$$

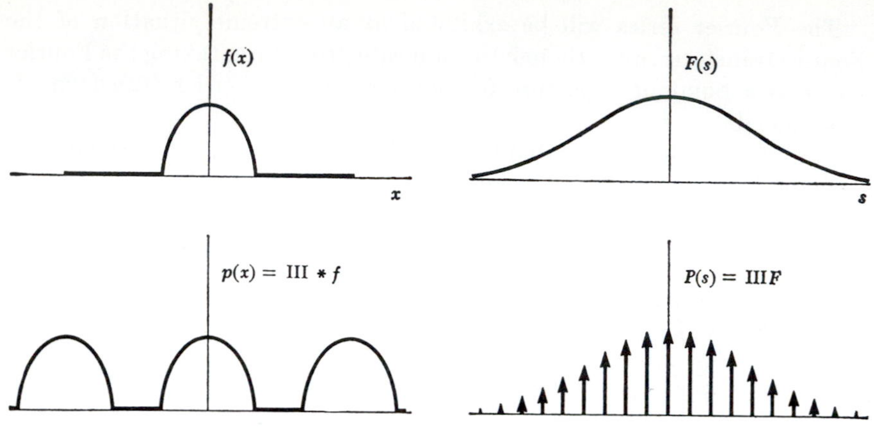

Fig. 10.10 The transform in the limit of a periodic function.

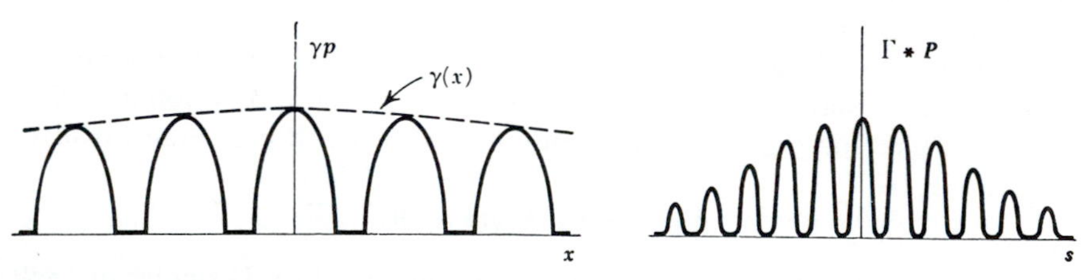

Fig. 10.11 The convergence factor γ applied to a periodic function $p(x)$ renders its infinite integral convergent while the convolvent Γ removes infinite discontinuities from a line spectrum $P(s)$.

We now recover this result by considering the Fourier transform of a periodic function.

Let $f(x)$ be a function that possesses a regular Fourier transform $F(s)$ (see Fig. 10.10). Then its convolution with the replicating symbol $\text{III}(x)$ will be the periodic function $p(x)$, defined by

$$p(x) = \text{III}(x) * f(x) = \sum_{-\infty}^{\infty} f(x - n) \qquad n \text{ integral.}$$

The convergence of the summation is guaranteed by the absolute integrability of $f(x)$, which was requisite for the possession of a Fourier transform. The period of $p(x)$ is unity; that is,

$$p(x + 1) = p(x)$$

for all x.

There will be no regular Fourier transform for $p(x)$ because

$$\int_{-\infty}^{\infty} |p(x)|\, dx$$

cannot converge, save in the trivial case where $p(x)$ is identically zero. Hence we multiply $p(x)$ by a factor $\gamma(x)$ that dies out to zero for large values of x, both positive and negative. In effect we are bringing the strictly periodic function $p(x)$ back to the realm of the physically possible, but only barely so (see Fig. 10.11). Let

$$\gamma(x) = e^{-\pi\tau^2 x^2};$$

then the Fourier transform $\Gamma(s)$ of $\gamma(x)$ will be given by

$$\Gamma(s) = \tau^{-1} e^{-\pi s^2/\tau^2}.$$

The function $\gamma(x)p(x)$ will possess a Fourier transform,

$$\begin{aligned} \gamma(x)p(x) &= \gamma(x) \sum_{-\infty}^{\infty} f(x-n) \\ &\supset \Gamma(s) * \sum_{-\infty}^{\infty} e^{-i2\pi ns} F(s) \\ &= \sum_{-\infty}^{\infty} F(n)\Gamma(s-n). \end{aligned}$$

In *shah*-symbol notation,

$$\sum_{-\infty}^{\infty} F(n)\Gamma(s-n) = \Gamma(s) * [\text{III}(s)F(s)].$$

Now let

$$P(s) = \text{III}(s)F(s).$$

This entity $P(s)$ is a whole set of equidistant impulses of various strengths, as given by samples of $F(s)$ at integral values of s, and has the property that

$$\gamma p \supset \Gamma * P.$$

By the convolution theorem we see that P is a suitable symbol to assign as the Fourier transform of $p(x)$, and indeed as $\tau \to 0$ and γp runs through a sequence of functions having p as limit, the sequence $\Gamma * P$ defines an entity P of such a type that, as agreed, we may call $p(x)$ and $P(s)$ a Fourier transform pair in the limit.

Taking $f(x)$ to be the one-period segment $g(x)\Pi(x)$, we see that the spectrum of a periodic function is a set of impulses whose strengths are given by equidistant samples of $F(s)$, the Fourier transform of the one-period segment. Now

$$\begin{aligned}\int_{-\infty}^{\infty} \text{III}(s)F(s)e^{+i2\pi sx}\,ds &= \sum_{-\infty}^{\infty} \int_{n-0}^{n+0} F(s)e^{+i2\pi sx}\,ds \\ &= \sum_{-\infty}^{\infty} F(n)e^{+i2\pi nx} \\ &= F(0) + \sum_{1}^{\infty} [F(n)e^{+i2\pi nx} + \text{conjugate}] \\ &= F(0) + 2\sum_{1}^{\infty} (\text{Re}\, F \cos 2\pi nx - \text{Im}\, F \sin 2\pi nx) \\ &= a_0 + \sum_{1}^{\infty} (a_n \cos 2\pi nx + b_n \sin 2\pi nx)\end{aligned}$$

if $a_n - ib_n = 2F(n)$. And this is precisely the value that was quoted earlier for the complex coefficient $a_n - ib_n$.

The fact that rigorous deliberations on Fourier series generally are more complex than those encountered with the Fourier integral is essentially connected with the infinite energy of periodic functions (the integrals of which are not absolutely convergent). It is therefore very natural physically to regard a periodic function as something to be approached through functions having finite energy, and to consider *line* spectra as something to be approached via continuous spectra with finite energy density. The strange thing is that the physically possible functions and spectra are often presented as elaborations of the physically impossible. Some people cannot see how a *line* spectrum, no matter how closely the lines are packed, can ultimately become a continuous spectrum. This order of presentation is inherited from the historical precedence of the theory of trigonometric series, and runs as follows (in our terminology). The periodic function $\text{III} * f$ has a line spectrum $\text{III}F$ (set of Fourier coefficients). If the repetition period is lengthened to τ, the lines of the spectrum are packed τ times more closely and are τ times weaker (note compensating change in ordinate scale in Fig. 10.12). Now let the period become infinite, so that the pulse f does not recur. Then the trigonometric sum which represented its periodic predecessors passes into an infinite integral, and the finite integral which specified the series coefficients does likewise. These two integrals are the plus-i and minus-i Fourier integrals.

On the view described here line spectra and periodic functions are regarded as included in the theory of Fourier transforms, to be handled

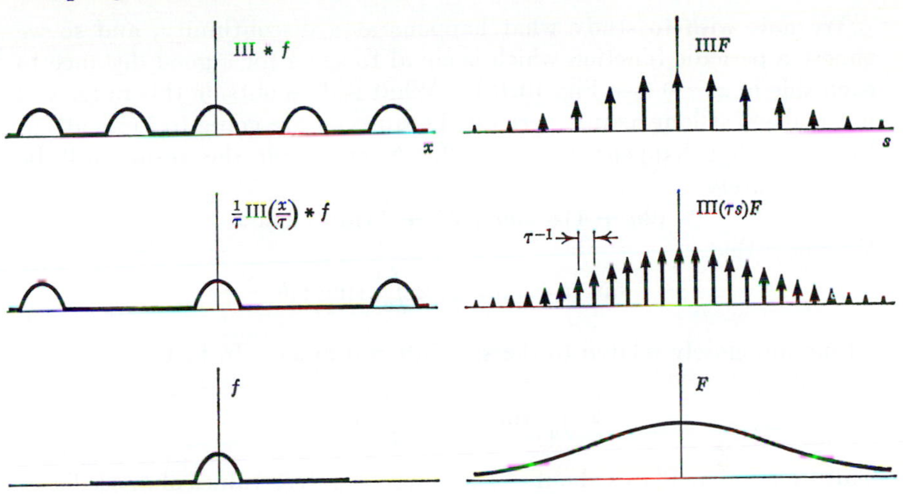

Fig. 10.12 Deriving the Fourier integral from the Fourier series.

exactly as other transforms by means of the III symbology, and with the same caution accorded to other transforms in the limit.

Gibbs phenomenon One of the classical topics of the theory of Fourier series may be studied profitably from the point of view that we have developed. In situations where periodic phenomena are analyzed to determine the coefficients of a Fourier series, which is then used to predict, it is a matter of practical importance to know how many terms of the series to retain. Various considerations enter into this, but one of them is the phenomenon of overshoot associated with discontinuities, or sharp changes, in the periodic function to be represented.

It is quite clear that by omitting terms beyond a certain limiting frequency we are subjecting the periodic function $p(x)$ to low-pass filtering. Thus, if the fundamental frequency is s_0, and frequencies up to ns_0 are retained, it is as though the spectrum had been multiplied by a rectangle function $\Pi(s/2s_c)$, where s_c is a cutoff frequency between ns_0 and $(n+1)s_0$. It makes no difference precisely where s_c is taken; for convenience it may be taken at $(n+\frac{1}{2})s_0$. Multiplication of the spectrum by $\Pi[s/(2n+1)s_0]$ corresponds to convolution of the original periodic function $p(x)$ with $(2n+1)s_0$ sinc $[(2n+1)s_0x]$. Therefore, when the series is summed for terms up to frequencies ns_0 only, the sum will be

$$p(x) * (2n+1)s_0 \operatorname{sinc} [(2n+1)s_0x].$$

The convolving function has unit area, so in places where $p(x)$ is slowly varying, the result will be in close agreement with $p(x)$.

We now wish to study what happens at a discontinuity, and so we choose a periodic function which is equal to sgn x for a good distance to each side of $x = 0$ (see Fig. 10.13). What it does outside this range will not matter, as long as it is periodic, because we are going to focus attention on what happens near $x = 0$. Near $x = 0$ the result will be approximately

$$(2n + 1)s_0 \operatorname{sinc}[(2n + 1)s_0 x] * \operatorname{sgn} x.$$

We know that

$$\operatorname{sinc} x * \operatorname{sgn} x = 2\int_0^x \operatorname{sinc} t \, dt,$$

a function closely related to the sine integral Si(x). In fact

$$2\int_0^x \operatorname{sinc} t \, dt = \frac{2}{\pi}\operatorname{Si}(\pi x).$$

This function oscillates about -1 for large negative values of x, oscillates with increasing amplitude as the origin is approached, passes through zero at $x = 0$, shoots up to a maximum value of 1.18, and then settles down to decaying oscillations about a value of $+1$. If we change the scale factors of the sinc function, compressing it by a factor $N = (2n + 1)s_0$, and

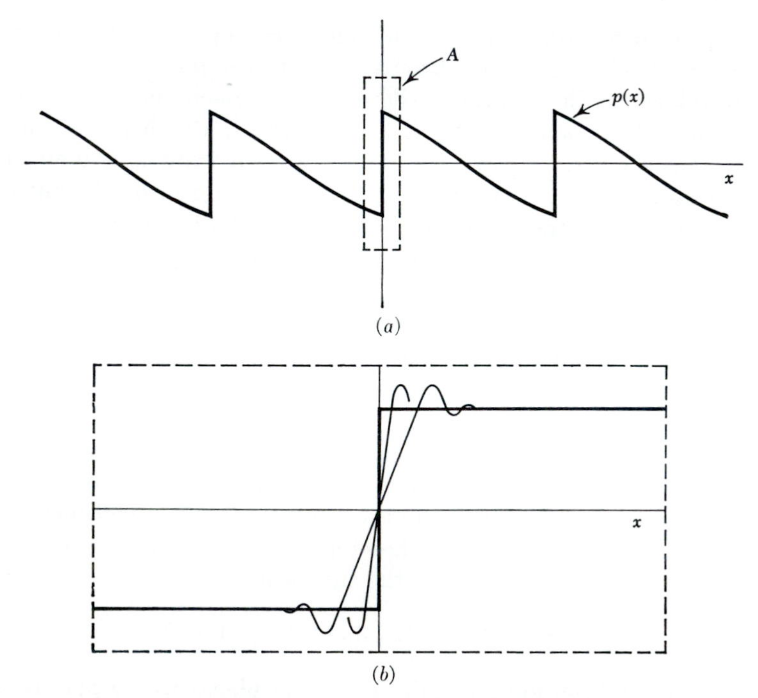

Fig. 10.13 (*a*) *A periodic function* $p(x)$; (*b*) *an enlargement of area A.*

strengthening it by a factor N, so as to retain its unit area, then convolution with sgn x will result in oscillations about -1 and then about $+1$ that are faster but have the same amplitude, that is,

$$N \operatorname{sinc} Nx * \operatorname{sgn} x = \frac{2}{\pi} \operatorname{Si}(N\pi x).$$

The overshoot, amounting to 9 per cent of the amount of the discontinuity, remains at 9 per cent, but the maximum is reached nearer to the discontinuity. The same applies to the minimum that occurs on the negative side.

Now we see precisely what happens when a Fourier series is truncated. There is overshoot on both sides of any discontinuity, amounting to about 9 per cent, regardless of the inclusion of more and more terms. At any given point to one side of the discontinuity the oscillations decrease indefinitely in amplitude as N increases, so that in the limit the sum of the series approaches the value of the function of which it is the Fourier series (and at the point of discontinuity, the sum of the series approaches the midpoint of the jump). In spite of this, the maximum departure of the sum of the series from the function remains different from zero, and as the maximum moves in close to the step, it approaches the precise value of 9 per cent that was derived for sgn x, because the parts of the function away from the discontinuity have indeed become irrelevant.

This behavior is reminiscent of that of $x\,\delta(x)$, which is zero for all x, even though sequences defining it have nonvanishing maxima and minima.

Finite Fourier transforms In problems where the range of the independent variable is not from $-\infty$ to ∞, advantages accrue from the introduction of finite transforms; for example,

$$F(s,a,b) = \int_a^b f(x)e^{-i2\pi xs}\,dx.$$

With such a definition one can work out an inversion formula, a convolution theorem, theorems for the finite transform of derivatives of functions, and so on. As a particular example of an inversion formula we have

$$F_s(s,0,\tfrac{1}{2}) = \int_0^{\frac{1}{2}} f(x) \sin 2\pi xs\,dx,$$

$$f(x) = 4 \sum_{s=1}^{\infty} F_s(s,0,\tfrac{1}{2}) \sin 2\pi xs.$$

The right-hand side of the last equation will be recognized as the Fourier series for a periodic function of which the segment in the interval $(0,\frac{1}{2})$ is identical with $f(x)$.

It will be evident that the theory of finite transforms will be the same as the theory of Fourier series and that the principal advantage of their use will lie in the approach. We have seen the convenience of embracing Fourier series within the scope of Fourier transforms through the concept of transforms in the limit, and we shall therefore also include finite transforms. Thus we may write the foregoing example in terms of ordinary sine transforms as follows:

$$\int_0^{\frac{1}{2}} f(x) \sin 2\pi xs \, dx = 2 \int_0^{\infty} [\tfrac{1}{2}\Pi(x) f(x)] \sin 2\pi xs \, dx$$

and in the general case

$$\int_a^b f(x) e^{-i2\pi xs} \, dx = \int_{-\infty}^{\infty} \left[\Pi\left(\frac{x - \frac{1}{2}(b+a)}{b-a} \right) f(x) \right] e^{-i2\pi xs} \, dx.$$

In other words, we substitute for integration over a finite range infinite integration of a function which is zero outside the old integration limits.

All the special properties of finite transforms then drop out. For example, the derivative of a function will (in general) be impulsive at the points a and b where it cuts off, and therefore the Fourier transform of the derivative will contain two special terms proportional to the jumps at a and b. It is not necessary to make explicit mention of this property of the transformation when stating the theorem that the Fourier transform of the derivative of a function is $i2\pi s$ times the transform of the function; for example, the Fourier transform of $\Pi'(x)$ is $i2\pi s.\text{sinc } s = 2i \sin \pi s$. However, the derivative theorem for finite transforms contains these additive terms explicitly. Thus

$$\int_a^b f'(x) e^{-i2\pi xs} \, dx = i2\pi s F(s,a,b) + f(a) e^{-i2\pi as} - f(b) e^{-i2\pi bs}.$$

Fourier coefficients If we consider the usual formula for the series coefficients a_n and b_n for a periodic function $p(x)$ of unit period, namely,

$$a_n - ib_n = 2 \int_{-\frac{1}{2}}^{\frac{1}{2}} p(x) e^{-i2\pi nx} \, dx,$$

we note that the integral has the form of a finite Fourier transform. Thus in spite of the fact that $p(x)$ is a function of a continuous variable x, whereas the Fourier series coefficients depend on a variable n which can assume only integral values, Fourier transforms as we have been studying them enter directly into the determination of series coefficients. The finite transform can, we know, be expressed as a standard transform of a slightly different function $\Pi(x)p(x)$ as follows:

$$\int_{-\infty}^{\infty} \Pi(x) p(x) e^{-i2\pi nx} \, dx.$$

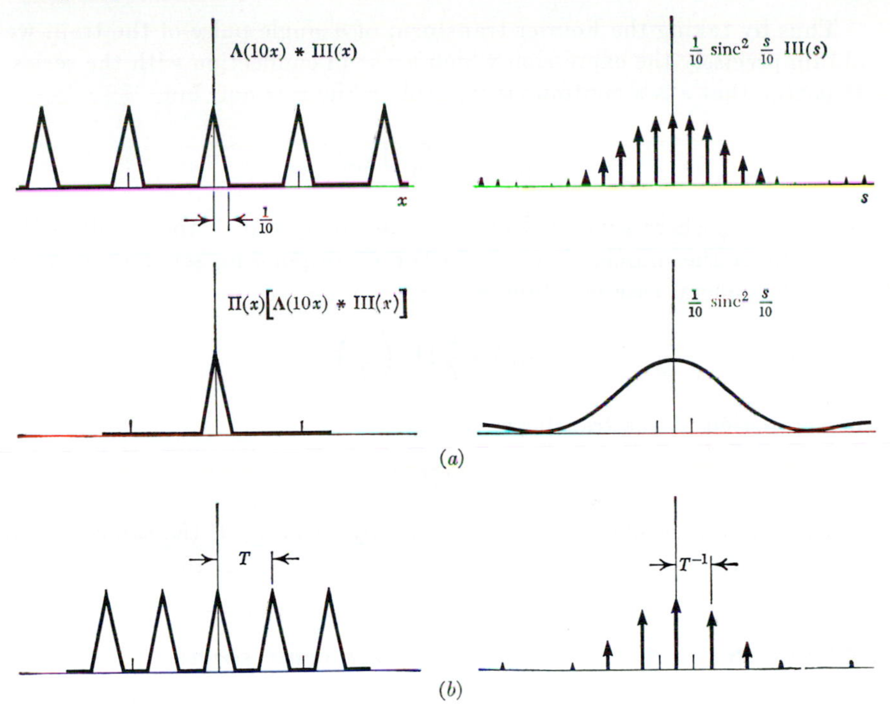

Fig. 10.14 Obtaining Fourier series coefficients.

Our ability to handle transforms can thus be freely applied to the determination of Fourier series coefficients.

As an example consider a periodic train of narrow triangular pulses shown in Fig. 10.14*a*,

$$\sum_{n=-\infty}^{\infty} \Lambda[10(x-n)].$$

With the aid of the *shah* notation this train can also be expressed in the form

$$\Lambda(10x) * \mathrm{III}(x),$$

whose transform is

$$\frac{1}{10}\,\mathrm{sinc}^2\,\frac{s}{10}\,\mathrm{III}(s).$$

We note that the $\frac{1}{10}$ sinc² $(s/10)$ part of this expression came from

$$\int_{-\infty}^{\infty} \Lambda(10x)e^{-i2\pi sx}\,dx,$$

which, in terms of the periodic train $\Lambda(10x) * \mathrm{III}(x)$, is the same as

$$\int_{-\infty}^{\infty} \Pi(x)[\Lambda(10x) * \mathrm{III}(x)]e^{-i2\pi sx}\,dx.$$

Thus by taking the Fourier transform of a single pulse of the train we obtain precisely the expression which arose in connection with the series. It is true that s is a continuous variable while n is not, but

$$\frac{1}{10}\,\mathrm{sinc}^2\,\frac{s}{10}\,\mathrm{III}(s)$$

is zero everywhere save at discrete values of s, and at these values the strength of the impulse is equal to the corresponding series coefficient.

In the general case of a function

$$f(x) * \frac{1}{T}\,\mathrm{III}\left(\frac{x}{T}\right)$$

of any period T, the transform

$$F(s)\mathrm{III}(Ts)$$

shows that the coefficients are obtained by reading off the same $F(s)$ at different intervals $s = T^{-1}$. This is illustrated in Fig. 10.14*b*.

The shah symbol is its own Fourier transform

The *shah* symbol $\mathrm{III}(x)$ is defined by

$$\mathrm{III}(x) = \sum_{-\infty}^{\infty} \delta(x - n)$$

and therefore, in accordance with the approach being adopted here, is to be considered in terms of defining sequences. If, as asserted, its Fourier transform proves to be $\mathrm{III}(s)$, then it too will be considered in terms of sequences.

We proceed therefore to construct a sequence of regular Fourier transform pairs of ordinary functions such that one sequence is suitable for defining $\mathrm{III}(x)$, and we then see whether the other sequence defines $\mathrm{III}(s)$.

Consider the function

$$f(x) = \tau^{-1} e^{-\pi\tau^2 x^2} \sum_{n=-\infty}^{\infty} e^{-\pi\tau^{-2}(x-n)^2}.$$

For a given small value of τ (which we shall later allow to vary to generate a sequence), the function $f(x)$ represents a row of narrow Gaussian spikes of width τ, the whole multiplied by a broad Gaussian envelope of width τ^{-1} (see Fig. 10.15). As $\tau \to 0$, each spike narrows in on an integral value of x and increases in height. For any value of x not equal to an integer, we can show that $f(x) \to 0$ as $\tau \to 0$ and, in addition, the area under each

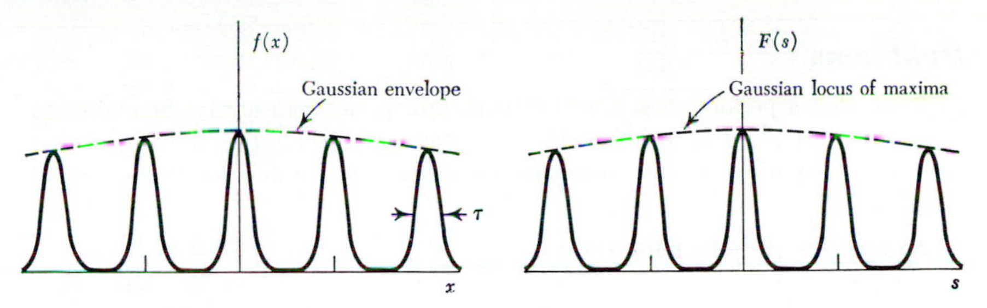

Fig. 10.15 A transform pair for discussing the shah *symbol.*

spike approaches unity. The sequence is therefore a suitable one for defining a set of equal unit impulses situated at integral values of x.

The function $f(x)$ possesses a regular Fourier transform, since $|f(x)|$ is integrable, and there are no discontinuities. The Fourier transform $F(s)$ is given by

$$F(s) = \tau^{-1} \sum_{m=-\infty}^{\infty} e^{-\pi\tau^2 m^2} e^{-\pi\tau^{-2}(s-m)^2}.$$

One way of establishing this is to note that the factor

$$\tau^{-1} \sum_{n=-\infty}^{\infty} e^{-\pi\tau^{-2}(x-n)^2}$$

is periodic, with unit period, and hence may be expressed as a Fourier series. The theory of Fourier series is a well-established branch of mathematics that we are entitled to draw upon here, as long as we do not attempt to make the self-transforming property of the *shah* symbol a basis for reestablishing the theory of Fourier series. The Fourier series is

$$\sum_{m=-\infty}^{\infty} e^{-\pi\tau^2 m^2} \cos 2\pi m x.$$

Therefore
$$\begin{aligned} f(x) &= \sum_{m=-\infty}^{\infty} e^{-\pi\tau^2 m^2} e^{-\pi\tau^2 x^2} \cos 2\pi m x \\ &= \sum_{m=-\infty}^{\infty} e^{-\pi\tau^2 m^2} e^{-\pi\tau^2 x^2} e^{2\pi i m x}. \end{aligned}$$

By applying the shift theorem term by term, we obtain $F(s)$ in the form quoted above.

The function $F(s)$ is a row of Gaussian spikes of width τ with maxima lying on a broad Gaussian curve of width τ^{-1}. As before, it may be verified that as $\tau \to 0$ a suitable defining sequence for the *shah* symbol results.

Problems

1 Show that a periodic function $p(x)$ with unit period can always be expressed in the form $\text{III}(x) * f(x)$ in infinitely many ways, and relate this to the fact that infinitely many different functions can share the same infinite set of equidistant samples.

2 Express the periodic pulse train

$$\sum_{n=-\infty}^{\infty} \Pi[10(x-n)]$$

in the form $\text{III}(x) * f(x)$ in three distinct ways.

3 Determine the Fourier series coefficients for the functions of period equal to unity which, in the interval $-\frac{1}{2} < x < \frac{1}{2}$, are defined as follows:

a. $\cos \pi x$, $\Lambda(2x)$, $\Pi(2x) - \frac{1}{2}$,
b. $\Lambda(8x - 1)$, $(1 - 4x)H(x)$,
c. $e^{-\pi x^2}$, $1 - 4x^2$, $e^{-|x|}$, $e^{-x}H(x)$.

4 The following sample set defines a band-limited function:

$$\ldots\ 0, 10, 30, 50, 50, -40, -35, -10, -5, 0, \ldots$$

All samples omitted are zero. Establish the form of the function by numerical interpolation. What is the minimum value assumed by the function?

5 A certain function is approximately band-limited, that is, a small fraction μ of its power spectrum in fact lies beyond the nominal cutoff frequency. It is sampled at the nominal sampling interval and reconstituted by the usual rules. Use the inequality

$$f(x) \leqslant \int_{-\infty}^{\infty} |F(s)|\, ds$$

to examine how great the discrepancy between the original and reconstituted functions can become.

6 The approximately band-limited function mentioned above is subjected to ordinate-and-slope sampling. Use the inequality

$$f'(x) \leqslant 2\pi \int_{-\infty}^{\infty} |sF(s)|\, ds$$

to show that the discrepancy between the original and reconstituted functions can be serious, even when μ is small.

7 A little noise is added to a band-limited function. Before sampling, one subjects the noisy function to filtering that eliminates the noise components beyond the original cutoff. However, the function is still contaminated with a little band-limited noise. Examine the relative susceptibilities of ordinate-and-slope sampling and of ordinary ordinate sampling to the presence of the noise.

8 A finite sequence of equispaced impulses of finite strength

$$a_0 \, \delta(x) + a_1 \, \delta(x - 1) + \ldots + a_n \, \delta(x - n)$$

is convolved with itself many times, and the result is another sequence of impulses

$$\Sigma \alpha_i \, \delta(x - i).$$

Combine the central-limit theorem with the sampling theorem to show that a graph of α_i against i will approach a normal distribution. State the simple condition that the coefficients $a_0, a_1, \ldots, a_n$ must satisfy.

9 In the previous problem, practice with simple cases that violate the condition, and develop the theory that suggests itself.

10 Show that the overshoot quoted as around 9 per cent in the discussion of the Gibbs phenomenon is given exactly by

$$-\int_1^\infty \operatorname{sinc} x \, dx.$$

11 A carrier telephony channel extends from a low-frequency limit $f_l = 95$ kilocycles per second to a high-frequency limit $f_h = 105$ kilocycles per second. The signal is sampled at a rate of 21 kilocycles per second, and a new waveform is generated consisting of very brief pulses at the sampling instants with strength proportional to the sample values. Make a graph showing the spectral bands occupied by the new waveform, and verify that the original waveform could be reconstructed. Show that in general the critical sampling rate is $2f_h$ divided by the largest integer not exceeding $f_h/(f_h - f_l)$.

12 In the previous problem, show that $2(f_h - f_l)$ is in general too slow a sampling rate to suffice to reconstitute a carrier signal. Show that by suitably interlacing two trains of equispaced samples the average sampling rate can, however, be brought down to twice the bandwidth.

13 A band-limited signal $X(t)$ passes through a filter whose impulse response is $I(t)$. The input and output signals $X(t)$ and $Y(t)$ can be represented by sample values X_i and Y_i. Show that

$$\{Y_i\} = \{I_i\} * \{X_i\}$$

and explain how to derive the sequence $\{I_i\}$.

14 In the previous problem the input signal samples are $\{X_i\} = \{1 \ 2 \ 3 \ 4 \ 5 \ \ldots\}$, and the sequence $\{I_i\}$ describing the filter is $\{1 \ 2 \ 1\}$. Show that the output sequence $\{Y_i\}$ is $\{1 \ 4 \ 8 \ 12 \ \ldots\}$ and that it is possible to work back from the known output and determine the input by evaluating

$$\{1 \ -2 \ 3 \ -4 \ 5 \ \ldots\} * \{1 \ 4 \ 8 \ 12 \ \ldots\}.$$

15 In the previous problem, verify numerically that a particular output signal sample can be expressed in terms of the history of the *output* plus a knowledge of the most recent input signal sample, that is,

$$Y_i = \alpha_1 Y_{i-1} + \alpha_2 Y_{i-2} + \dots + \beta X_i.$$

Show that the coefficients $\alpha_1, \alpha_2, \dots$ and β are given in terms of the reciprocal sequence

$$\{K_0 \ K_1 \ K_2 \dots\} = \{I_i\}^{-1}$$

by

$$\{\alpha_1 \ \alpha_2 \ \alpha_3 \dots\} = -\frac{1}{K_0}\{K_1 \ K_2 \ K_3 \dots\}$$

and

$$\beta = \frac{1}{K_0}.$$

16 The input signal to a filter ceases, but the output continues, with each new sample deducible from the previous ones by the relation

$$Y_i = \alpha_1 Y_{i-1} + \alpha_2 Y_{i-2} + \dots$$

Show that

$$Y(t) = [\alpha_1 \, \delta(t-1) + \alpha_2 \, \delta(t-2) + \dots] * Y(t).$$

In a particular case, there are only two nonzero coefficients:

$$Y_i = 1.65 Y_{i-1} - 0.9 Y_{i-2}.$$

The first two output samples immediately after the input ceased were each 100; calculate and graph enough subsequent output values to determine the general character of the behavior. Show that a damped oscillation $Y(t) = e^{-\sigma t} \cos [\omega(t-a)]$ satisfies the convolution relation given above when

$$\alpha_1 = 2e^{-\sigma} \cos \omega$$

and

$$\alpha_2 = -e^{-2\sigma}$$

Is this band-limited behavior?

17 In the previous problem, show that the series for Y_i contains only a finite number of terms if the filter is constructed of a finite number of inductors, capacitors, and resistors. Hence show that the output due to any input signal is deducible, for a filter whose internal construction is unknown, after a certain number of consecutive sample measurements have been made at its terminals. How would you know that enough samples had been taken?

18 The input voltage $X(t)$, and output voltage $Y(t)$ of an electrical system are sampled simultaneously at regular intervals with the following results.

$X(t)$	15	10	6	2	1	0	0	0	0	0
$Y(t)$	15	15	7.5	−2.75	−2.5					

Calculate the missing values of $Y(t)$, and also calculate what the output would be if, after some time had elapsed, $X(t)$ began to rise linearly.

Chapter 11 The Laplace transform

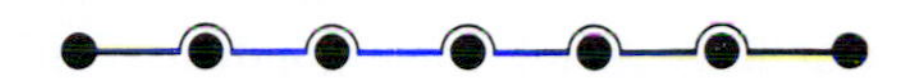

Hitherto we have taken the variables x and s to be real variables. Now, however, let t be a real variable and p a complex one, and consider the integral

$$\int_{-\infty}^{\infty} f(t)e^{-pt}\,dt.$$

This is known as the (two-sided) Laplace transform of $f(t)$ and will be seen to differ from the Fourier transform merely in notation. When the real part of p is zero, the identity with the Fourier transform (as interpreted with noncomplex variables) is complete. In spite of this, however, there is a profound difference in application between the two transforms.

Alternative definitions of the Laplace transform include

$$\int_{0+}^{\infty} f(t)e^{-pt}\,dt,$$

which may be referred to as the one-sided Laplace transform, and

$$p\int f(t)e^{-pt}\,dt,$$

which may be referred to as the p-multiplied form.

The one-sided transform arises in the analysis of transients, where $f(t)$ comes into existence following the throwing of a switch at $t = 0$. However, if we deem that $f(t) = 0$ for $t < 0$, such cases are seen to be included in the two-sided definition. It is not always stressed that the lower limit of the integral defining the one-sided Laplace transform is $0+$; indeed in practice it is normally written as 0. One must remember that

$$\int_{0}^{\infty} f(t)e^{-pt}\,dt \qquad \text{usually means} \qquad \lim_{h\to 0}\int_{|h|}^{\infty} f(t)e^{-pt}\,dt.$$

Many writers have adopted the p-multiplied form for the sake of continuity with the operational notation of Heaviside, whose early success with the differential equations of electrical transients stimulated still-continuing development in transform theory. The admittance operator $Y(p)$, where p is interpreted as the time-derivative operator, turns out to be the p-multiplied Laplace transform of the response to a unit voltage step.

The p-multiplied transform is now probably dying out, and so the connection with a good deal of electrotechnical and applied-mathematical literature will be disturbed. On the other hand, by omitting the p, one can have exact correspondence with the Fourier transform. Furthermore, one can preserve the compactness of routine operational solutions (which save a line or two), by dealing in terms of the *impulse response* instead of, as formerly, the step response. Thus in the operational relation

$$i(t) = Y(p)v(t)$$

between an applied voltage waveform $v(t)$ and a response current $i(t)$, where p is interpreted as the time-derivative operator, $Y(p)$ proves to be the not-p-multiplied two-sided Laplace transform of the response to a unit voltage impulse. This operational formula is often also true for the one-sided transform.

The inversion formula for regaining $f(t)$ from its Laplace transform has the sort of symmetry possessed by the inverse Fourier transformation. However, the variable p, unlike ω, is normally complex, so that a contour of integration in the complex plane of p must be specified. Thus

$$f(t) = \frac{1}{2\pi i} \int_{c-i\infty}^{c+i\infty} F_L(p) e^{pt}\, dp,$$

where

$$F_L(p) = \int_{-\infty}^{\infty} f(t) e^{-pt}\, dt,$$

and c is a suitably chosen positive constant.

Advantages of the Laplace transform over the Fourier transform for handling electrical transient and other problems are often quoted in books. The essential advantage of the Fourier transform is its physical interpretability—as a spectrum, a diffraction pattern, and so on. Laplace transforms are not so interpretable, and once we take Laplace transforms of an equation we retain only a mathematical, not a physical, grasp of its meaning.

In general, however, the Laplace transform is used for initial-value problems, such as transients in dynamical systems, and is not used at all in a broad class of problems included in the present work.

Convergence of the Laplace integral

The definition integral tells us that we are to multiply a given function $f(t)$ by e^{-pt}, where p is any complex number, and integrate from $-\infty$ to $+\infty$. If we consider for a moment only values of p which are real, we can graph the product $f(t)e^{-pt}$ for a sequence of values of p (see Fig. 11.1). When $p = 0$, the product is equal to $f(t)$ itself, hence the Laplace transform at $p = 0$ is equal to the infinite integral of $f(t)$; that is,

$$F_L(0) = \int_{-\infty}^{\infty} f(t)\,dt.$$

The areas under the successive curves labeled $p = -0.1$, -0.2 give $F_L(-0.1)$, $F_L(-0.2)$. Clearly, as p goes more and more negative, e^{-pt} represents an ever-larger factor by which the values of $f(t)$ for positive values of t have to be multiplied. There may be a limit, as suggested by the figure, which p must exceed if the integral is to remain finite. Let this

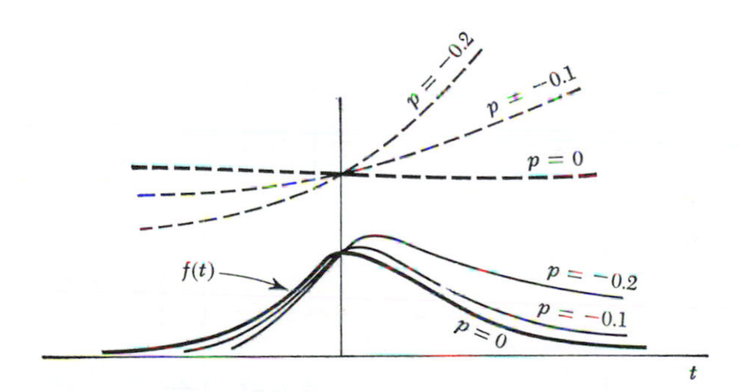

Fig. 11.1 The broken curves show e^{-pt}, and the full curves $e^{-pt}f(t)$, for certain real values of p.

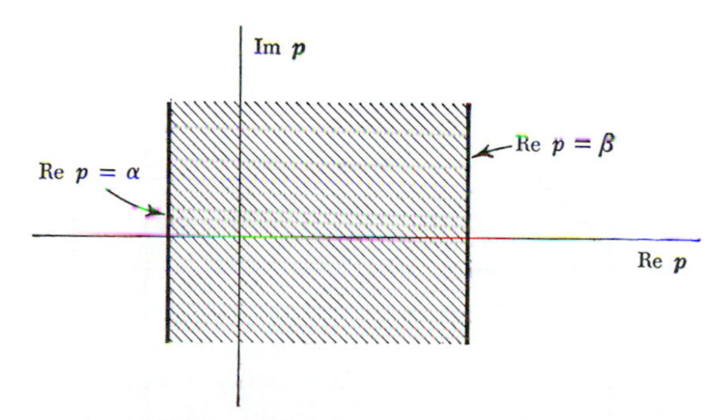

Fig. 11.2 Complex plane of p showing the strip of convergence.

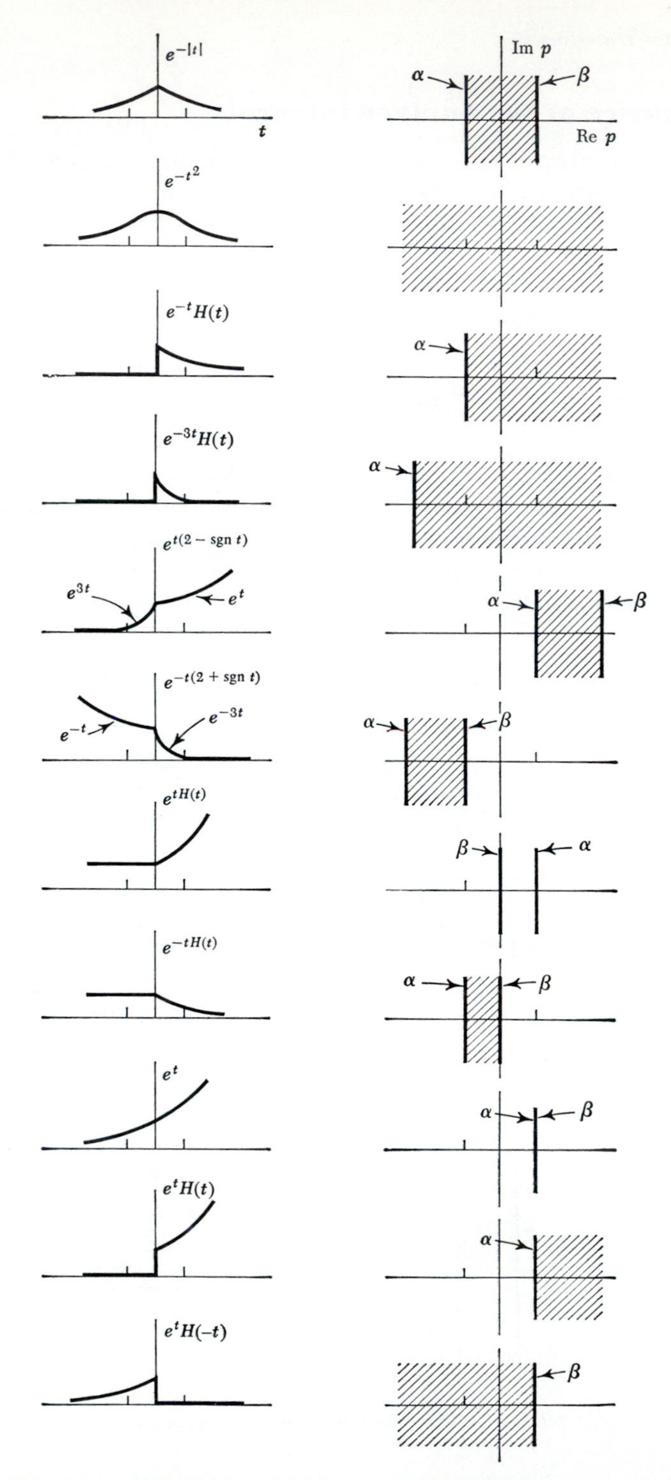

Fig. 11.3 Some simple functions and their strips of convergence.

limiting value of p be α. When p has an imaginary part, it can be shown that the integral will not be finite if the real part of p falls below this same real number α. Similarly, if positive real values of p are considered, it is possible that the area under the product may become infinite to the left of the origin. Of course, if $f(t) = 0$ for $t < 0$ this cannot happen, nor can it happen if the definition integral is one-sided. But in general there will be a real number β which p may not exceed, and as before it can be shown in the complex case that the real part of p is what counts.

Laplace transforms therefore may not exist for all values of p, and in the typical case there is a range specified by

$$\alpha < \text{Re } p < \beta$$

in which the definition integral converges. This means that in the complex plane of p, convergence occurs in a strip (see Fig. 11.2). It is not necessary that convergence occur on the imaginary axis of p, as happens in the figure, or that it should occur at all.

Consider, for example, $f(t) = e^t$. This elementary function, widely used in circuit analysis and other subjects studied by means of the Laplace transform, does not possess a convergent Laplace integral if $\text{Re } p > 1$, or if $\text{Re } p < 1$. In this case the strip of convergence has contracted to a line: the integral converges only where $\text{Re } p = 1$, and even then not exactly at $p = 1$.

The function $\exp [tH(t)]$ would require $\text{Re } p < 0$ for the left-hand part of the integral to converge; but convergence of the right-hand side requires $\text{Re } p > 1$; hence there can be convergence for no value of p. On the other hand, $\exp [-tH(t)]$ can converge on the right-hand side if $\text{Re } p > -1$, hence there is a strip of convergence such that $-1 < \text{Re } p < 0$.

To summarize the convergence situation let us note that the left-hand limit α is determined by the behavior of the right-hand part of $f(t)$. The more rapidly $f(t)$ dies away with increasing t, the further to the left α may be, and if $f(t)$ falls to zero and remains zero, or diminishes as $\exp(-t^2)$ then α recedes to $-\infty$.

Similarly, the right-hand limit β depends on the left-hand half of $f(t)$, receding to $+\infty$ if, for example, $f(t)$ is zero to the left of some point. Examples illustrating the various possibilities are given in Fig. 11.3.

Theorems for the Laplace transform

Most of the theorems given for the Fourier transform are restated in Table 11.1 for the Laplace transform. To a large extent the theorems are similar, the main additional item concerning the strip of convergence.

Table 11.1 Theorems for the Laplace transform

Theorem	$f(t)$	$F(p)$	*Strip of convergence*
Similarity	$f(at)$	$\frac{1}{\lvert a\rvert} F\left(\frac{p}{a}\right)$	$\alpha < a^{-1} \operatorname{Re} p < \beta$
Addition	$f_1(t) + f_2(t)$	$F_1(p) + F_2(p)$	$\max(\alpha_1,\alpha_2) < \operatorname{Re} p < \min(\beta_1,\beta_2)$
Shift	$f(t-\tau)$	$e^{-\tau p}F(p)$	$\alpha < \operatorname{Re} p < \beta$
Modulation	$f(t) \cos \omega t$	$\frac{1}{2}F(p - i\omega) + \frac{1}{2}F(p + i\omega)$	$\alpha < \operatorname{Re} p < \beta$
	$e^{-at}f(t)$	$F(p+a)$	$\alpha - \operatorname{Re} a < \operatorname{Re} p < \beta - \operatorname{Re} a$
Convolution	$f_1(t) * f_2(t)$	$F_1(p)F_2(p)$	$\max(\alpha_1,\alpha_2) < \operatorname{Re} p < \min(\beta_1,\beta_2)$
Product	$f_1(t)f_2(t)$	$\frac{1}{2\pi i}\int_{c-i\infty}^{c+i\infty} F_1(s)F_2(p-s)\,ds$	$\alpha_1 + \alpha_2 < \operatorname{Re} p < \beta_1 + \beta_2,\ \alpha_1 < c < \beta_1$
Autocorrelation	$f(t) * f(-t)$	$F(p)F(-p)$	$\lvert\operatorname{Re} p\rvert < \min(\lvert\alpha\rvert,\lvert\beta\rvert)$
Differentiation	$f'(t)$	$pF(p)$	$\alpha < \operatorname{Re} p < \beta$ (often)
Finite difference	$f(t + \frac{1}{2}\tau) - f(t - \frac{1}{2}\tau)$	$2 \sinh \frac{1}{2}\tau p\, F(p)$	$\alpha < \operatorname{Re} p < \beta$
Integration	$\int_{-\infty}^{t} f(u)\,du$	$p^{-1}F(p)$	$\max(\alpha,0) < \operatorname{Re} p < \beta$
	$\int_{t}^{\infty} f(u)\,du$	$p^{-1}F(p)$	$\alpha < \operatorname{Re} p < \min(\beta,0)$
Reversal	$f(-t)$	$F(-p)$	$-\beta < \operatorname{Re} p < -\alpha$

Transient-response problems

Let a time-varying stimulus $V_1(t)$ elicit a response $V_2(t)$ from a linear time-invariant system. For example, $V_1(t)$ might be the voltage applied to the input of a linear time-invariant electrical network, and $V_2(t)$ might be the output voltage caused by $V_1(t)$. Then the relation between the two will often be conveniently expressible in the form of a linear differential equation with constant coefficients:

$$a_0V_1(t) + a_1\frac{d}{dt}V_1(t) + a_2\frac{d^2}{dt^2}V_1(t) + \cdots$$
$$= b_0V_2(t) + b_1\frac{d}{dt}V_2(t) + b_2\frac{d^2}{dt^2}V_2(t) + \cdots.$$

Such a differential equation is suitable for solution by the Laplace transformation. Multiply each term by exp $(-pt)$ and integrate each term from $-\infty$ to ∞. Then

$$a_0\int_{-\infty}^{\infty} V_1(t)e^{-pt}\,dt + a_1\int_{-\infty}^{\infty} V'(t)e^{-pt}\,dt + \cdots$$
$$= b_0\int_{-\infty}^{\infty} V_2(t)e^{-pt}\,dt + b_1\int_{-\infty}^{\infty} V_2'(t)e^{-pt}\,dt + \cdots$$

Using bar notation for Laplace transforms, that is,

$$\bar{V}_1(p) = \int_{-\infty}^{\infty} V_1(t)e^{-pt}\,dt,$$

and recalling that the Laplace transform of the derivative $V_1'(t)$ is $p\bar{V}_1(p)$, we have

$$a_0\bar{V}_1(p) + a_1p\bar{V}_1(p) + a_2p^2\bar{V}_1(p) + \cdots$$
$$= b_0\bar{V}_2(p) + b_1p\bar{V}_2(p) + b_2p^2\bar{V}_2(p) + \cdots.$$

This equation relating the Laplace transforms of $V_1(t)$ and $V_2(t)$ is known as the *subsidiary equation*. It is not a differential equation; it is merely an algebraic equation and may readily be solved for $\bar{V}_2(p)$. Thus, where the denominator is not zero,

$$\bar{V}_2(p) = \frac{a_0 + a_1p + a_2p^2 + \cdots}{b_0 + b_1p + b_2p^2 + \cdots}\bar{V}_1(p).$$

This simple step typifies the essence of transform methods. By translating a problem to the transform domain one gets a simpler problem.

It will be found that partial differential equations in which there are two independent variables, such as x and t on a transmission line, are reduced to ordinary differential equations. Simultaneous differential equations are reduced to simultaneous algebraic equations, and so on.

In the present case, by the act of solving the trivial algebraic problem, we have solved the differential equation. It only remains to retranslate the result into the original domain.

If $V_1(t)$ is a simple conventional stimulus, and we shall refer later to cases in which it is not, then its Laplace transform $\bar{V}_1(p)$ may be recalled from the memory or from a small stock of transform pairs kept for reference. The reference list is eked out by theorems and one or two established procedures, for example, the expansion of ratios of polynomials in partial fractions. After calculating $\bar{V}_2(p)$ from $\bar{V}_1(p)$ by multiplication with the ratio of polynomials as indicated, one retransforms, by the same procedures, to find $V_2(t)$.

Laplace transform pairs

It is rarely necessary to evaluate a Laplace transform by integration. Most frequently, the method is used on systems with a finite number of degrees of freedom, for which a quite small stock of transform pairs suffices. The key pair is

$$e^{-\alpha t}H(t) \supset \frac{1}{p+\alpha} \qquad -\operatorname{Re}\alpha < \operatorname{Re} p.$$

Taking α to be zero, we have

$$H(t) \supset \frac{1}{p} \qquad 0 < \operatorname{Re} p.$$

Taking $\alpha = i\omega$, we have

$$e^{-i\omega t}H(t) \supset \frac{1}{p+i\omega} \qquad 0 < \operatorname{Re} p$$

and

$$e^{i\omega t}H(t) \supset \frac{1}{p-i\omega} \qquad 0 < \operatorname{Re} p.$$

Adding the last two pairs, we have

$$\begin{aligned}\cos\omega t\, H(t) &\supset \frac{1}{2}\frac{1}{p-i\omega} + \frac{1}{2}\frac{1}{p+i\omega} \\ &= \frac{p}{p^2+\omega^2} \qquad 0 < \operatorname{Re} p,\end{aligned}$$

and subtracting, we have

$$\begin{aligned}\sin\omega t\, H(t) &\supset \frac{1}{2i}\frac{1}{p-i\omega} - \frac{1}{2i}\frac{1}{p+i\omega} \\ &= \frac{\omega}{p^2+\omega^2} \qquad 0 < \operatorname{Re} p.\end{aligned}$$

Table 11.2 Some Laplace transforms

$f(t)$	$\int_{-\infty}^{\infty} f(t)e^{-pt}\,dt$	*Strip of convergence*
$e^{-\alpha t}H(t)$	$\dfrac{1}{p+a}$	$-\mathrm{Re}\,\alpha < \mathrm{Re}\,p$
$-e^{-\alpha t}H(t)$		$\mathrm{Re}\,p < -\mathrm{Re}\,\alpha$
$H(t)$	$\dfrac{1}{p}$	$0 < \mathrm{Re}\,p$
$tH(t)$	$\dfrac{1}{p^2}$	$0 < \mathrm{Re}\,p$
$\delta(t)$	1	all p
$\delta'(t)$	p	all p
$(1 - e^{-\alpha t})H(t)$	$\dfrac{\alpha}{p(p+\alpha)}$	$\max(0, -\mathrm{Re}\,\alpha) < \mathrm{Re}\,p$
$\cos \omega t\, H(t)$	$\dfrac{p}{p^2+\omega^2}$	$0 < \mathrm{Re}\,p$
$\sin \omega t\, H(t)$	$\dfrac{\omega}{p^2+\omega^2}$	$0 < \mathrm{Re}\,p$
$te^{-\alpha t}H(t)$	$\dfrac{1}{(p+\alpha)^2}$	$-\mathrm{Re}\,\alpha < \mathrm{Re}\,p$
$e^{-\sigma t}\cos \omega t\, H(t)$	$\dfrac{p+\sigma}{(p+\sigma)^2+\omega^2}$	$-\sigma < \mathrm{Re}\,p$
$e^{-\sigma t}\sin \omega t\, H(t)$	$\dfrac{\omega}{(p+\sigma)^2+\omega^2}$	$-\sigma < \mathrm{Re}\,p$
$e^{-\alpha\|t\|}$	$\dfrac{2\alpha}{\alpha^2-p^2}$	$-\mathrm{Re}\,\alpha < \mathrm{Re}\,p < \mathrm{Re}\,\alpha$
$\Pi(t)$	$\dfrac{\sinh \frac{1}{2}p}{\frac{1}{2}p}$	all p
$\Pi(t-\frac{1}{2})$	$\dfrac{1-e^{-p}}{p}$	all p
$\Lambda(t)$	$\left(\dfrac{\sinh \frac{1}{2}p}{\frac{1}{2}p}\right)^2$	all p
$\mathrm{III}(t)H(t+\frac{1}{2})$	$\dfrac{1}{1-e^{-p}}$	$0 < \mathrm{Re}\,p$
${}^{\mathrm{I}}{}_{\mathrm{I}}(t)$	$\sinh \frac{1}{2}p$	all p
$\mathrm{II}(t)$	$\cosh \frac{1}{2}p$	all p
$-H(-t)$	$\dfrac{1}{p}$	$\mathrm{Re}\,p < 0$
$\mathrm{sgn}\, t$	$\dfrac{2}{p}$	$\mathrm{Re}\,p = 0$
$\mathrm{III}(t)H(t)$	$\coth \frac{1}{2}p$	$0 < \mathrm{Re}\,p$

By applying the integration theorem to $H(t) \supset p^{-1}$ we find

$$tH(t) \supset \frac{1}{p^2} \qquad 0 < \text{Re } p,$$

and the derivative theorem gives

$$\delta(t) \supset 1 \qquad -\infty < \text{Re } p < \infty.$$

By the addition theorem, we have

$$(1 - e^{-\alpha t})H(t) \supset \frac{\alpha}{p(p + \alpha)} \qquad \max(0, -\text{Re } \alpha) < \text{Re } p.$$

By differentiating $\exp(-\alpha t)H(t) \supset (p + \alpha)^{-1}$ with respect to α, we have

$$te^{-\alpha t}H(t) \supset \frac{1}{(p + \alpha)^2} \qquad -\text{Re } \alpha < \text{Re } p.$$

Splitting α into real and imaginary parts, $\alpha = \sigma + i\omega$, we have

$$e^{-(\sigma \pm i\omega)t}H(t) \supset \frac{1}{p + \sigma \pm i\omega} \qquad -\sigma < \text{Re } p,$$

whence $$e^{-\sigma t} \cos \omega t \, H(t) \supset \frac{p + \sigma}{(p + \sigma)^2 + \omega^2} \qquad -\sigma < \text{Re } p,$$

and $$e^{-\sigma t} \sin \omega t \, H(t) \supset \frac{\omega}{(p + \sigma)^2 + \omega^2} \qquad -\sigma < \text{Re } p.$$

From the representative variety of familiar transient waveforms that are so readily derivable from

$$e^{-\alpha t}H(t) \supset \frac{1}{p + \alpha},$$

it is clear that the exponential waveform plays a fundamental role. As shown elsewhere, exponential response to exponential stimulus is synonymous with linearity plus time invariance. All the time functions listed are thus inherent in linear time-invariant systems and may be elicited by simple stimuli.

For reference, the examples are tabulated in Table 11.2, together with a few other examples that are characteristic of simple stimuli. These further examples may all be readily obtained by integration, or converted directly from the known Fourier transforms, by replacing $i2\pi f$ by p. The strip of convergence has to be considered separately.

Natural behavior

In the transient-response problem described above, the answer $V_2(t)$ is the response elicited by the stimulus $V_1(t)$. The system may also, however, be capable of natural behavior that requires no stimulus for its maintenance. Usually, some prior stimulus has been applied to the network to excite the natural behavior (but not necessarily; the system may have been assembled from elements containing stored energy).

Evidently any natural behavior associated with the circulation of energy not injected by the stimulus $V_1(t)$ has to be added to $V_2(t)$ to give the total behavior.

The well-known procedure for finding the natural behavior amounts to solving for a response $V_2(t)$ when the stimulus $V_1(t) = 0$. Thus putting $\bar{V}_1(p) = 0$ in the subsidiary equation, we have

$$b_0 + b_1 p + b_2 p^2 + \ . \ . \ . = 0.$$

This is known as the *characteristic equation*, and it furnishes the set of complex roots $\lambda_1, \lambda_2, \ldots$, that define the natural modes of the system.

A variation $V_2(t) = \exp(\lambda_k t)$ will be seen on substitution in the differential equation to be a solution when $V_1(t) = 0$. A simultaneous combination of all the modes, with arbitrary strengths, expresses the natural behavior in its most general form: $\sum_k a_k \exp(\lambda_k t)$, provided the roots $\lambda_1, \lambda_2, \ldots$ are all different.

If there is a repeated root, say for example that λ_3 equals λ_4, then non-exponential behavior $t \exp \lambda_3 t$ can occur. In a sense this is conceivable as a combination of exponential modes as follows. Consider a system in which λ_3 and λ_4 are almost equal, and then consider a sequence of systems in which λ_3 and λ_4 approach equality as attention progresses from one system to the next. The expression

$$a_3 e^{\lambda_3 t} + a_4 e^{\lambda_4 t}$$

denotes possible forms of behavior, where a_3 and a_4 are arbitrary strengths. Let a_3 be equal to a_4 in magnitude, but opposite in sign. Let $\lambda_4 = \lambda_3 + \delta\lambda$. Then the expression above becomes

$$a_3[e^{\lambda_3 t} - e^{(\lambda_3 + \delta\lambda)t}],$$

which describes a humped waveform, rising from zero and then falling back; but as $\delta\lambda$ approaches zero the expression also goes to zero. Since, however, the coefficients a_k are arbitrary, and the modes may be evoked in any strength, we are quite entitled to amplify the faint behavior as $\delta\lambda$ diminishes. Suppose that as we move from system to system we make the coefficients a_3 and a_4 increase in inverse proportion to $\delta\lambda$. Then a sequence of possible behaviors is

$$\frac{e^{(\lambda_3+\delta\lambda)t} - e^{\lambda_3 t}}{\delta\lambda},$$

an expression which in the limit is the derivative of exp $(\lambda_3 t)$ with respect to λ_3. Therefore, in the limiting system, where λ_4 is equal to λ_3,

$$te^{\lambda_3 t}$$

is a possible time variation and any multiple thereof. It might seem that the merging modes would have to be hit extremely hard in order to produce this result, but in fact only a finite amount of energy is required. Hence, if repeated roots occur, derivatives of exp $(\lambda_k t)$ with respect to λ_k, must be included in the expression for the most general natural behavior.

Impulse response and transfer function

Let $V_2(t) = I(t)$ when $V_1(t) = \delta(t)$, irrespective of the choice of the origin of t. Then $I(t)$ may be called the impulse response, and by the superposition property,

$$V_2(t) = \int_{-\infty}^{\infty} I(u)V_1(t-u)\,du.$$

Now if $V_1(t)$ is an exponential function,

$$V_1(t) = e^{pt},$$

where p is some complex constant, it follows that

$$\begin{aligned} V_2(t) &= \int_{-\infty}^{\infty} I(u)e^{p(t-u)}\,du \\ &= \left[\int_{-\infty}^{\infty} I(u)e^{-pu}\,du\right] e^{pt}. \end{aligned}$$

This result shows that the response to exponential stimulus is also exponential, with the same constant p, but with an amplitude given by the quantity in brackets. We call this quantity the *transfer function*, and it is seen to be the Laplace transform of the impulse response. From the subsidiary equation given earlier we see that the transfer function is

$$\frac{a_0 + a_1p + a_2p^2 + \cdots}{b_0 + b_1p + b_2p^2 + \cdots}.$$

For lumped electrical networks, one can write the transfer function readily from experience with calculating a-c impedance. Hence, through the Laplace transformation, the impulse response also becomes readily accessible. Thus in the circuit of Fig. 11.4, the ratio of $V_2(t)$ to $V_1(t)$ for the case where $V_1(t)$ varies as exp pt is, by inspection,

$$R\frac{\dfrac{\dfrac{1}{R+L_2p}}{\dfrac{1}{R+L_2p}+Cp}}{L_1p+\dfrac{(R+L_2p)\dfrac{1}{Cp}}{R+L_2p+\dfrac{1}{Cp}}} = R\frac{1}{R+(L_1+L_2)p+L_1RCp^2+L_1L_2Cp^3}.$$

By the method of partial fractions, this ratio of polynomials can be expressed in the form

$$\frac{A}{p+a}+\frac{B}{p+b}+\frac{C}{p+c},$$

and thus the impulse response is

$$(Ae^{-at}+Be^{-bt}+Ce^{-ct})H(t).$$

In problems where $V_2(t)$ must be found when $V_1(t)$ is given, two distinct courses must always be borne in mind: one is to proceed by multiplication with the transfer function in the transform domain; the other method is to proceed, by use of the impulse response, directly through the superposition integral.

A strange difference between the two procedures may be noted. The superposition integral

$$V_2(t_1) = \int_{-\infty}^{\infty} I(u)V_1(t_1-u)\,du$$

does not make use of values of $V_1(t)$ for $t > t_1$ during the course of calculating $V_2(t_1)$. (This is clear because of the fact that the integrand vanishes for negative values of u.) In other words, to calculate what is happening now, we need only know the history of excitation up to the present moment.

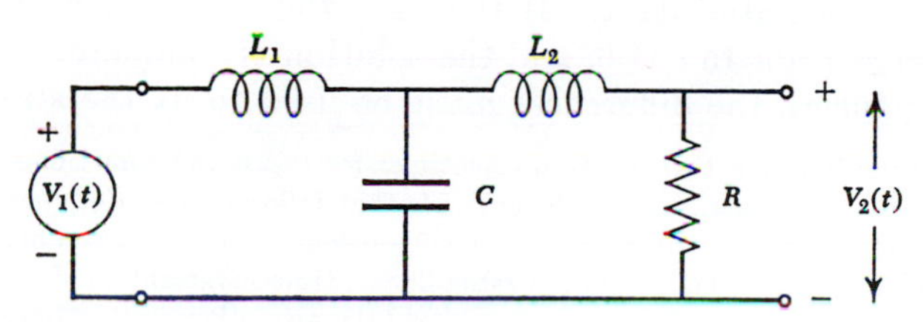

Fig. 11.4 Find $V_2(t)$ when $V_1(t)$ is given.

But the procedure in the transform domain normally makes use of information about the future excitation. Indeed, to calculate the Laplace transform of the excitation $V_1(t)$ we have to know its behavior indefinitely into the future.

As a rule, one obtains the same result for the response at $t = t_1$ irrespective of what $V_1(t)$ does after t_1 passes, but occasional trouble can be expected. For example, it is perfectly reasonable to ask what is the response of a system to $V_1(t) = \exp t^2[H(t)]$, and the superposition integral gives it readily, but the Laplace integral diverges because of troubles connected with the infinitely remote future. One way out of this dilemma is to assume that the excitation is perpetually on the point of ceasing; in other words, assume an excitation $\exp t^2[H(t) - H(t - t_1)]$.

Initial-value problems

Various types of initial-value problems may be distinguished. One type represents the classical problem of a differential equation plus boundary conditions. Thus the driving function is given for all[1] $t > 0$, but no information is provided about the prior excitation of the system. Consequently, some continuing response to prior excitation may still be present, and therefore extra facts must be furnished. The number of extra facts is the same as, or closely connected with, the number of modes of natural behavior, for the continuing response may be regarded as a mixture of natural modes excited by a stimulus that ceased at $t = 0$. Thus sufficient facts must be supplied to allow the strength of each natural mode to be found. These data often consist of the "initial" response and a sufficient number of "initial" derivatives. The word "initial" refers to $t = 0+$; the value at $t = 0$, which may seem to have more claim to the title, is frequently different. To deal with this problem, calculate the response due to a stimulus $V_1(t)$ that is equal to zero for $-\infty < t < 0$,[2] and equal to the given driving function for $t > 0$.[3] Of the many possible particular solutions of the differential equation, this one corresponds to the condition of no stored energy prior to $t = 0$. At $t = 0+$, this response function and its derivatives possess values that may be compared with the initial data. If they are the same, then there was no stored energy prior to $t = 0$, and the solution is complete. But if there are discrepancies, the differences must be used to fix the strengths a_k of

[1] The exciting function $V_1(t)$ is given sometimes for $t \geqslant 0$ and sometimes for $t > 0$, but the statement of the problem, as in the example that follows, often does not say which.

[2] In this statement $-\infty$ means a past epoch subsequent to the assembly of a physical system which has since remained in the same linear, time-invariant condition.

[3] At $t = 0$, take a value equal to $\frac{1}{2}V(0+)$; or take any other *finite* value, since, as previously seen, there is no response to a null function.

the natural modes by solving a small number of simultaneous equations of the form

$$V_2(0+) = \sum_k a_k e^{\lambda_k t}$$
$$V_2'(0+) = \sum_k a_k \lambda_k e^{\lambda_k t},$$
$$\cdots$$

A much better method, which injects the boundary conditions in advance, depends on assuming that the *response*, not the stimulus, is zero for $t < 0$. Let it be required to solve the differential equation

$$b_0 \hat{V}_2(t) + b_1 \hat{V}_2'(t) + b_2 \hat{V}_2''(t) + \cdots = \hat{V}_1(t)$$

for $\hat{V}_2(t)$, given $\hat{V}_1(t)$ for $t > 0$, and the initial values $\hat{V}_2(0+)$, $\hat{V}_2'(0+)$, ... The circumflex accents indicate that the functions are undefined save for positive t. We replace the differential equation by a new one for $V_2(t)$ where

$$V_2(t) = \begin{cases} \hat{V}_2(t) & t > 0 \\ 0 & t < 0. \end{cases}$$

The function $V_2(t)$ is thus defined for all t, and agrees with $\hat{V}_2(t)$, where the latter is defined. Since

$$V_2'(t) = \hat{V}_2(0+)\delta(t) + \begin{cases} \hat{V}_2'(t) & t > 0 \\ 0 & t < 0 \end{cases}$$

and $$V_2''(t) = \hat{V}_2'(0+)\delta(t) + \hat{V}_2(0+)\delta'(t) + \begin{cases} \hat{V}_2''(t) & t > 0 \\ 0 & t < 0, \end{cases}$$

the new differential equation is

$$b_0 V_2(t) + b_1 V_2'(t) + b_2 V_2''(t) + \cdots = V_1(t),$$

where $$V_1(t) = [b_1 \hat{V}_2(0+)\delta(t)] + [b_2 \hat{V}_2'(0+)\delta(t) + b_2 \hat{V}_2(0+)\delta'(t)] + \cdots + \begin{cases} \hat{V}_1(t) & t > 0 \\ 0 & t < 0. \end{cases}$$

This differential equation differs from the original one by the inclusion in the driving function of impulses necessary to produce the initial discontinuities of $V_2(t)$ and its derivatives. Taking Laplace transforms throughout, we have the subsidiary equation

$$b_0 \bar{V}_2(p) + b_1 p \bar{V}_2(p) + b_2 p^2 \bar{V}_2(p) + \cdots = \bar{V}_1(p).$$

Thus $$\bar{V}_2(p) = \frac{\bar{V}_1(p)}{b_0 + b_1 p + b_2 p^2 + \cdots}$$

$$= \frac{[b_1 \hat{V}_2(0+)] + [b_2 \hat{V}_2'(0+) + b_2 \hat{V}_2(0+)p] + \cdots + \int_{0+}^{\infty} \hat{V}_1(t) e^{-pt}\, dt}{b_0 + b_1 p + b_2 p^2 + \cdots}.$$

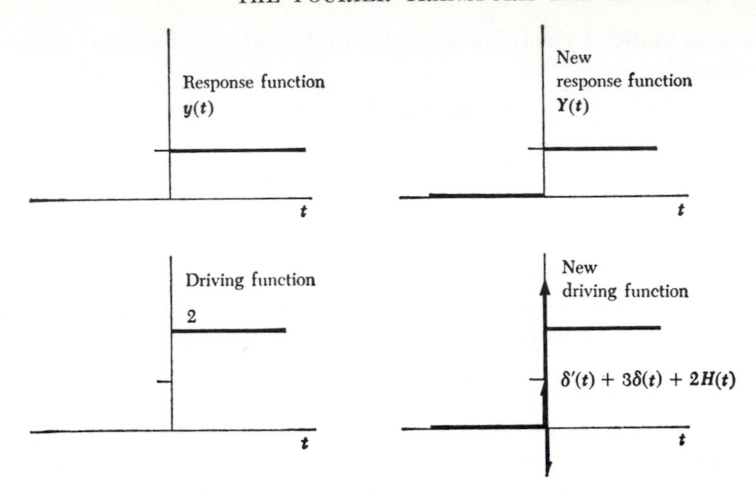

Fig. 11.5 An initial-value problem (left) and its equivalent (right).

Inversion of this transform to the time domain gives $V_2(t)$; and the part of it referring to $t > 0$ gives the desired solution $\hat{V}_2(t)$, with built-in satisfaction of the initial conditions.

Consider the following example: Given the differential equation

$$\frac{d^2y}{dt^2} + \frac{3\,dy}{dt} + 2y = 2,$$

solve for $y(t)$ $(t > 0)$, subject to the initial conditions $y(0) = 1, y'(0) = 0$. (Although this is customary notation, $y(0)$ really stands for $y(0+)$, as we have noted.) It may be observed that the driving function is 2, independent of time. However, it is not explicitly stated whether the driving function is equal to 2 for $t > 0$, or for $t \geqslant 0$. Let

$$Y(t) = \begin{cases} y(t) & t > 0 \\ 0 & t < 0, \end{cases}$$

as shown in Fig. 11.5. Then

$$Y'(t) = y(0+)\,\delta(t) + \begin{cases} y'(t) & t > 0 \\ 0 & t < 0 \end{cases}$$

and

$$Y''(t) = y'(0+)\,\delta(t) + y(0+)\,\delta'(t) + \begin{cases} y''(t) & t > 0 \\ 0 & t < 0, \end{cases}$$

and the new differential equation for $Y(t)$ is

$$Y''(t) + 3Y'(t) + 2Y = \delta'(t) + 3\delta(t) + 2H(t).$$

The subsidiary equation is

$$p^2\bar{Y}(p) + 3p\bar{Y}(p) + 2\bar{Y}(p) = p + 3 + (2/p),$$

whence

$$\bar{Y}(p) = \frac{p + 3 + (2/p)}{p^2 + 3p + 2} = \frac{1}{p}$$

and hence

$$Y(t) = H(t) \text{ for all } t.$$

Thus finally, for the limited region $t > 0$, the desired solution is

$$y(t) = 1.$$

It is now instructive to study this example from the standpoint of investigating the prior excitation, which was not divulged in the statement of the problem, and in lieu of which the initial conditions were given. For the sake of concreteness, let it be observed that

$$L\frac{d^2Q}{dt^2} + R\frac{dQ}{dt} + \frac{Q}{C} = V(t)$$

is the differential equation for the charge Q on the capacitor of a series LRC circuit across which there is a voltage $V(t)$. If $L = 1$, $R = 3$, $C^{-1} = 2$, we deduce from the previous work that a Q of one coulomb, constant for $t > 0$, is compatible with an applied voltage of 2 volts ($t > 0$), and with the conditions $Q(0+) = 1$ and $Q'(0+) = 0$.

Unless the circuit was assembled from elements with energy stored in them, the capacitor charge must have been drawn from the voltage source. We can easily find what the voltage source must have done by making some simple supposition. Suppose, for example, that the charge was placed instantaneously at $t = 0$; that is, take

$$Q(t) = H(t).$$

Then, from the differential equation,

$$V(t) = \delta'(t) + 3\delta(t) + 2H(t).$$

The impulsive parts force a current impulse through the inductance and resistance, placing the charge on the capacitor, and the voltage step holds it there.

We have just solved a problem in which a complete history of excitation is given and no supplementary initial conditions enter. By now suppressing the information about the drastically impulsive behavior of $V(t)$ at $t = 0$ and introducing in its stead the fact that at $t = 0+$ the capacitor was seen to be already charged, with zero rate of change, we return to the original problem.

There are many other ways in which the capacitor could have been given its initial charge. In particular, in this problem, the charging could have been done in the same way but earlier, as described by

$$\begin{aligned} Q(t) &= H(t+1) \\ V(t) &= \delta'(t+1) + 3\delta(t+1) + 2H(t+1). \end{aligned}$$

The positive-t part of these quantities would also satisfy the initial-condition problem, but $Q(0)$ and $V(0)$ would not be the same as before.

The peculiarity of the systematic method given for incorporating initial conditions is that impulsive terms are added to the driving function so as to inject all requisite initial energy instantaneously at $t = 0$.

Setting out initial-value problems

When the principle of the Laplace transform solution is understood, fewer steps in setting out are required. From the differential equation and initial conditions the subsidiary equation is written directly by inspection. Thus, given

$$\frac{d^2y}{dt^2} + \frac{3\,dy}{dt} + 2y = \varphi(t),$$

we write

$$p^2\,{}_0\bar{y}(p) + 3p\,{}_0\bar{y}(p) + 2\,{}_0\bar{y}(p) = [y'(0+) + y(0+)p] + [3y(0+)] + {}_0\bar{\varphi}(p),$$

where
$${}_0\bar{y}(p) \subset \begin{cases} y(t) & t > 0 \\ 0 & t < 0, \end{cases}$$

and
$${}_0\bar{\varphi}(p) \subset \begin{cases} \varphi(t) & t > 0 \\ 0 & t < 0. \end{cases}$$

Since these transforms refer to functions that are zero for $t < 0$, they are equal to the one-sided transforms of the functions $y(t)$, $\varphi(t)$. Early emphasis on this type of problem accounts for the prevalence of the one-sided transform.

Switching problems

Two types of problems have been described which are suited for handling by Laplace transforms:

1. Transient response of a linear, time-invariant system to a driving function given for all t (includes problem of natural behavior on taking driving function to be zero).

2. Initial-value problems where information on the driving function for $t > 0$ is supplemented by initial data regarding the response.

In a third type of problem the time invariance is violated in a simple way by switching. It is important to realize that throwing a switch changes an electric circuit to a different circuit, as regards application of the linear, time-invariant theory. For example, when switching places a battery voltage across a resistor, a new circuit comes into existence. This simple switching problem is quite different in kind from the transient-response problem in which a voltage source in series with a resistor develops a voltage $H(t)$. It is true that in each case a step function of current flows in the resistor, but the second case involves a time-invariant circuit whose resistance is the same before $t = 0$ and after $t = 0$. There is an equivalence between the two problems, but only the second is amenable to the treatment already given for linear, time-invariant systems.

Of course, after the switch is thrown, both systems are identical. Consequently, if initial values of response are furnished ("initial" means after the switch has closed), together with details of the subsequent excitation, if any, the problem can be handled as an initial-value problem; for in those problems one does not inquire into $t < 0$.

More commonly, however, preinitial conditions (at $t = 0-$) are furnished, or the previous history of excitation is given. The technique for handling switching problems is to convert to an equivalent problem concerning a time-invariant circuit, as in the simple case of the battery and resistor.

A voltage $V_1(t)$, given for all t, is applied to a circuit; at $t = 0$ an open switch somewhere in the circuit closes. To find the response $V_2(t)$, at some place, we can proceed as before for times preceding the throwing of the switch. We could also easily calculate what the response to $V_1(t)$ would have been if the switch had always been closed. In fact we expect that this response will become apparent after the switch has been closed long enough, and the circuit has "forgotten" the effects of the preswitching excitation. To find the response during the transition following $t = 0$ is the present problem.

Across the terminals of the open switch a certain voltage exists, and it could be calculated as a transient response to the given driving function $V_1(t)$. Let it be $V_s(t)$. If a voltage source having zero internal impedance were connected across the switch, and if its voltage varied in exact agreement with $V_s(t)$, no current would flow in it. Since no current was flowing through the open switch, the voltage source could thus be connected without disturbing the behavior of the circuit. An important change has, however, been effected, for the zero-impedance connection now introduced across the switch has brought the circuit, for negative values of time, into identity with the circuit which was to be created at $t = 0$ by throwing the switch. The throwing of the switch can be fully simulated by the introduction of a voltage source $V_s(t)H(-t)$. This can be split into two terms $V_s(t) - V_s(t)H(t)$. The response associated with the first term is already known; it is the response, for all t, of the open-switch system to $V_1(t)$. (The superposition of the responses of the closed-switch system to $V_1(t)$ and $V_s(t)$ would give the same result.) The response associated with the second term, $-V_s(t)H(t)$, gives the desired transition term that has to be added to the response of the open-switch system to get the answer to the problem.

In problems where a switch is opened and any current through it thereby interrupted, one applies the principle of duality and introduces a suitable current generator.

Problems

1 State the strip of convergence for the following functions:

$$H(t) \qquad \Pi(t) \qquad \exp(-t^{\frac{1}{2}})H(t) \qquad (1+t^2)^{-1}$$
$$(1+t^3)^{-1}H(t) \qquad (\log t)H(t) \qquad \operatorname{sgn} t \qquad t^{-1}\Pi(t)$$
$$t^{-1} \qquad \delta(t) \qquad (1+e^t)^{-1} \qquad (1+t^3)^{-1}$$

2 Find the Laplace transforms of the following functions, stating the strip of convergence:

$$(e^{-t} - e^t)H(t) \qquad e^{-|t|} \qquad (e^t - e^{-t})H(-t)$$

3 Derive the Laplace transforms of $\Pi(t)$, $\Pi(t - \frac{1}{2})$, and $\Lambda(t)$, stating the strip of convergence.

4 Find the current that flows in a series-LR circuit in response to a voltage $\exp(-|t|)$.

5 From one coil, one capacitor, and one resistor, 17 different two-terminal networks can be made. Tabulate the step and impulse responses of all these.

6 Find the current in a series-LR circuit in response to a voltage $\exp(t) \sin t\, H(t)$.

7 A current impulse passes through a series-LR circuit. What is the voltage across the circuit?

8 The input impedance to a short-circuited length of cable is $75 \tan(10^{-8}f)$. A voltage generator of zero internal impedance applies an impulse. What current flows in response?

9 A voltage $\sin \omega t\, H(t)$ is applied to a series-LR circuit. Show that the response current is

$$\frac{L\omega e^{-Rt/L} - L\omega \cos \omega t + R \sin \omega t}{R^2 + L^2\omega^2} H(t).$$

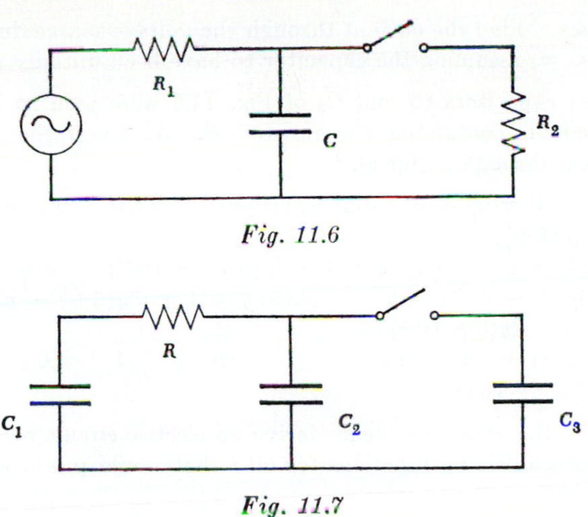

Fig. 11.6

Fig. 11.7

10 A voltage $\cos \omega t\, H(t)$ is applied to a series-RC circuit. Show that the response current is

$$\frac{R\omega^2C^2 \cos \omega t - \omega C \sin \omega t + R^{-1}e^{-t/RC}}{1 + \omega^2C^2R^2} H(t).$$

11 A voltage $\exp(-at)H(t)$ is applied to a series-LR circuit. Show that the response current is

$$\frac{e^{-at} - e^{-Rt/L}}{R - aL} H(t).$$

12 A series-LR circuit is shunted by a capacitor C, and a ramp voltage $tH(t)$ is applied. Show that the response current is

$$\left[C + \frac{t}{R} - \frac{L}{R^2} + \frac{L}{R^2} e^{-Rt/L}\right] H(t).$$

13 An alternating voltage $\sin \omega t\, H(t)$ is applied to a circuit, and direct current is observed. What is the circuit?

14 A steady voltage is applied to a circuit, and a current $\sin \omega t\, H(t)$ results. What is the circuit?

15 A capacitor C carries a charge Q. At $t = -1$ a switch is closed, allowing the capacitor to discharge through a resistor R. At $t = 0$ the switch is opened. Make a graph showing the current flow during the interval $-2 < t < 1$.

16 A resistor and an uncharged capacitor are connected in parallel to form a two-terminal network. At $t = -1$ a battery is switched across the network, and at $t = 0$ the switch is opened. Find the current in the resistor and capacitor for all t.

17 The voltage source applies a voltage $\exp(t)H(t)$ (see Fig. 11.6). At $t = 1$

the switch closes. Find the current through the voltage source during the interval $-\infty < t < \infty$, assuming the capacitor to have been initially uncharged.

18 At $t = -1$ capacitors C_1 and C_2 of Fig. 11.7 were seen to be discharging through the circuit containing the resistor R. At $t = 0$ the switch closed. Find the current through C_1 for all t.

19† Solve the following differential equations for $y(t)(t > 0)$, subject to initial values $y(0+)$, $y'(0+)$, ...

a. $y'(t) + y(t) = 1$ [Ans: $y = 1 + [y(0+) - 1]e^{-t}$]

b. $y''(t) + 5y'(t) + 6y(t) = 6$ [Ans: $y = 1 + [3y(0+) + y'(0+) - 3]e^{-2t} + [2 - y'(0+) - 2y(0+)]e^{-3t}$]

c. $y''(t) + 2y'(t) + y(t) = 2 \sin t$ [Ans: $y = [1 + y(0+) + y'(0+)]te^{-t} + [1 + y(0+)]e^{-t} - \cos t$

20 For each of the above problems derive an electric circuit obeying the same equation. Deduce driving functions for all t that would produce the responses quoted for $t > 0$.

21 Find four different time functions all having $[(p - 2)(p - 1)(p + 1)]^{-1}$ as their Laplace transform. For each time function state the strip of convergence.

22 Show that the Laplace transform of the staircase function $H(t) + H(t - T) + H(t - 2T) + \cdots$ is given by

$$\frac{1}{p(1 - e^{-Tp})} \qquad 0 < \text{Re } p.$$

23 A voltage $v(t)H(t)$ is applied to a circuit that already contained energy before $t = 0$. As no information about the stimulus has been suppressed, no compensating postinitial data are needed. However, in order to get the total behavior $f(t)$ it will be necessary to know about the natural behavior that is going on independently. It may be that the full history of earlier energy injection is unavailable, but it will suffice to know the situation immediately prior to the application of the voltage at $t = 0$. For example, the energy stored in each inductor and capacitor at $t = 0-$ might be given; but suppose here that a sufficient number of preinitial values of the behavior are given: $f(0-)$, $f'(0-)$, $f''(0-)$, ... Show that the *total* behavior can be conveniently calculated by means of a special form of the Laplace transform defined by

$$F_-(p) = \int_{0-}^{\infty} f(t)e^{-pt}\,dt.$$

Work out the theorems for this transform, deducing, for example, that the derivative theorem is $f'(t) \supset pF_-(p) - f(0-)$.

† For a well-organized list of transforms of ratios of polynomials with denominators up to degree 5, which is the practical tool for solving this sort of problem once a sound knowledge of transform methods is attained, see P. A. McCollum and B. F. Brown: "Laplace Transform Tables and Theorems," Holt, Rinehart and Winston, New York, 1965.

Chapter 12 Relatives of the Fourier transform

Many of the linear transforms in common use have a direct connection with either the Fourier or the Laplace transform. The closest relationship is with the generalizations of the Fourier transform to two or more dimensions, and with the Hankel transforms of the zero and higher orders, into which the multidimensional Fourier transforms degenerate under circumstances of symmetry. The Mellin transform is illuminated by previous study of the Laplace transform and is the tool by which the fundamental theory of Fourier kernels is constructed. Particularly impressive is the simplification of the Hilbert transform when it is studied by the Fourier transform, and finally there is the intimate relationship whereby the Abel, Fourier, and Hankel transformations, applied in succession, regenerate the original function.

The two-dimensional Fourier transform

The variable x may stand for some physical quantity such as time or frequency, which is essentially one-dimensional, or it may be the coordinate in a one-dimensional physical system such as a stretched string or an electrical transmission line. However, in cases which are two dimensional—stretched membranes, antennas and arrays of antennas, lenses and diffraction gratings, pictures on television screens, and so on—more general formulas apply.

A two-dimensional function $f(x,y)$ has a two-dimensional transform $F(u,v)$, and between the two the following relations exist:

$$F(u,v) = \int_{-\infty}^{\infty} \int_{-\infty}^{\infty} f(x,y) e^{-i2\pi(ux+vy)}\, dx\, dy$$

$$f(x,y) = \int_{-\infty}^{\infty} \int_{-\infty}^{\infty} F(u,v) e^{i2\pi(ux+vy)}\, du\, dv.$$

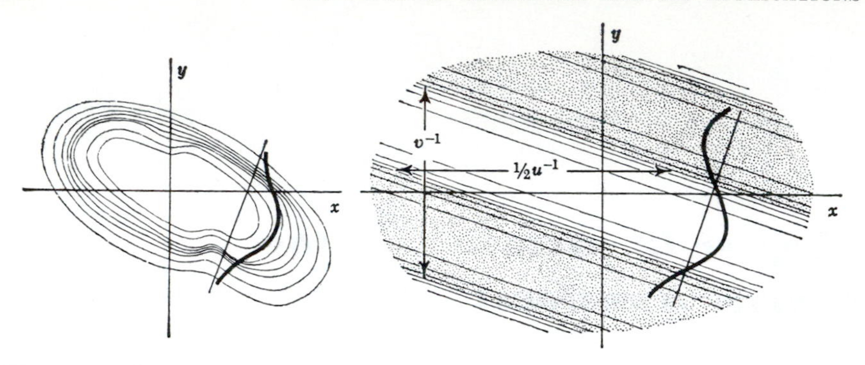

Fig. 12.1 A mountain (left) and a prominent Fourier component thereof (right).

These equations describe an analysis of the two-dimensional function $f(x,y)$ into components of the form $\exp[i2\pi(ux + vy)]$. Since any such component can be split into cosine and sine parts, we may begin by considering a cosine component $\cos[2\pi(ux + vy)]$.

As an example of a two-dimensional function consider the height of the ground at the geographical point (x,y), for example, over the area occupied by the mountain which is conventionally represented in Fig. 12.1 by contours of constant height. The function $\cos[2\pi(ux + vy)]$ represents a cosinusoidally corrugated land surface whose contours of constant height coincide with lines whose equation is

$$ux + vy = \text{const.}$$

The corrugations face in a direction that makes an angle $\arctan(v/u)$ with the x axis, and their wavelength is $(u^2 + v^2)^{-\frac{1}{2}}$. If a section is made through the corrugations, in the x direction, it will undulate with a frequency of u cycles per unit of x. Similarly, v may be interpreted as the number of cycles per unit of y, in the y direction.

In Fig. 12.1 a prominent Fourier component of the mountain is shown. In the transform domain the complex component is characterized in wavelength and orientation by the point (u,v) in the uv plane and its amplitude by $F(u,v)$. The interpretation of u and v as spatial frequencies is emphasized by dimensioning u^{-1} and v^{-1}, the wavelengths of sections taken in the x and y directions, respectively (see Fig. 12.1). The second of the Fourier relations quoted above asserts that a summation of corrugations of appropriate wavelengths and orientations, taken with suitable amplitudes, can reproduce the original mountain. The sinusoidal components, which must also be included, allow for the possibility that the corrugations may have to be slid into appropriate spatial phases.

Two-dimensional convolution

The convolution integral of two two-dimensional functions $f(x,y)$ and $g(x,y)$ is defined by

$$f ** g = \int_{-\infty}^{\infty} \int_{-\infty}^{\infty} f(x',y')g(x - x', y - y')\, dx'\, dy'.$$

Thus one of the functions is rotated half a turn about the origin by reversing the sign of both x and y, displaced, and multiplied with the other function, and the product is then integrated to obtain the value of the convolution integral for that particular displacement.

The two-dimensional autocorrelation function is formed in the same way save that the sign reversal is omitted; thus

$$f \star\star\, g = \int_{-\infty}^{\infty} \int_{-\infty}^{\infty} f(x',y')g(x + x', y + y')\, dx'\, dy'.$$

It is often convenient to be able to perceive ways in which a given function can be expressed as a convolution. For example, the two-dimensional function

$${}^2\Pi(x,y) = \begin{cases} 1 & |x| \text{ and } |y| < \frac{1}{2} \\ 0 & \text{elsewhere} \end{cases}$$

may be expressed as a product or as a convolution:

$${}^2\Pi(x,y) = \Pi(x)\Pi(y) = [\Pi(x)\ \delta(y)] * [\Pi(y)\ \delta(x)].$$

These two possibilities are illustrated in Fig. 12.2.

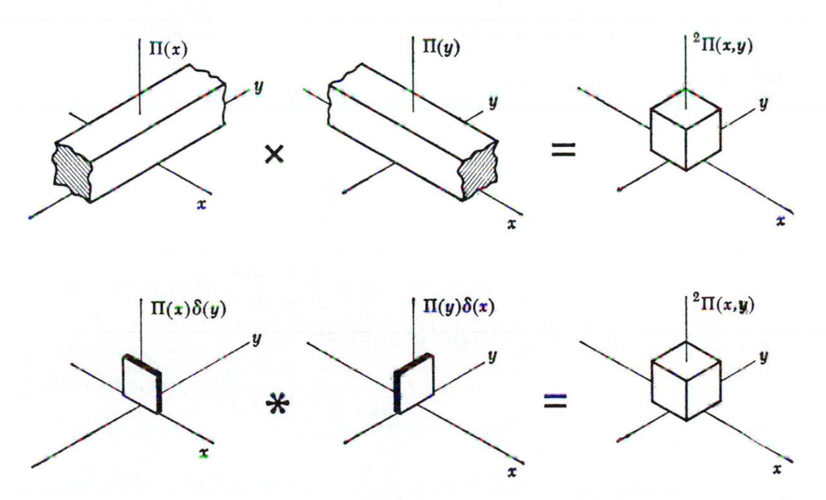

Fig. 1.22 Expressing a two diminsional function as a product and as a convolution.

The theorems pertaining to the one-dimensional transform generalize readily, as shown briefly in Table 12.1.

Figure 12.3 illustrates a number of two-dimensional Fourier transforms.

The Hankel transform

Two-dimensional systems may often show circular symmetry, for example, optical systems are often constructed from components that, in themselves, are circularly symmetrical. Then again, waves spreading out in two dimensions from a source of energy exhibit symmetry for natural reasons. It may be expected that in these cases a simplification will result, for one radial variable will suffice in place of the two independent variables x and y. The appropriate expression of such problems is in terms of the Hankel transform, a one-dimensional transform with Bessel function kernel.

When circular symmetry exists, that is, when

$$f(x,y) = \mathbf{f}(r),$$

where

$$r^2 = x^2 + y^2,$$

Table 12.1 Theorems for the two-dimensional Fourier transform

Theorem	$f(x,y)$	$F(u,v)$
Similarity	$f(ax,by)$	$\dfrac{1}{\lvert ab\rvert} F\left(\dfrac{u}{a}, \dfrac{v}{b}\right)$
Addition	$f(x,y) + g(x,y)$	$F(u,v) + G(u,v)$
Shift	$f(x-a, y-b)$	$e^{-2\pi i(au+bv)}F(u,v)$
Modulation	$f(x,y) \cos \omega x$	$\tfrac{1}{2}F\left(u + \dfrac{\omega}{2\pi}, v\right) + \tfrac{1}{2}F\left(u - \dfrac{\omega}{2\pi}, v\right)$
Convolution	$f(x,y) * g(x,y)$	$F(u,v)G(u,v)$
Autocorrelation	$f(x,y) * f^*(-x,-y)$	$\lvert F(u,v)\rvert^2$
Rayleigh	$\int_{-\infty}^{\infty} \int_{-\infty}^{\infty} \lvert f(x,y)\rvert^2 \, dx\, dy = \int_{-\infty}^{\infty} \int_{-\infty}^{\infty} \lvert F(u,v)\rvert^2 \, du\, dv$	
Power	$\int_{-\infty}^{\infty} \int_{-\infty}^{\infty} f(x,y)g^*(x,y)\, dx\, dy = \int_{-\infty}^{\infty} \int_{-\infty}^{\infty} F(u,v)G^*(u,v)\, du\, dv$	
Parseval	$\int_{-\frac{1}{2}}^{\frac{1}{2}} \int_{-\frac{1}{2}}^{\frac{1}{2}} \lvert f(x,y)\rvert^2 = \sum\sum a_{mn}{}^2,$ where $F(u,v) = \sum\sum a_{mn}[{}^2\delta(u-m, v-n)]$	

Differentiation	$\left(\frac{\partial}{\partial x}\right)^m \left(\frac{\partial}{\partial y}\right)^n f(x,y)$	$(2\pi iu)^m(2\pi iv)^n F(u,v)$
	$\frac{\partial}{\partial x} f(x,y) = f'_x(x,y)$	$2\pi iuF(u,v)$
	$\frac{\partial}{\partial y} f(x,y) = f'_y(x,y)$	$2\pi ivF(u,v)$
	$\frac{\partial^2}{\partial x^2} f(x,y) = f''_{xx}(x,y)$	$-4\pi^2u^2F(u,v)$
	$\frac{\partial^2}{\partial y^2} f(x,y) = f''_{yy}(x,y)$	$-4\pi^2v^2F(u,v)$
	$\frac{\partial^2}{\partial x\, \partial y} f(x,y) = f''_{xy}(x,y)$	$-4\pi^2uvF(u,v)$
	$\left(\frac{\partial^2}{\partial x^2} + \frac{\partial^2}{\partial y^2}\right) f(x,y)$	$-4\pi^2(u^2 + v^2)F(u,v)$
Definite integral	$\int_{-\infty}^{\infty} \int_{-\infty}^{\infty} f(x,y)\, dx\, dy = F(0,0)$	
First moments	$\int_{-\infty}^{\infty} \int_{-\infty}^{\infty} xf(x,y)\, dx\, dy = \frac{1}{-2\pi i} F'_u(0,0)$	
	$\int_{-\infty}^{\infty} \int_{-\infty}^{\infty} (x \cos \theta + y \sin \theta) f(x,y)\, dx\, dy = \frac{1}{-2\pi i} [\cos \theta\, F'_u(0,0) + \sin \theta\, F'_v(0,0)]$	
Center of gravity	$\langle x \rangle = \frac{F'_u(0,0)}{-2\pi i F(0,0)} \qquad \langle y \rangle = \frac{F'_v(0,0)}{-2\pi i F(0,0)}$	
Second moments	$\int_{-\infty}^{\infty} \int_{-\infty}^{\infty} x^2 f(x,y)\, dx\, dy = \frac{F''_{uu}(0,0)}{-4\pi^2}$	
	$\int_{-\infty}^{\infty} \int_{-\infty}^{\infty} xyf(x,y)\, dx\, dy = \frac{F''_{uv}(0,0)}{-4\pi^2}$	
	$\int_{-\infty}^{\infty} \int_{-\infty}^{\infty} (x^2 + y^2) f(x,y)\, dx\, dy = -\frac{1}{4\pi^2} [F''_{uu}(0,0) + F''_{vv}(0,0)]$	
Equivalent width	$\frac{\int_{-\infty}^{\infty} \int_{-\infty}^{\infty} f(x,y)\, dx\, dy}{f(0,0)} = \frac{F(0,0)}{\int_{-\infty}^{\infty} \int_{-\infty}^{\infty} F(u,v)\, du\, dv}$	
Finite differences†	$\Delta_x f(x,y)$	$i2 \sin \pi u\, F(u,v)$
	$\Delta_{xy}{}^2 f(x,y)$	$-4 \sin \pi u \sin \pi v\, F(u,v)$
	$\Delta_{xx}{}^2 f(x,y)$	$-4(\sin \pi u)^2 F(u,v)$
Running means	$\left[\Pi\left(\frac{x}{a}\right) \Pi\left(\frac{y}{b}\right)\right] * f(x,y)$	$ab \operatorname{sinc} au \operatorname{sinc} bv\, F(u,v)$
Separable product	$f(x)g(y)$	$F(u)G(v)$

† The finite differences in the table are defined as follows:

$\Delta_x f(x,y) = f(x + \tfrac{1}{2}, y) - f(x - \tfrac{1}{2}, y)$

$\Delta_{xy}{}^2 f(x,y) = f(x + \tfrac{1}{2}, y + \tfrac{1}{2}) - f(x - \tfrac{1}{2}, y + \tfrac{1}{2}) - f(x + \tfrac{1}{2}, y - \tfrac{1}{2}) + f(x - \tfrac{1}{2}, y - \tfrac{1}{2})$

$\Delta_{xx}{}^2 f(x,y) = f(x + 1, y) - 2f(x,y) + f(x - 1, y).$

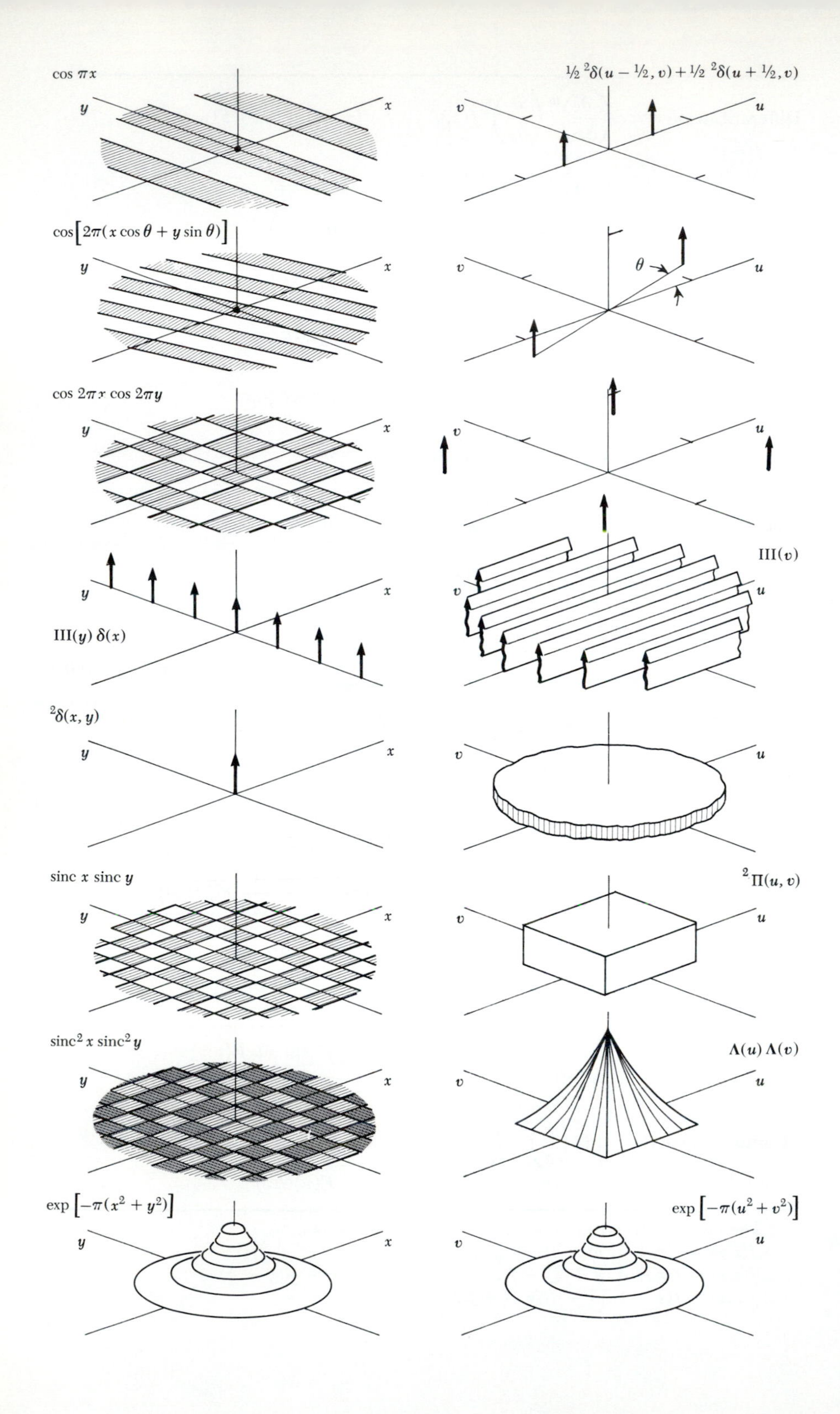

$\cos \pi x$
$\frac{1}{2}\,{}^2\delta(u-\frac{1}{2},v)+\frac{1}{2}\,{}^2\delta(u+\frac{1}{2},v)$
$\cos\left[2\pi(x\cos\theta+y\sin\theta)\right]$
θ
$\cos 2\pi x \cos 2\pi y$
$\mathrm{III}(v)$
$\mathrm{III}(y)\,\delta(x)$
${}^2\delta(x,y)$
$\operatorname{sinc} x \operatorname{sinc} y$
${}^2\Pi(u,v)$
$\operatorname{sinc}^2 x \operatorname{sinc}^2 y$
$\Lambda(u)\,\Lambda(v)$
$\exp\left[-\pi(x^2+y^2)\right]$
$\exp\left[-\pi(u^2+v^2)\right]$
x
y
u
v

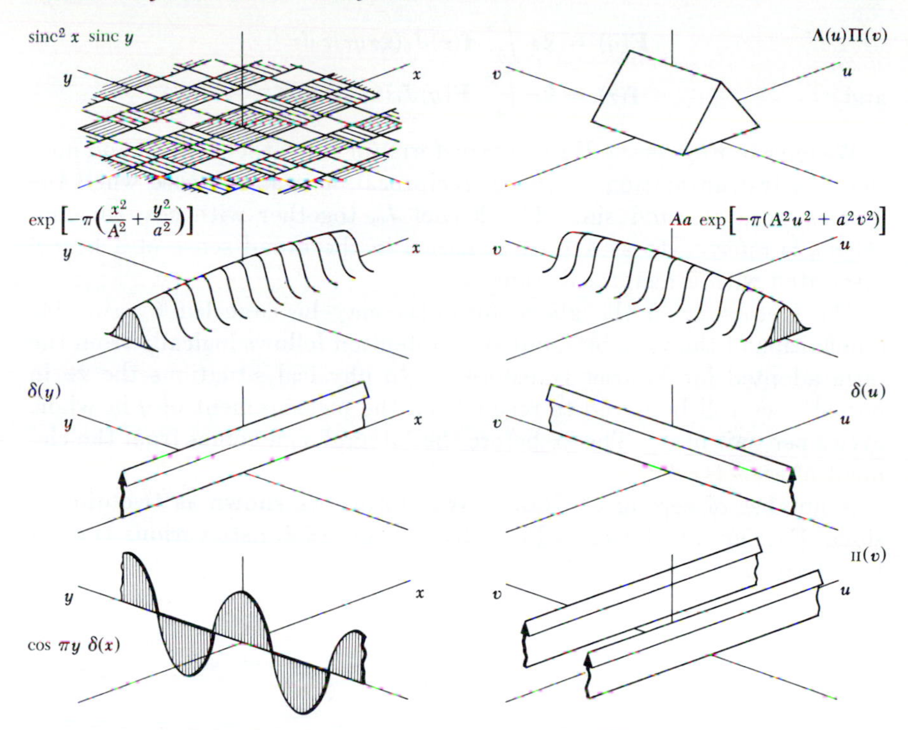

Fig. 12.3 Some two-dimensional Fourier transforms.

then $F(u,v)$ proves also to be circularly symmetrical; that is,

$$F(u,v) = \mathbf{F}(q),$$

where

$$q^2 = u^2 + v^2.$$

To show this, change the transform formula to polar coordinates and integrate over the angular variable.[7] Then the relations between the two one-dimensional functions $\mathbf{f}(r)$ and $\mathbf{F}(q)$ are

[7] That is,

$$\begin{aligned}\int_{-\infty}^{\infty}\int_{-\infty}^{\infty} f(x,y)e^{-i2\pi(xu+yv)}\,dx\,dy &= \int_0^{\infty}\int_0^{2\pi} \mathbf{f}(r)e^{-i2\pi qr\cos(\theta-\phi)}r\,dr\,d\theta \\ &= \int_0^{\infty} \mathbf{f}(r)\left[\int_0^{2\pi} e^{-i2\pi qr\cos\theta}\,d\theta\right]r\,dr \\ &= 2\pi\int_0^{\infty} \mathbf{f}(r)J_0(2\pi qr)r\,dr\end{aligned}$$

where $x + iy = re^{i\theta}$, $u + iv = qe^{i\phi}$ and we have used the relation

$$J_0(z) = \frac{1}{2\pi}\int_0^{2\pi} e^{-iz\cos\beta}\,d\beta.$$

$$\mathbf{F}(q) = 2\pi \int_0^\infty \mathbf{f}(r) J_0(2\pi qr) r \, dr$$

and
$$\mathbf{f}(r) = 2\pi \int_0^\infty \mathbf{F}(q) J_0(2\pi qr) q \, dq.$$

We refer to $\mathbf{F}(q)$ as the Hankel transform (of zero order) of $\mathbf{f}(r)$ and note that the transformation is strictly reciprocal, as was the case when the kernels were cos and sin. The kernel J_0, together with cos, sin, and others, is referred to as a Fourier kernel in the broad sense of a kernel associated with a reciprocal transform.

The factors 2π in the above formulas may be canceled by suitable redefinition of the variables, but their retention follows logically from the form adopted for Fourier transforms. In physical situations the 2π in parentheses will be found to result from the measurement of q in whole cycles per unit of r. The 2π before the integral sign comes from the element of area $2\pi r \, dr$.

A number of zero-order Hankel transforms are shown as two-dimensional Fourier transforms in Fig. 12.4. Table 12.2. lists various Hankel transforms for reference.

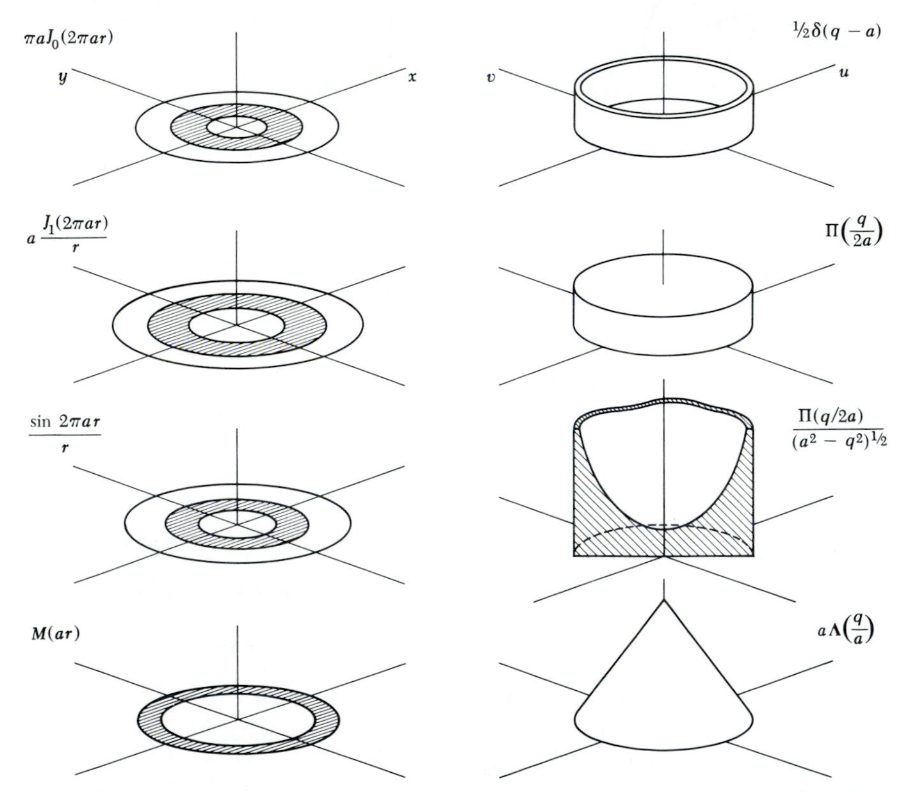

Fig. 12.4 Some zero-order Hankel transforms shown as two-dimensional Fourier transforms.

Table 12.2 Some Hankel transforms

$f(r)$	$F(q)$
$\Pi\left(\frac{r}{2a}\right)$	$\frac{aJ_1(2\pi aq)}{q}$
$\frac{\sin 2\pi ar}{r}$	$\frac{\Pi(q/2a)}{(a^2 - q^2)^{\frac{1}{2}}}$
$\frac{1}{2}\,\delta(r - a)$	$\pi aJ_0(2\pi aq)$
$M(ar)$†	$a^{-2}\Lambda\left(\frac{q}{a}\right)$
$e^{-\pi r^2}$	$e^{-\pi q^2}$
$\frac{1}{(a^2 + r^2)^{\frac{1}{2}}}$	$\frac{e^{-2\pi aq}}{q}$
$\frac{1}{(a^2 + r^2)^{\frac{3}{2}}}$	$\frac{2\pi e^{-2\pi aq}}{a}$
$\frac{1}{a^2 + r^2}$	$2\pi K_0(2\pi aq)$
$\frac{2a^2}{(a^2 + r^2)^2}$	$4\pi^2 aqK_1(2\pi aq)$
$\frac{4a^4}{(a^2 + r^2)^3}$	$4\pi^2 aqK_1(2\pi aq) + 4\pi^3 a^2 q^2 K_0(2\pi aq)$
$(a^2 - r^2)\,\Pi\left(\frac{r}{2a}\right)$	$\frac{a^2 J_2(2\pi aq)}{\pi q^2}$
$\frac{1}{r}$	$\frac{1}{q}$
e^{-ar}	$\frac{2\pi a}{(4\pi^2 q^2 + a^2)^{\frac{3}{2}}}$
$\frac{e^{-ar}}{r}$	$\frac{2\pi}{(4\pi^2 q^2 + a^2)^{\frac{1}{2}}}$
$\frac{\delta(r)}{\pi r} = {}^2\delta(x,y)$	1
$\frac{\Pi\left(\frac{r}{2a}\right) **_2 \Pi\left(\frac{r}{2a}\right)}{2a^2} = \left[\cos^{-1}\frac{r}{2a} - \frac{r}{2a}\left(1 - \frac{r^2}{4a^2}\right)^{\frac{1}{2}}\right]\Pi\left(\frac{r}{4a}\right)$	$\frac{[J_1(2\pi aq)]^2}{2q^2}$
$-4\pi^2 r^2 f(r)$	$\left(\frac{d^2F}{dq^2} + \frac{1}{q}\frac{dF}{dq}\right) = \nabla^2 F$
$r^2 e^{-\pi r^2}$	$\left(\frac{1}{\pi} - q^2\right)e^{-\pi q^2}$

† In this table $M(x) = 2\pi\left[x^{-3}\int_0^x J_0(x)\,dx - x^{-2}J_0(x)\right]$.

Table 12.3 Theorems for the Hankel transform

Theorem	f(r)	F(q)
Similarity	$\mathbf{f}(ar)$	$a^{-2}\mathbf{F}\left(\dfrac{q}{a}\right)$
Addition	$\mathbf{f}(r) + \mathbf{g}(r)$	$\mathbf{F}(q) + \mathbf{G}(q)$
Shift	Shift of origin destroys circular symmetry	
Convolution	$\int_0^\infty \int_0^{2\pi} \mathbf{f}(r')\mathbf{g}(R)r'\,dr'\,d\theta$ $(R^2 = r^2 + r'^2 - 2rr' \cos\theta)$	$\mathbf{F}(q)\mathbf{G}(q)$
Rayleigh	$\int_0^\infty \lvert\mathbf{f}(r)\rvert^2 r\,dr = \int_0^\infty \lvert\mathbf{F}(q)\rvert^2 q\,dq$	
Power	$\int_0^\infty \mathbf{f}(r)\mathbf{g}^*(r)r\,dr = \int_0^\infty \mathbf{F}(q)\mathbf{G}^*(q)q\,dq$	
Differentiation	Exercise for student	
Definite integral	$2\pi \int_0^\infty \mathbf{f}(r)r\,dr = \mathbf{F}(0)$	
Second moment	$2\pi \int_0^\infty r^2\mathbf{f}(r)r\,dr = \dfrac{\mathbf{F}''(0)}{-2\pi^2}$	
Equivalent width	$\dfrac{2\pi \int_0^\infty \mathbf{f}(r)r\,dr}{\mathbf{f}(0)} = \dfrac{\mathbf{F}(0)}{2\pi \int_0^\infty \mathbf{F}(q)q\,dq}$	

Many of the theorems for the two-dimensional Fourier transform can be restated in terms of the Hankel transform. The names of the corresponding Fourier theorems are listed in Table 12.3 to allow comparison.

Fourier kernels

Let two functions f and g be related through the following integral equation whose kernel is k:

$$g(s) = \int_0^\infty f(x)k(s,x)\,dx,$$

and let the kernel be such that a reciprocal relationship also holds; that is,

$$f(x) = \int_0^\infty g(s)k(s,x)\,ds.$$

We know that $2 \cos 2\pi ax$ and $2 \sin 2\pi ax$ are such kernels, and a whole further set is furnished by a theorem established by Hankel, namely,[2]

$$g(x) = \int_0^\infty ds\,(xs)^{\frac{1}{2}}J_\nu(xs) \int_0^\infty g(x)(xs)^{\frac{1}{2}}J_\nu(xs)\,dx,$$

whence $$G(s) = \int_0^\infty g(x)(xs)^{\frac{1}{2}}J_\nu(xs)\,dx,$$

[2] Where $f(x)$ is discontinuous, the left-hand side should be replaced by $\frac{1}{2}[f(x + 0) + f(x - 0)]$.

and conversely, $g(x) = \int_0^\infty G(s)(xs)^{\frac{1}{2}}J_\nu(xs)\,ds.$

By splitting off a factor $s^{\frac{1}{2}}$ from $G(s)$ and $x^{\frac{1}{2}}$ from $g(x)$, that is, by putting $G(s) = s^{\frac{1}{2}}F(s)$ and $g(x) = x^{\frac{1}{2}}f(x)$, we obtain the following alternative expressions of the above formulas:

$$F(s) = \int_0^\infty xf(x)J_\nu(xs)\,dx$$
$$f(x) = \int_0^\infty sF(s)J_\nu(xs)\,ds.$$

The case in which $\nu = 0$ was derived earlier from the two-dimensional Fourier transform under conditions of circular symmetry.

It is interesting that by taking $\nu = \pm\frac{1}{2}$ and using the relations

$$J_{\frac{1}{2}}(z) = \left(\frac{2}{\pi z}\right)^{\frac{1}{2}} \sin z, \qquad J_{-\frac{1}{2}}(z) = \left(\frac{2}{\pi z}\right)^{\frac{1}{2}} \cos z,$$

we recover the known kernels $2 \cos 2\pi ax$ and $2 \sin 2\pi ax$, which shows that the cosine and sine transform formulas are included in Hankel's theorem.

The three-dimensional Fourier transform

Undoubtedly physical systems have three dimensions, but for reasons of theoretical tractability, one seeks simplifications. The classical example in which Fourier analysis in three dimensions has nevertheless had to be faced is the diffraction of X rays by crystals. The formulas are

$$F(u,v,w) = \int_{-\infty}^\infty \int_{-\infty}^\infty \int_{-\infty}^\infty f(x,y,z)e^{-i2\pi(xu+yv+zw)}\,dx\,dy\,dz$$
$$f(x,y,z) = \int_{-\infty}^\infty \int_{-\infty}^\infty \int_{-\infty}^\infty F(u,v,w)e^{i2\pi(ux+vy+wz)}\,du\,dv\,dw.$$

Multidimensional transforms, should they be encountered, will be recognized without difficulty. By taking $\mathbf{x}$ and $\mathbf{s}$ to be vectors whose components are $(x_1,x_2, \ldots)$ and $(s_1,s_2, \ldots)$, we have the following convenient vector notation for n-dimensional transforms:

$$F(\mathbf{s}) = \iint \cdots \int_{-\infty}^\infty f(\mathbf{x})e^{-i2\pi\mathbf{x}\cdot\mathbf{s}}\,dx_1\,dx_2 \ldots dx_n.$$

In cylindrical coordinates r, θ, z where

$$x + iy = re^{i\theta},$$

the three-dimensional transform may be expressed in terms of the transform variables s, ϕ, w, where

$$u + iv = se^{i\phi},$$

by the formulas

$$G(s,\phi,w) = \int_0^\infty \int_0^{2\pi} \int_{-\infty}^\infty g(r,\theta,z) e^{-i2\pi[sr\cos(\theta-\phi)+wz]} r\,dr\,d\theta\,dz$$

$$g(r,\theta,z) = \int_0^\infty \int_0^{2\pi} \int_{-\infty}^\infty G(s,\phi,w) e^{i2\pi[sr\cos(\theta-\phi)+wz]} s\,ds\,d\phi\,dw.$$

These results are derivable directly from the basic formulas by substituting

$$g(r,\theta,z) = f(x,y,z)$$

and

$$G(s,\phi,w) = F(u,v,w).$$

Under circular symmetry, that is, when f is independent of θ (and hence F independent of ϕ), we find by writing

$$h(r,z) = f(x,y,z)$$

and

$$H(s,w) = F(u,v,w)$$

that

$$H(s,w) = 2\pi \int_0^\infty \int_{-\infty}^\infty h(r,z) J_0(2\pi sr) e^{-i2\pi wz} r\,dr\,dz$$

$$h(r,z) = 2\pi \int_0^\infty \int_{-\infty}^\infty H(s,w) J_0(2\pi sr) e^{i2\pi wz} s\,ds\,dw.$$

To obtain this result we use the formula derived earlier for the Hankel transform of zero order.

Under cylindrical symmetry, that is, when f is independent of both θ and z, being a function of r only, say $f(x,y,z) = \mathbf{k}(r)$ and $F(u,v,w) = K(s,w)$, then

$$K(s,w) = \mathbf{K}(s)\,\delta(w),$$

where

$$\mathbf{K}(s) = 2\pi \int_0^\infty \mathbf{k}(r) J_0(2\pi sr) r\,dr.$$

In spherical coordinates r,θ,ϕ, with transform variables s,Θ,Φ, we have

$$x = r\sin\theta\cos\phi \qquad y = r\sin\theta\sin\phi \qquad z = r\cos\theta$$

$$u = s\sin\Theta\cos\Phi \qquad v = s\sin\Theta\sin\Phi \qquad w = s\cos\Theta.$$

Writing $f(x,y,z) = g(r,\theta,\phi)$ and $F(u,v,w) = G(s,\Theta,\Phi)$, we find

$$G(s,\Theta,\Phi) = \int_0^\infty \int_0^\pi \int_0^{2\pi} g(r,\theta,\phi)\, e^{-i2\pi sr[\cos\Theta\cos\theta+\sin\Theta\sin\theta\cos(\phi-\Phi)]} r^2 \sin\theta\,dr\,d\theta\,d\phi.$$

$$g(r,\theta,\phi) = \int_0^\infty \int_0^\pi \int_0^{2\pi} G(s,\Theta,\Phi)\, e^{i2\pi sr[\cos\Theta\cos\theta+\sin\Theta\sin\theta\cos(\phi-\Phi)]} s^2 \sin\Theta\,ds\,d\Theta\,d\Phi.$$

With circular symmetry, that is, when $f(x,y,z)$ is independent of ϕ, we have, writing $f(x,y,z) = h(r,\theta)$ and $F(u,v,w) = H(s,\Theta)$,

$$H(s,\Theta) = 2\pi \int_0^\infty \int_0^\pi h(r,\theta) J_0(2\pi sr\sin\Theta\sin\theta) e^{-i2\pi sr\cos\Theta\cos\theta} r^2 \sin\theta\,dr\,d\theta.$$

With spherical symmetry, we have, writing $f(x,y,z) = k(r)$ and

$$F(u,v,w) = K(s),$$
$$K(s) = 4\pi \int_0^\infty k(r) \operatorname{sinc}(2sr)\, r^2\, dr,$$
$$k(r) = 4\pi \int_0^\infty K(s) \operatorname{sinc}(2sr)\, s^2\, ds.$$

A few examples of three-dimensional Fourier transforms are given in Table 12.4. Many more can be generated by noting that

$$f(x)g(y)h(z) \supset F(u)G(v)H(w),$$

where $f(x), F(u)$, and the like are one-dimensional Fourier transform pairs. This result is proved by expressing $f(x)g(y)h(z)$ in the form

$$f(x)\ \delta(y)\ \delta(z) * \delta(x)g(y)\ \delta(z) * \delta(x)\ \delta(y)h(z)$$

and then applying the convolution theorem in three dimensions. As various cases of this kind we have

$$\begin{aligned} f(x)g(y)h(z) &\supset F(u)G(v)H(w) \\ f(x)g(y) &\supset F(u)G(v)\ \delta(w) \\ f(x) &\supset F(u)\ \delta(v)\ \delta(w) \\ \mathbf{k}(r)h(z) &\supset \mathbf{K}(s)H(w) \\ \mathbf{k}(r) &\supset \mathbf{K}(s)\ \delta(w) \end{aligned}$$

Table 12.4 Some three-dimensional Fourier transforms†

$f(x,y,z)$		$F(u,v,w)$
${}^3\delta(x - a, y - b, z - c)$	point	$e^{i2\pi(au+bv+cw)}$
$e^{-\pi(x^2/a^2+y^2/b^2+z^2/c^2)}$	Gaussian	$abce^{-\pi(a^2u^2+b^2v^2+c^2w^2)}$
${}^3\Pi(x,y,z)$	cube	$\operatorname{sinc} u \operatorname{sinc} v \operatorname{sinc} w$
${}^2\Pi(x,y)$	bar	$\operatorname{sinc} u \operatorname{sinc} v\ \delta(w)$
$\Pi(x)$	slab	$\operatorname{sinc} u\ \delta(v)\ \delta(w)$
$\Pi(x)\Pi[(y^2 + z^2)^{\frac{1}{2}}]$	disk	$\operatorname{sinc} u \dfrac{J_1[\pi(v^2 + w^2)^{\frac{1}{2}}]}{2(v^2 + w^2)^{\frac{1}{2}}}$
$\Pi\left(\dfrac{r}{2}\right)$	ball	$\dfrac{\sin 2\pi s - 2\pi s \cos 2\pi s}{2\pi^2 s^3}$
$(1 - \|r\|)\Pi\left(\dfrac{r}{2}\right)$		$\dfrac{\pi}{3}\dfrac{12}{(2\pi)^4}\dfrac{2(1 - \cos 2\pi s) - 2\pi s \sin 2\pi s}{s^4}$
$(1 - r^2)\Pi\left(\dfrac{r}{2}\right)$		$\dfrac{8\pi}{(2\pi)^5}\dfrac{[3 - (2\pi s)^2] \sin 2\pi s - 3(2\pi s) \cos 2\pi s}{s^5}$
$\dfrac{e^{-r/R}}{\frac{4}{3}\pi R^3}$		$\dfrac{6}{(1 + 4\pi^2R^2s^2)^2}$
$e^{-\pi r^2}$		$e^{-\pi s^2}$

† In this table $r^2 = x^2 + y^2 + z^2$ and $s^2 = u^2 + v^2 + w^2$.

The Hankel transform in n dimensions

When symmetry in n dimensions exists, the resulting one-dimensional transform is

$$\mathfrak{F}(q) = \frac{2\pi}{q^{\frac{1}{2}n-1}} \int_0^\infty f(r)\, J_{\frac{1}{2}n-1}(2\pi qr) r^{\frac{1}{2}n}\, dr.$$

By putting $n = 1, 2, 3$, we recover the Fourier and Hankel transforms and the case of spherical symmetry previously stated (bearing in mind that $J_{\frac{1}{2}}(x) = (2/\pi x)^{\frac{1}{2}} \sin x$ and $J_{-\frac{1}{2}}(x) = (2/\pi x)^{\frac{1}{2}} \cos x$).

The Mellin transform

The transform defined by

$$F_M(s) = \int_0^\infty f(x) x^{s-1}\, dx$$

proves to be equivalent to the Laplace transform when a simple change of variable is made. Putting

$$x = e^{-t}$$

we have

$$dx = -e^{-t}\, dt$$
$$x^{s-1} = e^{-t(s-1)},$$

whence

$$F_M(s) = \int_{-\infty}^{\infty} f(e^{-t}) e^{-st}\, dt.$$

The Laplace transform of the t functions, and the Mellin transform of the x function, shown in Fig. 12.5, are the same. When we replot a function of time as a function of e^{-t}, we compress all positive time into the range from 1 to 0.

Several Mellin transforms are listed in Table 12.5.

The inversion formula for the Mellin transform is

$$f(x) = \frac{1}{2\pi i} \int_{c-i\infty}^{c+i\infty} F_M(s) x^{-s}\, ds.$$

Just as the Laplace transform approaches the Fourier series as a special

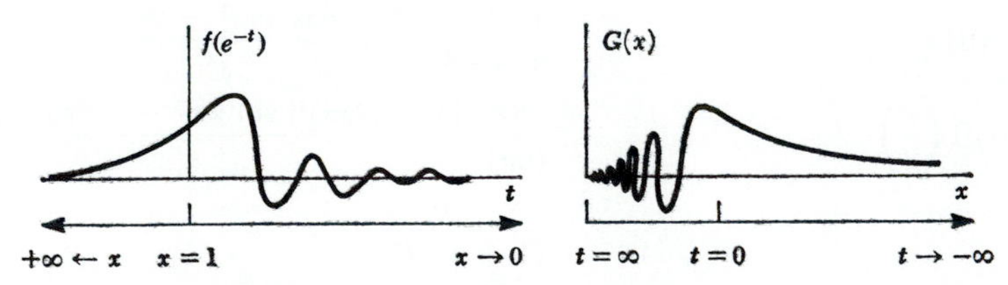

Fig. 12.5 Relation between Mellin and Laplace transforms.

Table 12.5 Some Mellin transforms

$f(x)$	$F_M(s)$	Convergence condition
$\delta(x-a)$	a^{s-1}	
$H(x-a)$	$-\dfrac{a^s}{s}$	
$H(a-x)$	$\dfrac{a^s}{s}$	
$x^n H(x-a)$	$-\dfrac{a^{s+n}}{s+n}$	
$x^n H(a-x)$	$\dfrac{a^{s+n}}{s+n}$	
e^{-ax}	$a^{-s}\Gamma(s)$	Re $a > 0$, Re $s > 0$
e^{-x^2}	$\frac{1}{2}\Gamma(\frac{1}{2}s)$	
$\sin x$	$\Gamma(s)\sin\frac{1}{2}\pi s$	$-1 <$ Re $s < 1$
$\cos x$	$\Gamma(s)\cos\frac{1}{2}\pi s$	$0 <$ Re $s < 1$
$\dfrac{1}{1+x}$	$\pi \operatorname{cosec} \pi s$	
$\dfrac{1}{1-x}$	$\pi \cot \pi s$	
$\dfrac{1}{(1+x)^a}$	$\dfrac{\Gamma(s)\Gamma(a-s)}{\Gamma(a)}$	Re $a > 0$
$\dfrac{1}{1+x^2}$	$\frac{1}{2}\pi \operatorname{cosec} \frac{1}{2}\pi s$	
$(1-x)^{a-1}H(1-x)$	$\dfrac{\Gamma(s)\Gamma(a)}{\Gamma(s+a)}$	Re $a > 0$
$(x-1)^a H(x-1)$	$\dfrac{\Gamma(a-s)\Gamma(1-a)}{\Gamma(1-s)}$	$0 <$ Re $a < 1$
$\ln(1+x)$	$\dfrac{\pi}{s}\operatorname{cosec}\pi s$	$-1 <$ Re $s < 0$
$\frac{1}{2}\pi - \tan^{-1} x$	$\frac{1}{2}\pi s^{-1} \sec \frac{1}{2}s$	
$\Lambda(x-1)$	$\begin{cases} \dfrac{2(2^s-1)}{s(s+1)} & s \neq 0 \\ 2\ln 2 & s = 0 \end{cases}$	Re $s > -1$
erfc x	$\dfrac{\Gamma(\frac{1}{2}s+\frac{1}{2})}{\pi^{\frac{1}{2}}s}$	Re $s > 0$
Si x	$-\dfrac{\Gamma(s)\sin\frac{1}{2}\pi s}{s}$	$-1 <$ Re $s < 0$

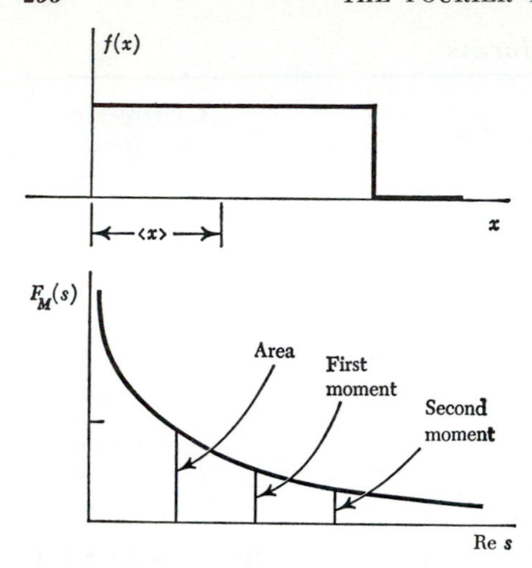

Fig. 12.6 Mellin transform pair.

case when the nonzero parts of $F_L(s)$ concentrate around imaginary integral values of s, so is the Mellin transform related to power series.

We may regard $F_M(s)$ as the $(s-1)$th moment of $f(x)$. For example, if $f(x) = \Pi(x - \frac{1}{2})$, then $F_M(s) = s^{-1}$ (provided Re s is positive). This Mellin transform pair is illustrated in Fig. 12.6. In general,

$$\begin{aligned}
&\text{Area of } f(x) &&= F_M(1)\\
&\text{First moment of } f(x) &&= F_M(2)\\
&\text{Second moment of } f(x) &&= F_M(3)\\
&\text{Abscissa of centroid, } \langle x\rangle &&= \frac{F_M(2)}{F_M(1)}\\
&\text{Radius of gyration} &&= \left[\frac{F_M(3)}{F_M(1)}\right]^{\frac{1}{2}}\\
&\text{Second moment about centroid} &&= F_M(3) - \frac{[F_M(2)]^2}{F_M(1)}
\end{aligned}$$

Theorems for the Mellin transform are listed in Table 12.6. These theorems are closely related to those discussed earlier in respect to Fourier transforms.

By taking Mellin transforms of each of the equations

$$g(\alpha) = \int_0^\infty f(x)k(\alpha x)\,dx$$

and

$$f(x) = \int_0^\infty g(\alpha)h(\alpha x)\,d\alpha,$$

the relation between the Mellin transforms $K_M(s)$ and $H_M(s)$ is found to be

$$K_M(s)H_M(1-s) = 1.$$

Table 12.6 Theorems for the Mellin transform

$f(x)$	$F_M(s)$	
$f(ax)$	$a^{-s}F_M(s)$	$a > 0$
$f_1 + f_2$	$F_{1M} + F_{2M}$	
$xf(x)$	$F_M(s+1)$	
$x^a f(x)$	$F_M(s+a)$	
$f\left(\dfrac{1}{x}\right)$	$F_M(-s)$	
$f(x^a)$	$a^{-1}F_M\left(\dfrac{s}{a}\right)$	$a > 0$
$f(x^{-a})$	$a^{-1}F_M\left(-\dfrac{s}{a}\right)$	$a > 0$
$f'(x)$	$-(s-1)F_M(s-1)$	
$xf'(x)$	$-sF_M(s)$	
$x^a f'(x)$	$-(s+a-1)F_M(s+a^{-1})$	
$f''(x)$	$(s-1)(s-2)F_M(s-2)$	
$x^2f''(x)$	$s(s+1)F_M(s)$	
$f^{(n)}(x)$	$(-1)^n(s-n)\ .\ .\ .\ (s-1)F_M(s-n)$	
$\left(x\dfrac{d}{dx}\right)^n f(x)$	$(-1)^n s^n F_M(s)$	
$\displaystyle\int_0^\infty f\left(\frac{x}{u}\right)g(u)\,\frac{du}{u}$	$F_M(s)G_M(s+1)$	
$\displaystyle\int_0^\infty f\left(\frac{x}{u}\right)g(u)\,du$	$F_M(s)G_M(s)$	
$f(x)g(x)$	$\displaystyle\frac{1}{2\pi i}\int_{c-i\infty}^{c+i\infty} F(\tau)G(s-\tau)\,d\tau$	

This relation leads to the solving kernel $h(\alpha x)$ for integral equations with kernel $k(\alpha x)$ and also, when $h(\alpha x) = k(\alpha x)$, leads to the condition for Fourier kernels of the form $k(\alpha x)$, namely,

$$K_M(s)K_M(1-s) = 1.$$

The z transform

A good deal of interest attaches to signals whose sample values at equispaced intervals of time constitute the full available information about the signal. Such signals arise in communications systems using pulse modulation, where several different messages are interlaced on a time-sharing basis, and another important sphere of interest includes control systems in which feedback is applied on a basis of samples, taken at regular intervals, of some quantity which is to be controlled. Many other examples come to mind, including topics involving probability theory and periodic structures in space, but pulse modulation and sampled-data control systems are the main fields where the z transform is customarily introduced.

It is usual to suppose that the time interval between samples is T, but here the interval will be taken to be unity. If it is necessary to deal with some other interval, one can always introduce a new independent variable differing by a suitable factor.

Let the function $f(t)$ be known to us only through its sample values at $t = 0, 1, 2, \ldots, n$, namely,

$$f(0) \quad f(1) \quad f(2) \quad \ldots \quad f(n).$$

Then the polynomial

$$F(z) = f(0) + f(1)z^{-1} + f(2)z^{-2} + \ldots + f(n)z^{-n}$$

is referred to as the z transform of $f(t)$. The function $f(t)$ is real and is taken to be zero for $t < 0$; the variable z is taken to be complex.

The impulse string

$$\phi(t) = f(0)\,\delta(t) + f(1)\,\delta(t-1) + f(2)\,\delta(t-2) + \ldots + f(n)\,\delta(t-n),$$

though quite different in character from the original function $f(t)$, is fully equivalent to the set of samples, as has been pointed out in connection with the sampling theorem. In fact it is often convenient to deal directly with the sequence

$$\{f(0) \quad f(1) \quad f(2) \quad \ldots \quad f(n)\}$$

without specifying the kind of signal.

If such signals are applied to the input of a system whose impulse response is

$$\{h(0) \quad h(1) \quad h(2) \quad \ldots \quad h(n)\},$$

then the output of the system is represented by the serial product

$$\{f(0) \quad f(1) \quad f(2) \quad \ldots \quad f(n)\} * \{h(0) \quad h(1) \quad h(2) \quad \ldots \quad h(n)\}.$$

But we have seen in Chapter 3 that the serial-product rule for convolving sequences is the same as that for multiplying polynomials. Hence the z transform of the output is the product

$$F(z)H(z),$$

where $H(z)$ is the z transform of the impulse response.

For example, if a signal $\{2 \quad 1\}$ is applied to a system whose impulse response is {8 4 2 1}, then

$$F(z) = 2 + z^{-1}$$
$$H(z) = 8 + 4z^{-1} + 2z^{-2} + z^{-3}$$

and the z transform of the output is given by

$$\begin{aligned} F(z)H(z) &= (2 + z^{-1})(8 + 4z^{-1} + 2z^{-2} + z^{-3}) \\ &= 16 + 16z^{-1} + 8z^{-2} + 4z^{-3} + z^{-4}. \end{aligned}$$

Hence the output sequence is

$$\{16 \; 16 \; 8 \; 4 \; 1\}.$$

Conversely, if the input and output are given and it is desired to find what impulse response the system must have, one simply divides the z transform of the output by the z transform of the input and expands the result as a polynomial in z^{-1}. The coefficients then give the answer.

Serial multiplication and its inversion, as introduced in Chapter 3, constitute an alternative and quite direct approach to the theory of such numerical problems, and present the essential arithmetic of multiplying and dividing z transforms which remains when the numerous z's are omitted.

The possibility of factoring polynomials entering into products or quotients makes for theoretical convenience of the z transform and allows one's familiarity with algebraic manipulation to be brought into play. For this reason, a table of z transforms of some common impulse signals is useful for reference. If the signal is specified by a sequence of numerical data, it is a trivial matter to write down the z transform. We consider here cases in which the strengths of the successive impulses are given instead by some simple formula. Let

$$\phi(t) = f(0)\,\delta(t) + f(1)\,\delta(t-1) + f(2)\,\delta(t-2) + \cdots + f(n)\,\delta(t-n).$$

Taking the Laplace transform of $\phi(t)$ and calling it $F_L(p)$, we have

$$F_L(p) = \int_{-\infty}^{\infty} \phi(t)e^{-pt}\,dt = \sum_{n=0}^{\infty} f(n)e^{-np}.$$

This is a simple polynomial in $\exp(-p)$. Putting

$$z = e^p,$$

we find

$$F_L(p) = \sum_{n=0}^{\infty} f(n)z^{-n},$$

but this polynomial is by definition the z transform of $f(t)$.

Table 12.7 Some z transforms

$\{f(n)\}$	$f(n)$	$F(z)$
$\{a_0\ a_1\ a_2\ .\ .\ .\}$	a_n	$a_0 + a_1z^{-1} + a_2z^{-2} + \ldots$
$\{1\ 1\ 1\ 1\ .\ .\ .\}$	1	$\dfrac{1}{1 - z^{-1}}$
$\{0\ 1\ 1\ 1\ .\ .\ .\}$		$\dfrac{z^{-1}}{1 - z^{-1}}$
$\{0\ 1\ 2\ 3\ .\ .\ .\}$	n	$\dfrac{z^{-1}}{(1 - z^{-1})^2}$
$\{0\ 1\ 4\ 9\ .\ .\ .\}$	n^2	$\dfrac{z^{-1}(1 + z^{-1})}{(1 - z^{-1})^3}$
$\{0\ 1\ 8\ 27\ .\ .\ .\}$	n^3	$\dfrac{3z^{-2}(1 + z^{-1})}{(1 - z^{-1})^4} + \dfrac{z^{-1}(1 + 2z^{-1})}{(1 - z^{-1})^3}$
$\{1\ e^{-\alpha}\ e^{-2\alpha}\ .\ .\ .\}$	$e^{-\alpha n}$	$\dfrac{1}{1 - e^{-\alpha}z^{-1}}$
$\{0\ 1 - e^{-\alpha}\ 1 - e^{-2\alpha}\ .\ .\ .\}$	$1 - e^{-\alpha n}$	$\dfrac{(1 - e^{-\alpha})z^{-1}}{(1 - z^{-1})(1 - e^{-\alpha}z^{-1})}$
$\{0\ e^{-\alpha}\ 2e^{-2\alpha}\ .\ .\ .\}$	$ne^{-\alpha n}$	$\dfrac{e^{-\alpha}z^{-1}}{(1 - e^{-\alpha}z^{-1})^2}$
$\{0\ e^{-\alpha}\ 4e^{-2\alpha}\ .\ .\ .\}$	$n^2e^{-\alpha n}$	$\dfrac{e^{-\alpha}z^{-1}(1 + e^{-\alpha}z^{-1})}{(1 - e^{-\alpha}z^{-1})^3}$
	$\dfrac{1}{\alpha}(\alpha n - 1 + e^{-\alpha n})$	$\dfrac{z^{-1}}{(1 - z^{-1})^2} - \dfrac{(1 - e^{-\alpha})z^{-1}}{\alpha(1 - z^{-1})(1 - e^{-\alpha}z^{-1})}$
	$e^{-\alpha n} - e^{-\beta n}$	$\dfrac{(e^{-\alpha} - e^{-\beta})z^{-1}}{(1 - e^{-\alpha}z^{-1})(1 - e^{-\beta}z^{-1})}$
	$\sin \alpha n$	$\dfrac{z^{-1}\sin \alpha}{1 - z^{-1}2\cos \alpha + z^{-2}}$
	$\cos \alpha n$	$\dfrac{1 - z^{-1}\cos \alpha}{1 - z^{-1}2\cos \alpha + z^{-2}}$
	$\sinh \alpha n$	$\dfrac{z^{-1}\sinh \alpha}{1 - z^{-1}2\cosh \alpha + z^{-2}}$
	$\cosh \alpha n$	$\dfrac{1 - z^{-1}\cosh \alpha}{1 - z^{-1}2\cosh \alpha + z^{-2}}$
	$e^{-\alpha n}\cos \omega n$	$\dfrac{1 - z^{-1}e^{-\alpha}\cos \omega}{1 - z^{-1}2e^{-\alpha}\cos \omega + e^{-2\alpha}z^{-2}}$
	$e^{-\alpha n}\sin \omega n$	$\dfrac{z^{-1}e^{-\alpha}\sin \omega}{1 - z^{-1}2e^{-\alpha}\cos \omega + e^{-2\alpha}z^{-2}}$

$f(n+1)$	$zF(z) - f(0)z$
$f(n+2)$	$z^2F(z) - f(0)z^2 - f(1)z$
$n^a f(n)$	$-z(d/dz)^a F(z)$

As an example, consider the sequence of pulses $\delta(t) + \delta(t-1) + \delta(t-2) + \ldots$. Its Laplace transform is given by $1 + e^{-p} + e^{-2p} + \ldots$, and its z transform is $1 + z^{-1} + z^{-2} + \ldots$. Thus the z transform is directly deducible from the Laplace transform. In this case the polynomial can be expressed in finite form,

$$\frac{1}{1-e^{-p}} = \frac{1}{1-z^{-1}},$$

and makes a compact entry in the table.

Operations carried out on $f(n)$ will correspond with operations on the z transform; pairs of such operations are listed in Table 12.8.

A comparison of this table with the theorems quoted for the Mellin transform will reveal that the two are virtually identical. This is because the Mellin transform and the inverse z transform are related to the Laplace transform in much the same way. Thus in the Laplace integral

$$\int_{-\infty}^{\infty} f(t)e^{-pt}\,dt$$

we put $x = \exp(-t)$ to derive the Mellin transform, while we put $z = \exp(-p)$ to derive the z transform. The three transforms and their inversion integrals are listed below for reference and comparison.

Table 12.8 Theorems for the z transform

$f(n)$	$F(z)$	
$f_1(n) + f_2(n)$	$F_1(z) + F_2(z)$	
$f(n-1)$	$z^{-1}F(z)$	
$f(n-a)$	$z^{-a}F(z)$	$a \geqslant 0$†
$nf(n)$	$-zF'(z)$	
$-(n-1)f(n-1)$	$F'(z)$	
$a^{-n}f(n)$	$F(az)$	$a > 0$
$f_1(n) * f_2(n)$	$F_1(z)F_2(z)$	

† This restriction is a consequence of assuming $f(n)$ to be zero for $n < 0$.

$$F_L(p) = \int_{-\infty}^{\infty} f(t)e^{-pt}\,dt \qquad f(t) = \frac{1}{2\pi i}\int_{c-i\infty}^{c+i\infty} F_L(p)e^{pt}\,dp$$

$$F_M(s) = \int_0^{\infty} f(x)x^{s-1}\,dx \qquad f(x) = \frac{1}{2\pi i}\int_{c-i\infty}^{c+i\infty} F_M(s)x^{-s}\,ds$$

$$f(n) = \frac{1}{2\pi i}\int_\Gamma F(z)z^{n-1}\,dz \qquad F(z) = \int_{-\infty}^{\infty} \phi(t)z^{-t}\,dt$$

$$= \sum_0^{\infty} f(n)z^{-n}$$

It is clear that the z transform is like the inverse Mellin transform except that t must assume real values whereas s may be complex, and conversely, x is real whereas z may be complex. The contour Γ on the z plane may be understood as follows. It must enclose the poles of the integrand. If the contour $c - i\infty$ to $c + i\infty$ for inverting the Laplace transformation is chosen to the right of all poles, then the circle into which it is transformed by the transformation $z = \exp(-p)$ will enclose all poles. In the common case where $c = 0$ is suitable (all poles of $F_L(p)$ in the left half-plane), the contour Γ becomes the circle $|z| = 1$.

The Abel transform

As soon as one goes beyond the one-dimensional applications of Fourier transforms and into optical-image formation, television-raster display, mapping by radar or passive detection, and so on, one encounters phenomena which invite the use of the Abel transform for their neatest treatment. These phenomena arise when circularly symmetrical distributions in two dimensions are projected in one dimension. A typical example is the electrical response of a television camera as it scans across a narrow line; another is the electrical response of a microdensitometer whose slit scans over a circularly symmetrical density distribution on a photographic plate.

Fractional-order derivatives are also closely connected with the Abel transform, which therefore also arises in fields, such as conduction of heat in solids or transmission of electrical signals through cables, where fractional-order derivatives are encountered.

The Abel transform $f_A(x)$ of the function $f(r)$ is commonly defined as

$$f_A(x) = 2\int_x^{\infty} \frac{f(r)r\,dr}{(r^2 - x^2)^{\frac{1}{2}}}.$$

The choice of the symbols x and r is suggested by the many applications in which they represent an abscissa and a radius, respectively, in the same plane.

The above formula may be written

$$f_A(x) = \int_0^\infty k(r,x)f(r)\,dr,$$

where
$$k(r,x) = \begin{cases} 2r(r^2 - x^2)^{-\frac{1}{2}} & r > x \\ 0 & r < x. \end{cases}$$

The kernel $k(r,x)$, regarded as a function of r in which x is a parameter, shifts to the right as x increases, and it also changes its form. A slight change of variable leads to a kernel which simply shifts without change of form. Thus putting $\xi = x^2$ and $\rho = r^2$, and letting $f_A(x) = F_A(x^2)$ and $f(r) = F(r^2)$, we have

$$F_A(\xi) = \int_0^\infty K(\xi - \rho)F(\rho)\,d\rho,$$

where
$$K(\xi) = \begin{cases} (-\xi)^{-\frac{1}{2}} & \xi < 0 \\ 0 & \xi \geqslant 0; \end{cases}$$

alternatively,
$$F_A(\xi) = \int_\rho^\infty \frac{F(\rho)\,d\rho}{(\rho - \xi)^{\frac{1}{2}}},$$

or again,
$$F_A = K * F.$$

When necessary, F_A will be referred to as the "modified Abel transform of F." Having reduced the formula to a convolution integral, we may take Fourier transforms and write

$$\bar{F}_A = \bar{K}\bar{F}.$$

Since
$$\bar{K}(s) = \frac{1}{(-2is)^{\frac{1}{2}}},$$

if follows that
$$\bar{F} = (-2is)^{\frac{1}{2}}\bar{F}_A$$
$$= -\frac{1}{\pi}\frac{1}{(-2is)^{\frac{1}{2}}}\,i2\pi s\bar{F}_A$$

whence
$$F = -\frac{1}{\pi}K * F'_A;$$

that is,
$$F(\rho) = -\frac{1}{\pi}\int_\rho^\infty \frac{F'_A(\xi)\,d\xi}{(\xi - \rho)^{\frac{1}{2}}},$$

The solution of the modified Abel integral equation enables F to be expressed in terms of the derivative of F_A. Integrating the solution by parts, or choosing different factors for the transform of F, we obtain a solution in terms of the second derivative of F_A:

$$F = \frac{2}{\pi}\mathcal{K} * F''_A,$$

where
$$\mathcal{K}(\xi) = \begin{cases} (-\xi)^{\frac{1}{2}} & \xi < 0 \\ 0 & \xi \geqslant 0. \end{cases}$$

Table 12.9 Some Abel transforms

$f(r)$		$f_A(x)$	
$\Pi(r/2a)$	Disk	$2(a^2 - x^2)^{\frac{1}{2}}\Pi(x/2a)$	Semiellipse
$(a^2 - r^2)^{-\frac{1}{2}}\Pi(r/2a)$		$\pi\Pi(x/2a)$	Rectangle
$(a^2 - r^2)^{\frac{1}{2}}\Pi(r/2a)$	Hemisphere	$\frac{1}{2}\pi(a^2 - x^2)\Pi(x/2a)$	Parabola
$(a^2 - r^2)\Pi(r/2a)$	Paraboloid	$\frac{4}{3}(a^2 - x^2)^{\frac{3}{2}}\Pi(x/2a)$	
$(a^2 - r^2)^{\frac{3}{2}}\Pi(r/2a)$		$(3\pi/8)(a^2 - x^2)^2\Pi(x/2a)$	
$a\Lambda(r/a)$	Cone	$[a(a^2 - x^2)^{\frac{1}{2}} - x^2 \cosh^{-1}(a/x)]\Pi(x/2a)$	
$\pi^{-1} \cosh^{-1}(a/r)\Pi(r/2a)$		$a\Lambda(x/a)$	Triangle
$\delta(r - a)$	Ring impulse	$2a(a^2 - x^2)^{-\frac{1}{2}}\Pi(x/2a)$	
$\exp(-r^2/2\sigma^2)$	Gaussian	$(2\pi)^{\frac{1}{2}}\sigma \exp(-x^2/2\sigma^2)$	Gaussian
$r^2 \exp(-r^2/2\sigma^2)$		$(2\pi)^{\frac{1}{2}}\sigma(x^2 + \sigma^2) \exp(-x^2/2\sigma^2)$	
$(r^2 - \sigma^2) \exp(-r^2/2\sigma^2)$		$(2\pi)^{\frac{1}{2}}\sigma x^2 \exp(-x^2/2\sigma^2)$	
$(a^2 + r^2)^{-1}$		$\pi(a^2 + x^2)^{-\frac{1}{2}}$	
$J_0(2\pi ar)$		$(\pi a)^{-1} \cos 2\pi ax$	
$2\pi\left[r^{-3}\int_0^r J_0(r)\,dr - r^{-2}J_0(r)\right] = M(r)$		$\operatorname{sinc}^2 x$	
$\delta(r)/\pi\lvert r\rvert$		$\delta(x)$	
$2a \operatorname{sinc} 2ar$		$J_0(2\pi ax)$	
$\frac{1}{2}r^{-1}J_1(2\pi ar)$		$\operatorname{sinc} 2ax$	

Since $\bar{K}$ is nowhere zero, the solution is unique (except for additive null functions).

Reverting to f and f_A, we may write the solutions as

$$f(r) = -\frac{1}{\pi}\int_r^\infty \frac{f'_A(x)\,dx}{(x^2 - r^2)^{\frac{1}{2}}} = -\frac{1}{\pi}\int_r^\infty (x^2 - r^2)^{\frac{1}{2}} \frac{d}{dx}\left[\frac{f'_A(x)}{x}\right] dx,$$

or, if the integral is zero beyond $x = r_0$, and allowing for the possibility that the integrand may behave impulsively at r_0, we have

$$\begin{aligned} f(r) &= -\frac{1}{\pi}\int_r^{r_0} \frac{f'_A(x)\,dx}{(x^2 - r^2)^{\frac{1}{2}}} + \frac{f_A(r_0)}{\pi(r_0{}^2 - r^2)^{\frac{1}{2}}} \\ &= -\frac{1}{\pi}\int_r^{r_0} (x^2 - r^2) \frac{d}{dx}\left[\frac{f'_A(x)}{x}\right] dx - \frac{f'_A(r_0)}{\pi r_0} (r_0{}^2 - r^2)^{\frac{1}{2}}. \end{aligned}$$

Useful relations for checking Abel transforms are

$$\int_{-\infty}^{\infty} f_A(x)\,dx = 2\pi \int_0^{\infty} f(r)r\,dr$$

and

$$f_A(0) = 2\int_0^{\infty} f(r)\,dr.$$

Another property is that

$$K * K * F' = -\pi F;$$

that is, the operation $K\,*$ applied twice in succession annuls differentiation; then F_A is the half-order integral of F, and conversely, F is the half-order differential coefficient of F_A. To prove this, note that if $F_A = K * F$ implies that $F = -\pi^{-1}K * F'_A$, then it follows further that $F'_A = K * F'$; whence

$$K * K * F' = K * F'_A = -\pi F.$$

In Table 12.9 the first eight examples are to be taken as zero for r and x greater than a.

Numerical evaluation of Abel transforms is comparatively simple in view of the possibility of conversion to a convolution integral. One first makes the change of variable, then evaluates sums of products of $K(\rho)$ and $f(\xi - \rho)$ at discrete intervals of ρ. The values of K turn out to be the same, however fine an interval is chosen, save for a normalizing factor; consequently, a universal table of values (see Table 12.10) can be set up for permanent reference. The table shows coefficients for immediate use with values of F read off at $\rho = \frac{1}{2}, 1\frac{1}{2}, \ldots, 9\frac{1}{2}$, the scale of ρ being such that F becomes zero or negligible at $\rho = 10$. The table gives mean values of K over the intervals $0 - 1, 1 - 2, \ldots$. Thus at $\rho = n + \frac{1}{2}$ the value is

$$\int_n^{n+1} K(-\rho)\,d\rho = 2(n+1)^{\frac{1}{2}} - 2n^{\frac{1}{2}}.$$

Table 12.10 Coefficients for performing or inverting the Abel transformation

ρ	K	ρ	K	ρ	K	ρ	K
$\frac{1}{2}$	2.000	$5\frac{1}{2}$	0.427	$10\frac{1}{2}$	0.309	$15\frac{1}{2}$	0.254
$1\frac{1}{2}$	0.828	$6\frac{1}{2}$	0.393	$11\frac{1}{2}$	0.295	$16\frac{1}{2}$	0.246
$2\frac{1}{2}$	0.636	$7\frac{1}{2}$	0.364	$12\frac{1}{2}$	0.283	$17\frac{1}{2}$	0.239
$3\frac{1}{2}$	0.536	$8\frac{1}{2}$	0.343	$13\frac{1}{2}$	0.272	$18\frac{1}{2}$	0.233
$4\frac{1}{2}$	0.472	$9\frac{1}{2}$	0.325	$14\frac{1}{2}$	0.263	$19\frac{1}{2}$	0.226

When N points of subdivision are used, the scale of ρ is arranged so that F becomes zero at $\rho = N$. The coefficients may then all be multiplied by $(10/N)^{\frac{1}{2}}$, or the coefficients may be left unchanged and the answers multiplied by $(10/N)^{\frac{1}{2}}$.

As an example consider $F(\rho) = (10 - \rho)^{\frac{1}{2}}$, for which the modified Abel transform is known to be $F_A(\xi) = \frac{1}{2}\pi(10 - \xi)$. We work at unit intervals and copy the coefficients on a movable strip. The calculation in progress is shown in Fig. 12.7. The movable strip is in position for calculating $F_A(\xi)$ as the sum of products of corresponding values of F and K:

$$7.78 = 2.12 \times 2.000 + 1.87 \times 0.828 + \ . \ . \ . \ + 0.71 \times 0.472.$$

The inverse problem, that of calculating F from F_A, can be handled by means of the relation $F = -\pi^{-1}K * F_A'$ if F_A is first differentiated. However, it will be perceived that the calculation just described can be done in reverse, using the values of F_A, and working the movable strip upward from the bottom. The strip is shown in position for calculating $F(5 - \frac{1}{2})$, let us say by means of a pocket calculator. Form the products 0.71×0.472, . . . , 1.87×0.828, allowing them to accumulate in the memory. Subtract this sum of products from 7.78 and divide by 2.000 to obtain the next wanted value, $F(5 - \frac{1}{2}) = 2.12$. The inverse transformation can be performed quickly in this way.

ρ	F	K	F_A
			15.65
	3.08		
1			14.08
	2.91		
2			12.52
	2.74		
3			10.94
	2.55		
4			9.37
	2.35		
5			7.78
	2.12	2.000	
6			6.20
	1.87	0.828	
7			4.62
	1.58	0.636	
8			3.03
	1.22	0.536	
9			1.42
	0.71	0.472	
10			0
		0.427	

Fig. 12.7 Calculating modified Abel transforms.

The Hilbert transform

The Hilbert transform has a variety of applications; the ones to be mentioned here concern causality and the generalization of the phasor idea beyond pure alternating current.

We define the Hilbert transform of $f(x)$ by

$$F_{\mathrm{Hi}}(x) = \frac{1}{\pi}\int_{-\infty}^{\infty}\frac{f(x')\,dx'}{x'-x}.$$

The divergence at $x = x'$ is allowed for by taking the Cauchy principal value of the integral. It will be seen that $F_{\mathrm{Hi}}(x)$ is a linear functional of $f(x)$; in fact it is obtainable from $f(x)$ by convolution with $(-\pi x)^{-1}$. To emphasize this relationship, we may write

$$F_{\mathrm{Hi}} = \frac{-1}{\pi x} * f(x).$$

From the convolution theorem we can now say how the spectrum of $F_{\mathrm{Hi}}(x)$ is related to that of $f(x)$.

The Fourier transform of $(-\pi x)^{-1}$ is $i \operatorname{sgn} s$ (Fig. 12.8), which is equal to $+i$ for positive s and $-i$ for negative s; hence Hilbert transformation is equivalent to a curious kind of filtering, in which the amplitudes of the spectral components are left unchanged, but their phases are altered by $\pi/2$, positively or negatively according to the sign of s (Fig. 12.9).

Since the application of two Hilbert transformations in succession reverses the phases of all components, it follows that the result will be the negative of the original function. Hence

$$f(x) = -\left(\frac{-1}{\pi x}\right) * F_{\mathrm{Hi}},$$

or

$$f(x) = -\frac{1}{\pi}\int_{-\infty}^{\infty}\frac{F_{\mathrm{Hi}}(x')\,dx'}{x'-x}.$$

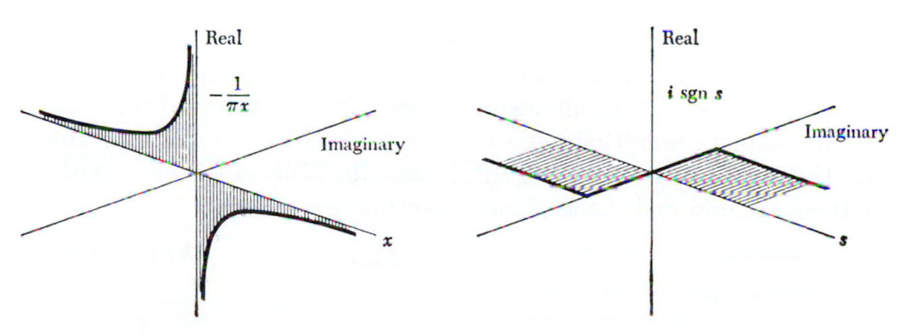

Fig. 12.8 The kernel $(-\pi x)^{-1}$ and its Fourier transform i sgn s.

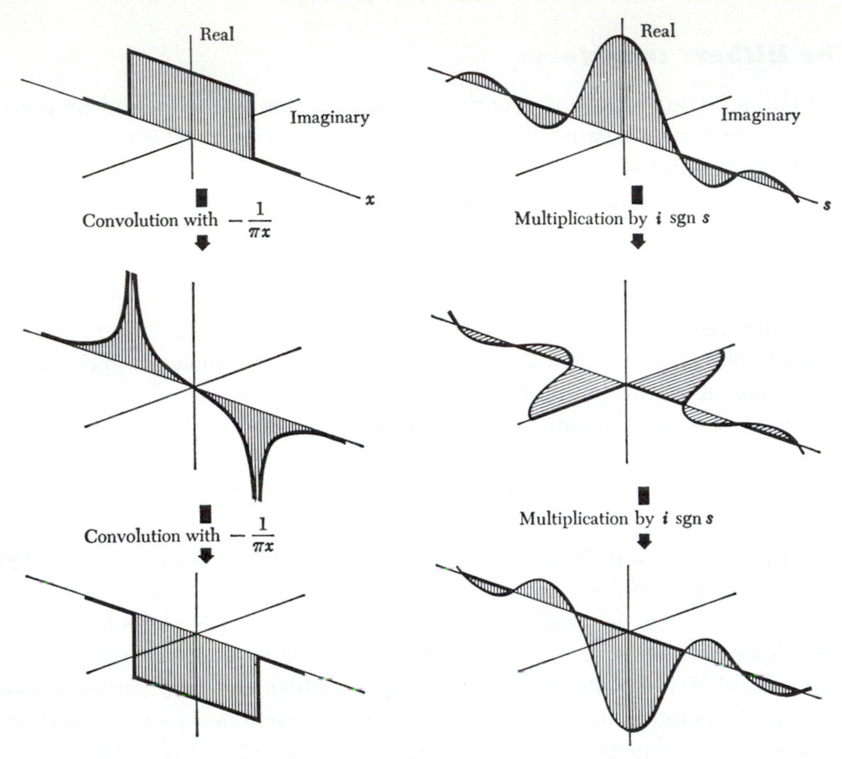

Fig. 12.9 The left-hand column shows a function and the result of two successive Hilbert transformations, while the right-hand column shows the corresponding Fourier transforms.

Had the kernel been chosen as $[i\pi(x' - x)]^{-1}$ instead of $[\pi(x' - x)]^{-1}$, the transformation would have been strictly reciprocal, for then the effect would have been to multiply the spectrum by sgn s, and two such multiplications produce no net change. The custom is to sacrifice the symmetry which would be gained by this procedure in favor of the property that the Hilbert transform of a real function should also be a real function.

It will be noticed that all cosine components transform into negative sine components and that all sine components transform into cosines (Fig. 12.10). A consequence of this is that the Hilbert transforms of even functions are odd and those of odd functions even.

The analytic signal Consider a real function $f(t)$. With it we may associate a complex function

$$f(t) - iF_{\mathrm{Hi}}(t),$$

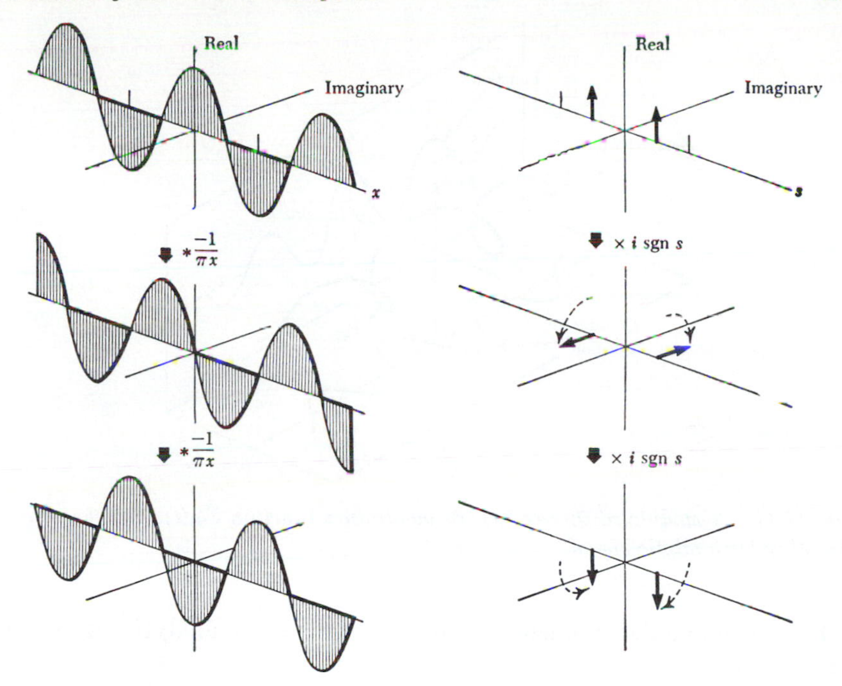

Fig. 12.10 Explaining the Hilbert transform of cosine and sine functions (left) in terms of their spectra (right).

whose real part is $f(t)$. In signal analysis and optics, where the independent variable is time, this associated complex function is known as the analytic signal, and the Hilbert transform is referred to as the quadrature function of $f(t)$. As an example, the quadrature function of $\cos t$ is $-\sin t$ and the analytic signal corresponding to $\cos t$ is $\exp(it)$. One might say that the analytic signal bears the same relationship to $f(t)$ as $\exp(it)$ does to $\cos t$.

Just as phasors simplify manipulations in a-c theory, so the analytic signal is useful in some situations in which departure from absolutely monochromatic behavior prevents use of phasors. Modulated carriers furnish an example. Figure 12.11 shows a function $f(t)$ that could be regarded as a modulated carrier; that is, it is a quasi-monochromatic pulse that slowly builds up and dies away. Its Hilbert transform is also shown, and together the two constitute the complex function of time represented by a helix that slowly dilates and contracts. The original signal $f(t)$ is the projection of this twisted curve on the plane defined by the time axis and the axis of reals (the vertical plane shown), and the quadrature function $F_{\text{Hi}}(t)$ is the projection on the horizontal plane.

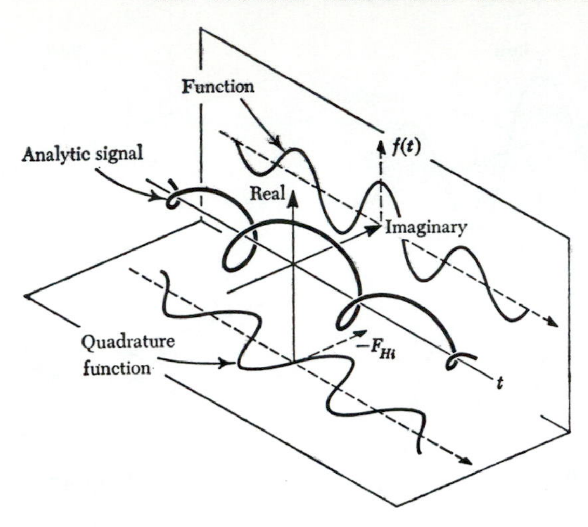

Fig. 12.11 A modulated carrier $f(t)$, its quadrature function $F_{\mathrm{Hi}}(t)$, and the associated complex analytic signal.

If a mean angular frequency $\bar{\omega}$ can be assigned, the analytic signal can be written

$$V(t)e^{i\bar{\omega}t},$$

where the complex coefficient $V(t)$, since it is a generalization of the phasor, can be described as a "time-varying phasor." In the case illustrated in Fig. 12.11, $V(t)$ is real and slowly rises and falls in value. If there were some frequency modulation, $V(t)$ would also rock backward and forward in phase.

From the explanation it is clear that the analytic signal contains no negative-frequency components; in fact it is obtainable from $f(t)$ by suppressing the negative frequencies. For example, noting that

$$\cos \omega t = \frac{e^{i\omega t} + e^{-i\omega t}}{2},$$

we obtain the analytic signal by suppressing the negative-frequency term $\exp(-i\omega t)$. It is also necessary to double the result. To show this directly, let $f(t) \supset F(f)$ and let $\hat{f}(t)$ be derived by suppressing the negative frequencies and doubling:

$$\hat{f}(t) \supset 2H(f)F(f).$$

Since

$$H(f) \subset \tfrac{1}{2}\delta(t) + \frac{i}{2\pi t},$$

it follows that

$$\acute{f}(t) = 2\left[\tfrac{1}{2}\delta(t) + \frac{i}{2\pi t}\right] * f(t)$$
$$= f(t) - i\left(\frac{-1}{\pi t}\right) * f(t)$$
$$= f(t) - iF_{\mathrm{Hi}}(t).$$

Thus $\acute{f}(t)$ is indeed the analytic signal, as originally defined.

Instantaneous frequency and envelope There is some question how one might define the envelope of a signal, since it only touches the signal waveform from time to time and might do anything in between. Similarly, the frequency of a quasi-monochromatic signal is not obviously defined from moment to moment. However, one does not hesitate to say what the amplitude of a sine wave is between peaks; furthermore, there is equal certainty about the moment-to-moment frequency. A generalization that provides a definite answer comes from the analytic signal $V(t) \exp i\bar{\omega}t$. We may say that $|V(t)|$ is the instantaneous amplitude, or envelope, and that the time rate of change of the phase of the analytic signal is the instantaneous frequency.

Causality It is well known that effects never precede their causes, and so $I(t)$, the response of a physical system to an impulse applied at $t = 0$, must be zero for negative values of t:

$$I(t) = 0 \qquad t < 0.$$

Impulse responses satisfying this condition are said to be "causal." This condition is sometimes called the condition of "physical realizability," which is a bad term because it ignores the fact that a system may be impossible to construct for quite other reasons.[3]

Consider some real function $J(t)$. The most general causal impulse response has the form

$$I(t) = H(t)J(t).$$

The spectrum of $I(t)$, namely the transfer function $T(f)$, where

$$T(f) = \int_{-\infty}^{\infty} I(t)e^{-i2\pi ft}\,dt,$$

must have a special property corresponding to the causal character of $I(t)$, and from the previous discussion of the analytic signal this property must be that the real and imaginary parts of $T(f)$ are a Hilbert trans-

[3] In the literature the term "physical realizability" may include a second condition having nothing to do with causality, namely that $I(t)$ should die away with time in such a way as to eliminate circuits, such as transmission-line segments, that do not have a finite number of degrees of freedom.

Table 12.11 Some Hilbert transforms

$f(x)$	$F_{\mathrm{Hi}}(x)$
$\cos x$	$-\sin x$
$\sin x$	$\cos x$
$\dfrac{\sin x}{x}$	$\dfrac{\cos x - 1}{x}$
$\Pi(x)$	$\dfrac{1}{\pi}\ln\left\lvert\dfrac{x-\frac{1}{2}}{x+\frac{1}{2}}\right\rvert.$
$\dfrac{1}{1+x^2}$	$-\dfrac{x}{1+x^2}$
$\operatorname{sinc}' x$	$-\pi \operatorname{sinc} x - \frac{1}{2}\pi \operatorname{sinc}^2 \frac{1}{2}x$
$\delta(x)$	$-\dfrac{1}{\pi x}$
$\mathrm{II}(x)$	$\dfrac{x}{\pi(\frac{1}{4}-x^2)}$
$\mathrm{I_I}(x)$	$-\dfrac{1}{2\pi(\frac{1}{4}-x^2)}$

form pair. Split $I(t)$ into its even and odd parts:

$$\begin{aligned} I(t) &= E(t) + O(t) \\ &= \tfrac{1}{2}[I(t) + I(-t)] + \tfrac{1}{2}[I(t) - I(-t)]. \end{aligned}$$

Now since $O(t) = \operatorname{sgn} tE(t)$, and $\operatorname{sgn} t \supset -i/\pi f$,

$$\begin{aligned} I(t) &= (1 + \operatorname{sgn} t)E(t) \\ &\supset G(f) + i\left(\frac{-1}{\pi f}\right) * G(f). \end{aligned}$$

Hence all causal transfer functions $T(f)$ are of the form

$$T(f) = G(f) + iB(f),$$

where $B(f)$ is the Hilbert transform of $G(f)$; that is,

$$B(f) = \frac{1}{\pi}\int_{-\infty}^{\infty} \frac{G(f')}{f'-f}\,df',$$

and

$$G(f) = -\frac{1}{\pi}\int_{-\infty}^{\infty} \frac{B(f')}{f'-f}\,df'.$$

Several Hilbert transforms are listed in Table 12.11.

Problems

1 Demonstrate the following theorems relating to a function $f(x)$ and its Hilbert transform $F_{\text{Hi}}(x)$.

a. $$f(ax) \supset F_{\text{Hi}}(ax) \qquad \text{(similarity)}$$

b. $$f(x) + g(x) \supset F_{\text{Hi}}(x) + G_{\text{Hi}}(x) \qquad \text{(addition)}$$

c. $$f(x - a) \supset F_{\text{Hi}}(x - a) \qquad \text{(shift)}$$

d. $$\int_{-\infty}^{\infty} f(x)f^*(x)\,dx \supset \int_{-\infty}^{\infty} F_{\text{Hi}}(x)F_{\text{Hi}}^*(x)\,dx \qquad \text{(power)}$$

e. $$\int_{-\infty}^{\infty} f^*(x)f(x-u)\,dx \supset \int_{-\infty}^{\infty} F_{\text{Hi}}^*(x)F_{\text{Hi}}(x-u)\,dx \qquad \text{(autocorrelation)}$$

f. $$f * g \supset -F_{\text{Hi}} * G_{\text{Hi}} \qquad \text{(convolution)}$$

2 Let $\mathcal{H}f$ represent the Hilbert transform of $f(x)$. Show that the Hilbert transform of a convolution can be expressed as follows:

$$\mathcal{H}(f * g) = \mathcal{H}f * g = f * \mathcal{H}g.$$

3 Explain why a function and its Hilbert transform have the same autocorrelation function.

4 It is said that the Hilbert transformation is not unique because the transform of unity is zero. Consequently, the transform of zero would be any constant, and therefore the Hilbert transform of any function would be uncertain to the extent of an additive constant. Examine this argument critically and make an authoritative report on the question of uniqueness.

5 What is the analytic signal corresponding to the following waveforms: sinc t, $\exp[-(t - t_0)^2] \cos \omega t$, $(1 + M \cos \Omega t) \cos \omega t$? Make three-dimensional sketches showing the analytic signal and its two projections.

6 Generate Hilbert transform pairs by separating the real and imaginary parts of the Fourier transforms of the following physically realizable impulse-response functions: $\exp(-t)H(t)$, $t \exp(-t)H(t)$, $\Lambda(t - 1)$, $\exp(-t) \cos \omega t\, H(t)$.

7 Generate Hilbert transform pairs by taking the Fourier transform of the odd and even parts of the following physically realizable impulse-response functions:

$$\Lambda(t)H(t),\ \Pi(t - \tfrac{1}{2}).$$

8 A real function of t is zero for $t < 1$. Prove that the real and imaginary parts of its Fourier transform form a Hilbert transform pair.

9 What can be said about the Fourier transform of a real function of t that is zero for $t < -1$?

10 Let the transfer function $T(f)$ of a filter be expressed as

$$T(f) = e^{\Theta(f)},$$

where $\Theta(f)$ is the complex electrical length. Show that $\Theta(f)$ is hermitian.

11 In the preceding problem, let

$$\Theta(f) = \alpha(f) + i\beta(f),$$

where $\alpha(f)$ is the gain and $\beta(f)$ the phase change of the filter. Thus if

$$T(f) = G(f) + iB(f),$$

then
$$\alpha(f) = \log (G^2 + B^2)^{\frac{1}{2}} = \log |T|$$

and
$$\beta(f) = \tan^{-1}\frac{B}{G} = \text{pha } T.$$

It has been shown that $\Theta(f)$ is hermitian, that is, that $\Theta(-f) = \Theta^*(f)$, or that $\alpha(f)$ is even and $\beta(f)$ is odd. To prove this, was it only necessary to assume that $T(f)$ was hermitian [$I(t)$ real]? If so, what further property must $\alpha(f)$ and $\beta(f)$ exhibit if $T(f)$ is, in addition, causal?

12 If $G(f) + iB(f)$ is the transfer function of a causal filter, show that

$$B(f) = -\frac{2f}{\pi}\int_0^\infty \frac{G(u)}{f^2 - u^2}\,du$$

and
$$G(f) = \frac{2}{\pi}\int_0^\infty \frac{uB(u)}{f^2 - u^2}\,du.$$

13 What can be said about the Hilbert transform of a function that is hermitian?

14 Show that the Hilbert transform of a band-limited function is band-limited.

15 A function $f(t)$ is band-limited. It is said that from a knowledge of $f(t)H(-t)$ only, the whole of $f(t)$ can be deduced. Examine this statement.

16 Show that the Hankel transformation is equivalent to an Abel transformation followed by a one-dimensional Fourier transformation, or that

$$2\pi \int_0^\infty dr\, J_0(2\pi \xi r)r \int_{-\infty}^{\infty} ds\, e^{i2\pi rs} \int_s^\infty dx\, \frac{2xf(x)}{(x^2 - s^2)^{\frac{1}{2}}} = f(\xi).$$

17 Let the impulse train

$$\sum_0^\infty f(n)\,\delta(t - n)$$

possess a Laplace transform where $\alpha < \text{Re } p < \beta$. Show that the z transform $F(z)$ of the sequence $f(n)$ exists in an annulus of the z plane where $e^{-\beta} < |z| < e^{-\alpha}$.

18 Establish the following Hankel transform pairs.

$J_0(2\pi ar)$	$(2\pi a)^{-1}\,\delta(q - a)$
$J_0^2(2\pi ar)$	$\pi^{-2}q^{-1}(\pi^{-2} - q^2)^{-\frac{1}{2}}\Pi(\pi q/2)$
$\text{jinc } r \equiv (2r)^{-1}J_1(\pi r)$	$\Pi(q)$
$\text{jinc}^2\, r$	$\frac{1}{2}[\cos^{-1} q - q(1 - q^2)^{\frac{1}{2}}]\Pi(q/2)$
$r^{-n}J_n(r)$	$2\pi(1 - 4\pi^2q^2)^{n-1}\Pi(\pi q)/2^{n-1}(n - 1)!$
$\exp (ir^2)$	$i\pi \exp (-i\pi^2q^2)$
$\text{sinc}^2\, ar$	$(\pi a^2)^{-1} \cosh^{-1} (a/q)$

Chapter 13 Antennas

Antenna theory is sufficiently analogous to the theory of waveforms and spectra that familiarity with one field can be carried over to the other, once one has the key. The first step is to establish a Fourier transform relation in this subject and then to see what the physical quantities are that enter into it. Since our purpose is to improve the versatility of the Fourier transform as a tool, only one aspect of antenna theory is touched on, but it is of fundamental and far-reaching importance in thinking about antennas.

We approach the subject by considering one-dimensional apertures. This may seem inappropriate, because the essence of antenna theory is that a higher dimensionality is required than suffices for the study of waveforms, which are functions of only one independent variable. The one-dimensional treatment is adequate, however, for discussing a great many antennas whose directivity is separable into a product of directivities of one-dimensional apertures, and it is sufficient for setting up the analogy. Features of the two-dimensional theory are introduced later.

Much of the material of this chapter is directly applicable, with changes in terminology, at optical as well as at radio wavelengths; thus the parallel light incident on an aperture is diffracted into space in much the same way as radio emission from an aperture which is of the same shape (but larger, in proportion to its wavelength). In one class of problem an irradiating field, optical or radio, is given, but the aperture illumination is not specified explicitly. It is often sufficient to assume that the illumination that an incident wave produces in an aperture in a screen is the same as would have been produced if the screen had not been there, but since these more subtle questions of diffraction theory are not to be considered here, our point of departure is the aperture distribution itself. In radio the excitation of an aperture, both in amplitude and in phase, may be directly imposed by separate transmission lines leading to an array of closely spaced dipoles, a possibility that does not exist at optical wavelengths. For this reason it seems appropriate for a discussion of diffraction from given aperture distributions to be couched in terms of antennas.

One-dimensional apertures

An electromagnetic horn feeding energy into semi-infinite space, as shown in Fig. 13.1, is an example of a type of antenna which can conveniently be replaced by a plane-aperture model. As far as the flow of energy into the right-hand half-plane is concerned, we may say that the source is equivalent to a certain distribution of electric field over an aperture plane AA'. The field is zero over the part of the plane occupied by conductor, but some definite distribution is maintained over the horn opening by the source of electromagnetic energy.

Suppose that all conditions are independent of y, the coordinate perpendicular to the plane of the paper. Then the antenna will be called a one-dimensional aperture. It is also supposed in most of what follows that the opening is at least several wavelengths across. This proves to be equivalent to assuming antennas of high directivity.

Since any antenna may be replaced, as far as its distant effects are concerned, by a plane distribution of field, restriction to plane apertures is not in principle destructive of generality. But some antennas (particularly end-fire arrays such as the Yagi antenna) are so nonplanar that they are normally considered by other methods.

Two-dimensional apertures are readily handled as a generalization of the one-dimensional case.

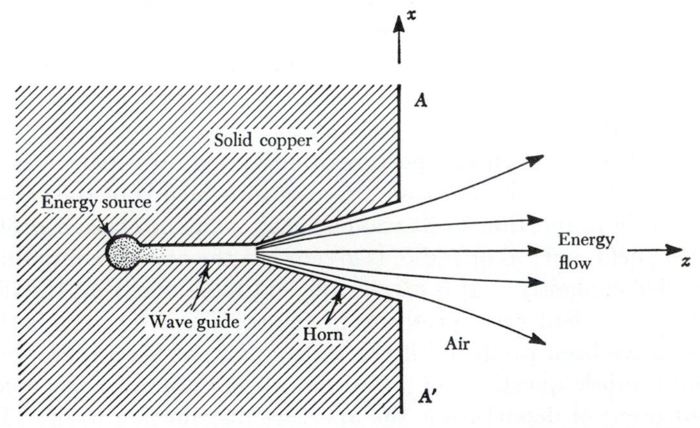

Fig. 13.1 An antenna which can be regarded as a plane aperture.

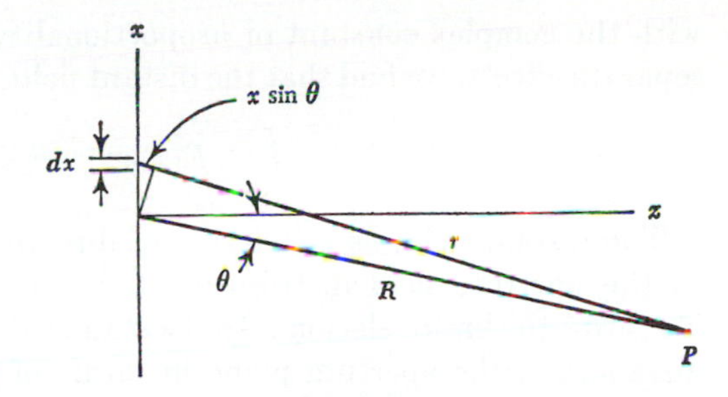

Fig. 13.2 Construction for applying Huyghens' principle.

Let the aperture distribution of electric field at the point x at time t be $F(x) \cos [\omega t - \phi(x)]$, where ω is the angular frequency of the (monochromatic) oscillation and $F(x)$ and $\phi(x)$ are the amplitude and phase, respectively. The phasor $E(x)$, that is, the complex quantity given by

$$E(x) = F(x)e^{-i\phi(x)},$$

will be understood wherever electric fields are dealt with below. We note that the instantaneous field is given by the real part of $E(x) \exp (i\omega t)$; that is,

$$F(x) \cos [\omega t - \phi(x)] = \text{Re } [E(x)e^{i\omega t}].$$

By Huyghens' principle, the electric field produced at a distant point will be the sum of the effects of the elements of the aperture. Each element produces an effect proportional in amplitude to the field at the element and retarded in phase by the number of cycles contained in the path to the distant point. Thus at a distance r the field due to the element between x and $x + dx$ is proportional to

$$E(x)\, dx e^{-i2\pi r/\lambda},$$

where λ is the wavelength.

Consider the field at a distant point P in a direction making an angle θ with the z axis (Fig. 13.2). Let the point P be at a distance R from the origin $x = 0$, $z = 0$. Then to an approximation,

$$r = R + x \sin \theta$$
$$= R + xs$$

where
$$s = \sin \theta.$$

Then
$$E(x)e^{-i2\pi r/\lambda}\, dx$$

becomes
$$E(x)e^{-i2\pi R/\lambda}e^{-i2\pi xs/\lambda}\, dx.$$

Let the factor $\exp (-i2\pi R/\lambda)$, which expresses the average phase retardation of the effects of the aperture elements at a distance R, be lumped with the complex constant of proportionality. Then, integrating all the separate effects, we find that the distant field, in the direction s, is given by

$$\int_{-\infty}^{\infty} E(x)e^{-i2\pi xs/\lambda}\,dx.$$

The distant field, as a function of direction, is thus seen to be related to the aperture field distribution through the Fourier transformation. To throw the basic relation into the standard form, we henceforth measure distances in the aperture plane in terms of the wavelength; that is, we deal in x/λ instead of x. We then define $E(x/\lambda)$ to be equal to the old $E(x)$ and introduce $P(s)$ such that

$$P(s) = \int_{-\infty}^{\infty} E\left(\frac{x}{\lambda}\right) e^{-i2\pi(x/\lambda)s}\, d\left(\frac{x}{\lambda}\right).$$

The quantity $P(s)$, which is proportional to the distant field produced in the direction s, is known as the field radiation pattern or angular spectrum. Usually it is normalized in some way, but as we have introduced it, it depends not only on the character of the aperture excitation $E(x)$ but also on the level of excitation. When normalized, it is characteristic of the antenna alone, irrespective of the strength of the energy source.

By the reciprocal property of the Fourier transformation, it then follows that

$$E\left(\frac{x}{\lambda}\right) = \int_{-\infty}^{\infty} P(s)e^{i2\pi(x/\lambda)s}\,ds.$$

Since s is the sine of an angle, the meaning of an integral taken between infinite limits of s may be obscure. However, the restriction to high directivity ensures that the integrand falls to zero while s is still well within the range -1 to $+1$. The precise limits then do not matter, provided only that they are large enough.

As an example consider an aperture for which

$$E\left(\frac{x}{\lambda}\right) = \Pi\left(\frac{x/\lambda}{w}\right);$$

that is, a uniform field is maintained over a distance of w wavelengths, outside which the field is zero. Then

$$P(s) = w \operatorname{sinc} ws,$$

as shown in Fig. 13.3.

In the one-dimensional theory that follows, we shall restrict the discussion to highly directive antennas. In the more general case the follow-

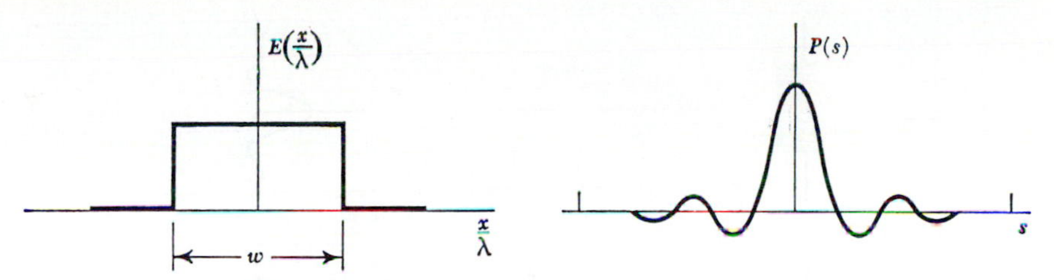

Fig. 13.3 An aperture distribution and its angular spectrum.

ing phenomena enter the discussion. First, the polarization of the field must be taken into account; in directions around the normal to the aperture, the radiated energy is approximately the same whether the aperture field is in the x or the y direction, but at wide angles a cosine factor enters in one case and not in the other. Another factor allows for the difference which sets in between the radian and the unit of s. Then sharp edges in the aperture distribution must be allowed for; they set up evanescent fields which die away with increasing z and do not contribute to the radiated energy. In a highly directive antenna the energy of the evanescent fields is negligible in comparison with the radiation.

Analogy with waveforms and spectra

Having established a Fourier transform relation between a pair of physical quantities in the subject of antenna theory, we are in a position to translate statements about antennas into statements about waveforms and spectra, and vice versa. It only remains to set up a table of corresponding terms as follows:

Waveforms and spectra	*Antennas*
Time t	Aperture distance in wavelengths x/λ
Frequency f	Direction sine s
Waveform $V(t)$	Aperture field distribution $E(x/\lambda)$
Spectrum $S(f)$	Field radiation pattern or angular spectrum $P(s)$

In view of the reciprocity of the Fourier transformation, it is possible to set up the correspondence in two different ways. The one given here is customary, as evidenced by terminology such as "angular spectrum." However, there are two separate and distinct antenna analogues to any one waveform-spectrum pair, and we draw on both as needed.

It will be noticed that $P(s)$ is the minus-i transform of $E(x/\lambda)$, just as

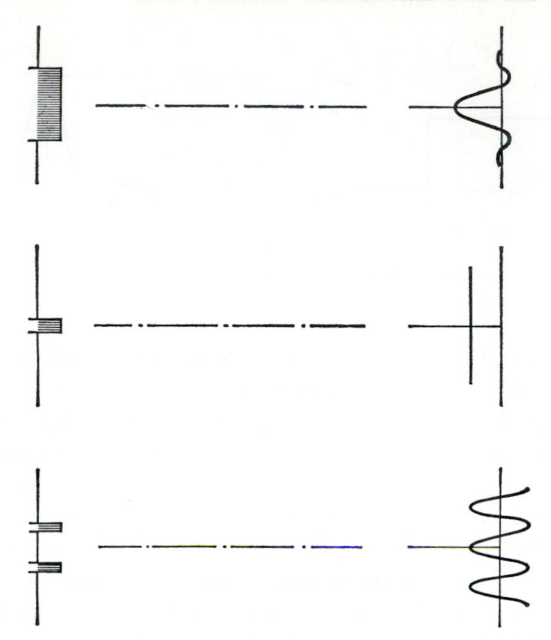

Fig. 13.4 Some aperture distributions and the field that they produce on a distant plane.

$S(f)$ is the minus-i transform of $V(t)$. Clearly this is a convenience in the present context, but the opposite choice might just as well have been made. For the present it is desirable to bear in mind that x and θ are measured in opposite directions.

If we know that a rectangular pulse has a spectrum which is a sinc function of frequency, it follows that a uniformly excited finite aperture produces an angular spectrum of radiation which is a sinc function of direction.

Knowing that an impulse has a flat spectrum tells us that a very small aperture radiates uniformly, and the fact that an impulse pair has a cosinusoidal spectrum means that two infinitesimal antennas produce a cosinusoidal variation of field on a distant plane. These cases are illustrated in Fig. 13.4, and any of the Fourier transform pairs which have been encountered may be so interpreted.

In addition to this possibility of reinterpreting particular transform pairs, there are other classes of information which may be transferred from our body of knowledge about waveforms and their spectra. For example, all the theorems given earlier may be endowed with new meaning in the new physical field. Furthermore, all the corresponding properties investigated previously may be translated into corresponding properties of aperture distributions and their angular spectra. As a first example, consider the similarity theorem.

Beam width and aperture width

The similarity theorem tells us that if the dimensions of an aperture are scaled up, the form of the aperture distribution being kept the same, then the field radiation pattern retains its form and is compressed by the scaling factor, for if

$$E\left(\frac{x}{\lambda}\right) \qquad \text{gives rise to} \qquad P(s),$$

then

$$E\left(\frac{ax}{\lambda}\right) \qquad \text{gives rise to} \qquad |a|^{-1}P\left(\frac{s}{a}\right).$$

Thus the wider the aperture, the narrower the beam.

The quantity $|P(s)|^2$ will be referred to as the angular power spectrum. Its Fourier transform is the autocorrelation function of the aperture distribution. From the reciprocal relation between the equivalent widths of a function and its transform it follows that, given an aperture distribution, we can calculate the effective beam width of its radiation by taking the reciprocal of the equivalent width of the autocorrelation function of the aperture distribution. For example, if we have a uniform aperture 10 wavelengths wide, the equivalent width of the (triangular) autocorrelation function will be 10 wavelengths, hence the equivalent width of the angular power spectrum will be one-tenth of a radian, or 5.7 degrees. This example is illustrated in Fig. 13.5.

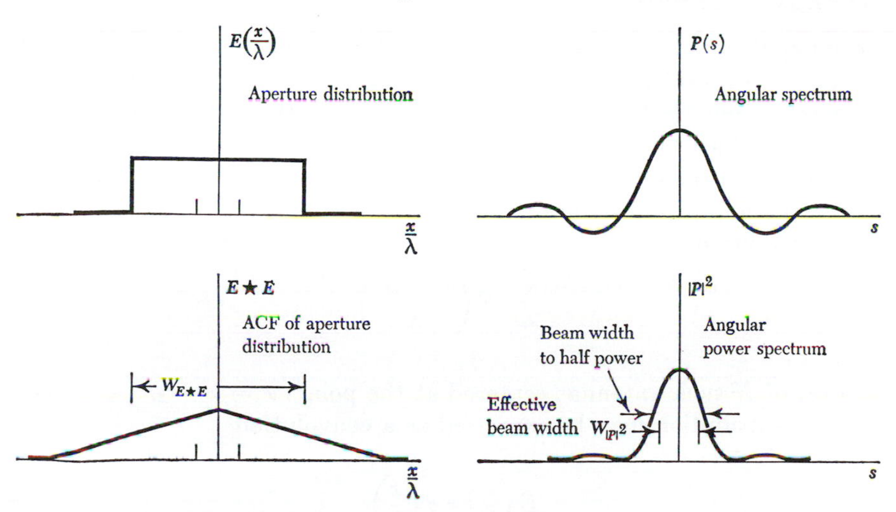

Fig. 13.5 Relations between some antenna quantities.

Beam swinging

According to the shift theorem, if a waveform is shifted with respect to $t = 0$, the result is a progressive phase delay increasing with increasing frequency f. One must therefore be able to shift the beam of an antenna (with respect to $s = 0$) by introducing a linear gradient of phase along the aperture coordinate x. In fact the shift theorem states that

$$E\left(\frac{x}{\lambda}\right) e^{-i2\pi Sx/\lambda} \supset P(s + S).$$

Hence, to shift the beam by an amount S it is necessary to introduce a phase gradient of S cycles per wavelength along the aperture.

Various methods of beam swinging are in use which depend on this principle. A direct approach is to insert phase shifters or to introduce extra lengths of transmission line. Extra air paths may be introduced, as in the method of swinging the beam of a parabolic reflector by displacing the feed point from the focus. Small shifts in frequency can also be used to introduce a suitable phase gradient. If a thin dielectric prism were placed on the aperture, it would introduce a progressive phase delay depending linearly on x, and the beam would consequently be shifted. The amount of shift would be precisely that which we would calculate on the basis of the prismatic refraction.

Arrays of arrays

In its application to antennas, the convolution theorem parades as the array-of-arrays rule. According to this well-known rule, the field pattern of a set of identical antennas is the product of the pattern of a single antenna and an "array factor." This factor is the pattern which would be obtained if the set of antennas were replaced by a set of point sources.

Let the aperture distribution of a certain antenna be $E(x/\lambda)$. Then the distribution

$$\sum_{n=1}^{N} E\left(\frac{x - x_n}{\lambda}\right)$$

is a set of N such antennas centered at the points $x = x_1, x_2, \ldots, x_N$. Such a distribution may be expressed as a convolution

$$E\left(\frac{x}{\lambda}\right) * q\left(\frac{x}{\lambda}\right),$$